Topics in Applied Physics Volume 40

Topics in Applied Physics Founded by Helmut K. V. Lotsch

1 **Dye Lasers** 2nd Edition
Editor: F. P. Schäfer

2 **Laser Spectroscopy** of Atoms and Molecules. Editor: H. Walther

3 **Numerical and Asymptotic Techniques in Electromagnetics** Editor: R. Mittra

4 **Interactions on Metal Surfaces**
Editor: R. Gomer

5 **Mössbauer Spectroscopy**
Editor: U. Gonser

6 **Picture Processing and Digital Filtering**
2nd Edition. Editor: T. S. Huang

7 **Integrated Optics**
2nd Edition. Editor: T. Tamir

8 **Light Scattering in Solids**
Editor: M. Cardona

9 **Laser Speckle** and Related Phenomena
Editor: J. C. Dainty

10 **Transient Electromagnetic Fields**
Editor: L. B. Felsen

11 **Digital Picture Analysis**
Editor: A. Rosenfeld

12 **Turbulence** 2nd Edition
Editor: P. Bradshaw

13 **High-Resolution Laser Spectroscopy**
Editor: K. Shimoda

14 **Laser Monitoring of the Atmosphere**
Editor: E. D. Hinkley

15 **Radiationless Processes** in Molecules and Condensed Phases. Editor: F. K. Fong

16 **Nonlinear Infrared Generation**
Editor: Y.-R. Shen

17 **Electroluminescence** Editor: J. I. Pankove

18 **Ultrashort Light Pulses**
Picosecond Techniques and Applications
Editor: S. L. Shapiro

19 **Optical and Infrared Detectors**
Editor: R. J. Keyes

20 **Holographic Recording Materials**
Editor: H. M. Smith

21 **Solid Electrolytes** Editor: S. Geller

22 **X-Ray Optics.** Applications to Solids
Editor: H.-J. Queisser

23 **Optical Data Processing.** Applications
Editor: D. Casasent

24 **Acoustic Surface Waves**
Editor: A. A. Oliner

25 **Laser Beam Propagation in the Atmosphere**
Editor: J. W. Strohbehn

26 **Photoemission in Solids I**
General Principles
Editors: M. Cardona and L. Ley

27 **Photoemission in Solids II.** Case Studies
Editors: L. Ley and M. Cardona

28 **Hydrogen in Metals I.** Basic Properties
Editors: G. Alefeld and J. Völkl

29 **Hydrogen in Metals II**
Application-Oriented Properties
Editors: G. Alefeld and J. Völkl

30 **Excimer Lasers** Editor: Ch. K. Rhodes

31 **Solar Energy Conversion.** Solid-State Physics Aspects. Editor: B. O. Seraphin

32 **Image Reconstruction from Projections**
Implementation and Applications
Editor: G. T. Herman

33 **Electrets** Editor: G. M. Sessler

34 **Nonlinear Methods of Spectral Analysis**
Editor: S. Haykin

35 **Uranium Enrichment**
Editor: S. Villani

36 **Amorphous Semiconductors**
Editor: M. H. Brodsky

37 **Thermally Stimulated Relaxation in Solids**
Editor: P. Bräunlich

38 **Charge-Coupled Devices**
Editor: D. F. Barbe

39 **Semiconductor Devices** for Optical Communication
Editor: H. Kressel

40 **Display Devices**
Editor: J. I. Pankove

41 **Computer Application in Optical Research**
Editor: B. R. Frieden

42 **Two-Dimensional Digital Signal Processing I.** Linear Filters
Editor: T. S. Huang

43 **Two-Dimensional Digital Signal Processing II.** Transforms and Median Filters. Editor: T. S. Huang

Display Devices

Edited by J. I. Pankove

With Contributions by

D. J. Channin R. S. Crandall T. N. Criscimagna
A. L. Dalisa B. W. Faughnan E. O. Johnson
C. J. Nuese J. I. Pankove P. Pleshko A. Sussman

With 140 Figures

Springer-Verlag Berlin Heidelberg New York 1980

Dr. *Jacques I. Pankove*

RCA Laboratories, David Sarnoff Research Center, Princeton, NJ 08540, USA

ISBN 3-540-09868-2 Springer-Verlag Berlin Heidelberg New York
ISBN 0-387-09868-2 Springer-Verlag New York Heidelberg Berlin

Library of Congress Cataloging in Publication Data. Main entry under title: Display devices. (Topics in applied physics; v. 40). Includes bibliographies and index. 1. Information display systems. I. Pankove, Jacques I., 1922–. II. Channin, D. J. TK7882.I6D563 621.3815'42 79-27608

Monophoto typesetting, offset printing and bookbinding: Brühlsche Universitätsdruckerei, Gießen
2153/3130-543210

Preface

Display devices produce the ultimate output of an information system. They are an interface between man and machine. Therefore, their appearance must be dictated by how pleasing and comfortable it is to the viewer. Hand-held personal systems such as calculators and watches need small displays, while advertising panels which must be viewed at a distance by large crowds need large display devices. There is a range of intermediate applications where neither large nor small devices are satisfactory. Ideally, all displays should be viewable in the dark and also in full ambient light. The esthetics of color and psychophysical characteristics of vision are important guidelines in choosing display devices.

Although display systems can be very complex and may involve very sophisticated driving circuits, most of the research is focused on materials. In fact, any innovation in display devices will be due either to improvements in materials or to discovery of new properties of materials. The circuitry to drive the new display will be readily modified to meet the new requirements.

A large variety of physical or chemical phenomena can be utilized in the operation of display devices. To dwell deeply on all the possible methods would require several volumes. The aim of this book is to present a good coverage of some of the most important display devices and to make the reader aware of the other devices by a more superficial treatment in the introductory chapter.

The various devices have been grouped according to whether they emit light or modulate it. Thus, we start with light-emitting diodes (Chap. 2) and plasma displays (Chap. 3) – two chapters had been planned on dc and ac EL but their authors reneged on their promise (may it weigh on their conscience!) – then we treat liquid-crystal displays (Chap. 4), electrochromic displays (Chap. 5) and electrophoretic displays (Chap. 6). Finally, a more general philosophical discussion of all types of displays is presented in Chap. 7.

This book is intended for the practicing engineer and the graduate student who wish a compact education in a rather specialized but important area of technology. This field is still evolving, and new breakthroughs may suddenly render very attractive some of the presently less promising display devices. It is our hope that this book will provide insights that will help future progress.

The authors have been chosen for their outstanding contributions and well-deserved reputations within each specialty.

I am most grateful to Donna Bailey who typed a substantial portion of this book.

Princeton, NJ., January, 1980 *Jacques I. Pankove*

Contents

Contributors

Channin, Donald J.
RCA Laboratories, David Sarnoff Research Center,
Princeton, NJ 08540, USA, and
Instituto de Fisica e Quimica de Sao Carlos, Sao Carlos (SP), Brasil

Crandall, Richard S.
RCA Laboratories, David Sarnoff Research Center,
Princeton, NJ 08540, USA

Criscimagna, Tony N.
IBM, System Communications Division, Kingston, NY 12401, USA

Dalisa, Andrew L.
EXXON Enterprises Inc., Information Systems,
Sunnyvale, CA 94086, USA

Faughnan, Brian W.
RCA Laboratories, David Sarnoff Research Center,
Princeton, NJ 08540, USA

Johnson, Edward O.
RCA Research Laboratories, Inc. 971-2 Aza 4-go,
Zushi-machi Machida City, Tokyo 194-02, Japan

Nuese, Charles J.
RCA Laboratories, David Sarnoff Research Center,
Princeton, NJ 08540, USA

Pankove, Jacques I.
RCA Laboratories, David Sarnoff Research Center,
Princeton, NJ 08540, USA

Pleshko, Peter
IBM, System Communications Division, Kingston, NY 12401, USA

Sussman, Alan
RCA Laboratories, David Sarnoff Research Center,
Princeton, NJ 08540, USA

1. Introduction

J. I. Pankove

With 19 Figures

In this chapter, we shall first identify the important parameters that characterize a display. This will give us a basis for comparing the various devices.

The ideal device would modulate the ambient light when it is abundant, but would emit bright light in the dark; it would be capable of producing saturated colors at will, be visible from all angles, have high resolution, respond in microseconds but retain the image indefinitely if so desired, have a contrast of 50:1 and 64 levels of gray, consume negligible power at a low voltage. There are devices possessing several of these qualities, but none combines all these virtues. Therefore, the search for perfection will continue.

The relevant parameters of the detector, the average human eye, will also be reviewed since these parameters guide the design of the display.

Then, we shall consider the various mechanisms that either generate or modulate light, and show how they are utilized in actual devices. The performance of each type of display will be mentioned and summarized in Table 1.1. The advantages, as well as the problems presented by each approach will be exposed.

1.1 Characterization of Displays

1.1.1 Luminance or Brightness

In the case of light-emitting displays, the most important property is the luminance or brightness. It determines how well the display can compete with ambient light – a very severe problem in sunlit places. Brightness is usually expressed in foot-Lamberts (fL), or else in nits or candelas per square meter (1 fL = 3.426 nt). These are units that take into account the spectral sensitivity of the average human eye. They are refered to as photometric units. The simpler radiometric unit is the W/cm^2. With radiometric units one can readily calculate the power efficiency of a device. However, radiated watts do not tell how visible the radiated power is unless one specifies the color of the radiation. This aspect of the eye response is treated next, and in more detail in Sect. 1.2.

Table 1.1. Summarized performance of display devices

Type of display	LED	EL		Plasma		LCD	ECD	EPID	Light valve	Electro plating	CL	PLZT
Properties		dc	ac	dc	ac							
Spectral output	*Red* to blue	Yellow	Yellow	Red	Red	B/W[a]	Blue/W		B/W[a]	B/W[a]	B/W[a]	B/W[a]
Luminance	50,000 fL	100 fL	1500 fL	20	700						200 fL	150 fL
Gray scale	>64		16		64		8			Yes	64	Yes
Contrast	$>50:1$		50:1	36:1	30:1	10:1	3:1	40:1	$10^3:1$	5:1	50:1	10:1
Electrical												
operating voltage	2–20 V	~ 100 V	~ 100 V	250 V	250 V	2–20 V	0.5 V	>40 V		0.5 V	5 V–20 kV	100 V
Power, current consumption	10 mW/mm²	mA/cm²	mA/cm²	mA/cm²	mA/cm²	μA/cm²	mC/cm²	μA/cm²		100 mA/cm²	mA/cm²	0.3 mW
Max.-power efficiency	10 %	0.1	0.1							1 %	30 %	
Luminous efficiency 1/w	2000	0.5	0.5	0.07	0.3							
Response time on/off	μs	0.5 ms	0.1 ms	10 μs		ms/0.1 s	0.1 s	10 ms	0.1 s	0.1 s	μs	ms
Storage, memory	No	Yes	Yes	No	Yes		Yes	Yes	Yes	Yes		Yes
Degradation lifetime	10^6 h	10^3 h	$>2\times10^4$ h			$>2\times10^4$ h	$>10^7$ cycles	$>10^7$ cycles	$>10^{10}$ cycles	$\sim 10^3$ cycles		$>10^9$ cycles
Geometrical												
Element size min/max	50 μm/10 mm		0.5 mm	0.5 mm	0.4 mm	0.25 mm	~ 0.2 mm	0.2 mm			30 μm	50 μm
Max. panel size practical	30 cm²	100 cm²		~ 200 lines					m²		0.1 m²	

[a] B/W: Black and White (all colors possible).

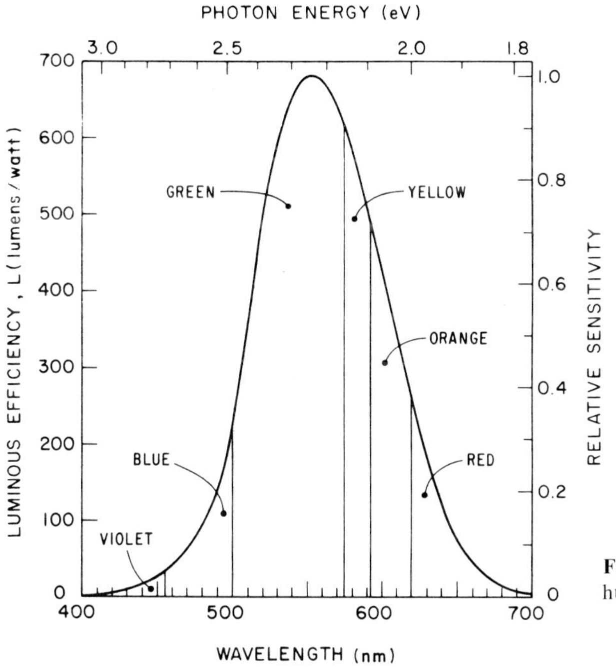

Fig. 1.1. Spectral response of human eye

1.1.2 Color

The spectral distribution of the emitted light must be compared to the spectral sensitivity of the eye, which is shown in Fig. 1.1. From this diagram, one can readily see that 100 μW in a narrow line of red light at 700 nm will be much less bright (by two orders of magnitude) than 100 μW at 556 nm, or less bright than a broad spectrum peaking at 700 nm but extending into shorter wavelengths. Note that 1 mW of red light is painful to look at, and it is also the maximum allowable safe limit.

As will be discussed in Sect. 1.2, some colors, especially red, can be fatiguing under prolonged viewing because the convergence of the human eye depends on the spectrum.

1.1.3 Contrast and Gradation

Contrast is a property of a boundary between light and dark regions of a display. Contrast is the brightness ratio of (maximum − minimum)/minimum. Sometimes, contrast is expressed as the normalized ratio (maximum − minimum)/(maximum + minimum), the highest value of which is one. The minimum brightness is the brightness of the display in the off state. It is the residual brightness due to the ambient light. The minimum brightness can be reduced by providing shade and a filter that is transparent only over the brightest portion of the emitted spectrum.

The gray scale is defined as the number of steps giving a $\sqrt{2}$ change in intensity.

1.1.4 Directional Visibility

The ability to view a display from all directions is an advantage. However, this advantage is often sacrificed in applications intended for personal viewing (e.g., watches and calculators). Plastic lenses enhance the apparent brightness by concentrating the emitted light into a narrow beam. The contrast of liquid-crystal displays fades as the angle of view deviates far from normal incidence because of the consequent longer path length through the birefringent liquid crystal; this produces eliptically polarized light which cannot be totally extinguished by plane polarizers.

1.1.5 Driving Power

Although the emitted power is an important parameter in light-emitting devices (see Sect. 1.1.1), the driving power is the most important quantity to the circuit engineer who must design the circuit that will energize the device. One must specify the operating voltage and current, and whether it will be dc, pulsed, or a pulse superimposed on a dc bias, or whether the pulses will be amplitude modulated or modulated by the repetition rate (pulse-position modulation) or by the duration of the pulse (pulse-width modulation). One must state if the pulses are unipolar or of alternating polarity as in ac EL. Some displays operate at high voltage (e.g., electroluminescence and cathodo-luminescence), other displays operate at high current density (e.g., LEDs or electroplating).

All future development will search for lower operating voltage, which eases the requirements on the driving circuit. Practical systems must be compatible with integrated circuits, i.e., with switching voltages under about 30 V. In many cases, the driving circuit is the most expensive part of the display system. Hence, the economics of the display materials or device may not be a major cost consideration.

1.1.6 Efficiency

The practicality of a device is eventually determined by its power efficiency. This is especially a limitation in light-emitting displays of large area. For example, a flat-panel display 1 m^2 in area emitting 1 W of yellow light (100 $\mu W/cm^2$) will have a total brightness of $\sim$1700 fL and will require $\sim$1 kW of driving power if an EL display is used with a power efficiency of 10^{-3}.

Light modulators such as liquid-crystal displays, electrophoretic displays, electrochromic displays, light valves, etc. can be much more efficient than light-emitting displays since the ambient light is available for free. An excellent discussion of efficiency will be found in Chap. 7.

1.1.7 Speed

The speed of response determines the applications for which the display device is best suited. Obviously flat-panel TV requires a response time on the order of 1 μs to address sequentially every spot of a display. However, since the eye cannot resolve a change faster than 0.1 s, for many applications a response of 0.1 s is adequate.

One consequence of faster operation is higher power dissipation and, therefore, lower efficiency.

1.1.8 Memory and Storage

The long persistence of a display can be an advantage from the point of view of energy conservation. The ability to store the display in the ON state with no (or low) sustaining power greatly increases the efficiency of the display. In many applications, the display need not change for a long time, and at most it need not change faster than the response time of the eye (~ 0.1 s). If a display element is turned on in one μs and can remain in that state for 0.1 s without further power consumption, the apparent efficiency is increased by 10^5. Also, storage frees the driving circuit to access other elements in a display panel without loss of information. The above considerations are applicable to electrochromic, electrophoretic, and light-valve types of displays.

In the memory-endowed scanned display of Sharp's EL TV panel (see Sect. 1.3.3), once the brightness level has been set, only a sustaining voltage (independent of brightness level) needs to be supplied.

1.1.9 Degradation

The life of a device is a parameter of obvious importance. Its control requires a profound knowledge of the physics of device operation. One must study the nature of defects and the mechanism of their formation in the various display materials, the migration of impurities, whether the defect formation is thermal or driven by a local electric field, the chemical and electrochemical interactions between the display material and the electrodes, the sensitivity to ambient moisture, etc. Even photochemical effects are possible, for example the uv bleaching of pigments.

1.1.10 Resolution and Size

The information content of a display increases with the number of picture elements or pixels. Therefore, the packing density of information depends on how small each element can be made. The fineness and visibility of detail has been extensively studied by the television industry, leading to a definition of "resolution" as the number of distinguishable lines per unit length. Thus, 20 lines/mm is judged adequate for normal viewing at a distance of 20 cm. This corresponds to a square picture element 50 μm on the side. However, the perception of an image depends in a very complex way on the light level and on the distance to the viewer. *Carlson* and *Cohen* [1.1] have developed a mathematical model of the visual system which successfully predicts the visual response under various conditions.

As the size of each picture element increases, the driving power increases and its speed tends to decrease.

1.1.11 Addressability

Arraying many elements opens an additional problem: how to address individual elements in the array. From a circuit point of view, the most economical form of addressing is the self-scan mode, whereby the address of the element propagates at a uniform rate along a line at a time. Then, the information is derived from a time-dependent modulation of the potential applied to the panel (like the grid modulation of a kinescope).

The next best method is the x–y addressing of elements that may be arranged in rows and columns. This presents a serious problem since a whole row and a whole column of elements are energized simultaneously. In order to turn ON only the intersection of the addressed row and column, the element must have a nonlinear response to the applied voltage. For example, the element could have a threshold that is exceeded only at the addressed intersection, the voltage across other elements in the row and column remaining below threshold. But what can one do if the display material has a linear dependence on the applied bias? Then the solution is to place a nonlinear device in series with the display material. Such devices can be a rectifying diode or the thin-film transistor (TFT) of *Weimer* [1.2] or the MOS transistor with similar characteristics of *Hofstein* and *Heiman* [1.3]. An x–y-addressed panel with a layer of evaporated circuity under each element in the array has been developed by *Brody* et al. [1.4]. One should note that the TFT has an inherent high-voltage capability (up to ~300 V).

1.2 Psychophysical Factors

Since display devices are meant to interact with the eye, one must be aware of the psychophysical characteristics of the eye. Hence, the response of the average human eye will be reviewed presently.

Color vision is provided by the cones in the retina, whereas sensitivity to low light levels is provided by the rods. At low light levels, there is a loss of color perception.

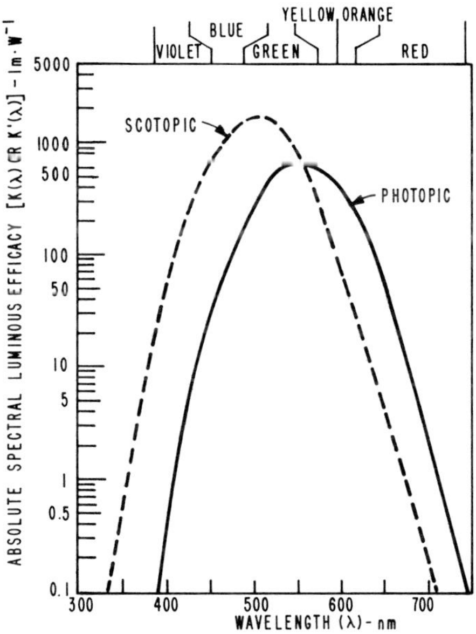

Fig. 1.2. Absolute spectral luminous efficacy of the human eye as a function of wavelength

The threshold of response to visual stimuli depends on the background illumination. The minimum detectable stimulus corresponds to about 60 quanta of blue-green light at 510 nm impinging on the cornea [1.5].

The rods in the retina provide the "scotopic response". It occurs after the eye has been dark adapted to 3×10^{-5} nt (cd/m^2) at most. Complete dark adaption requires about 45 min. The spectrum of luminous efficiency as a function of wavelength is shown in Fig. 1.2.

The "photopic" response is due to the cones. It occurs after the eye has been adapted to a background illumination of at least 3 nt, i.e., in the light-adapted state. This adaption requires about 2 min when the luminance is increased. Figure 1.2 shows the spectral response of photopic vision.

Between the light-adapted (photopic) and dark-adapted (scotopic) states, the response is continuously variable; the spectral response shifts towards the blue when the illumination decreases. This condition is known as the mesopic state, which occurs in the range 3 nt to 3×10^{-5} nt. For most practical applications it is the photopic state which is involved.

Color perception is characterized by three subjective attributes:

1) luminance or brightness;

2) hue – the distinguishable color (pure colors are directly related to wavelength);

3) saturation or chroma – which measures how intense the color is (or how undiluted it is by white light, which would render this color pale).

Any color sensation can be reproduced by the judicious combination of three monochromatic components (for example, red, green, and blue lights).

The Commision Internationale de l'Eclairage (CIE) has adopted a standard of colorimetry which represent the attributes of color by a three-dimensional diagram. The Cartesian vectors of this tridimensional diagram are derived from three idealized primaries which are nonmonochromatic as shown in Fig. 1.3. The coordinates to be used to represent a color are calculated by a procedure to be described later on. The values of x, y, and z are calculated in such a way that their sum is unity ($x+y+z=1$). This property makes a three-dimensional plot unnecessary. It is sufficient to identify x and y to automatically imply z. In other words, the color map may be expressed as a two-dimensional projection into the xy plane. This is the standard chromaticity diagram shown in Fig. 1.4.

The upper arc is the locus of saturated colors. All the colors and shades that the eye can resolve are enclosed between the area of saturated colors and the straight line labeled "magenta". The central region appears white. A black body assumes at different temperatures various shades of whiteness as shown by the arc labeled "black body locus".

Now returning to the tristimulus values of Fig. 1.3, the curve labeled $\bar{y}$ is the apparent intensity of a constant power of a monochromatic light shown on the abscissa; curve $\bar{y}$ is normalized at a peak at 550 nm. The reader will readily recognize that curve $\bar{y}$ is the photopic response of the eye. If a color is generated by a monochromatic source, one calculates its chromatic coordinates by the ratios:

$$x = X/(X+Y+Z)$$
$$y = Y/(X+Y+Z)$$
$$z = Z/(X+Y+Z)$$

where the tristimulus values X, Y, and Z are read at the chosen wavelength on Fig. 1.3. The ratios are the means of normalizing x, y, and z so that their sum is unity.

If the color is generated by nonmonochromatic sources, the three tristimulus values are calculated by the following integrations over the entire visible spectrum:

$$X = \int \phi_\lambda x(\lambda)\, d\lambda$$
$$Y = \int \phi_\lambda y(\lambda)\, d\lambda$$
$$Z = \int \phi_\lambda z(\lambda)\, d\lambda$$

where ϕ_λ is the total spectral radiant flux. The chroma coordinates are determined by the previously defined ratios.

Color discrimination by the eye is much better in the lower portion of the chromaticity diagram than in the upper portion; for example, the sensation of a given shade of green can be obtained from a larger area of the chromaticity diagram than can the sensation of a given shade of blue or of pink. In order to equalize this error in color discrimination, various new diagrams have been devised which stretch the x axis and compress the upper part of the y axis.

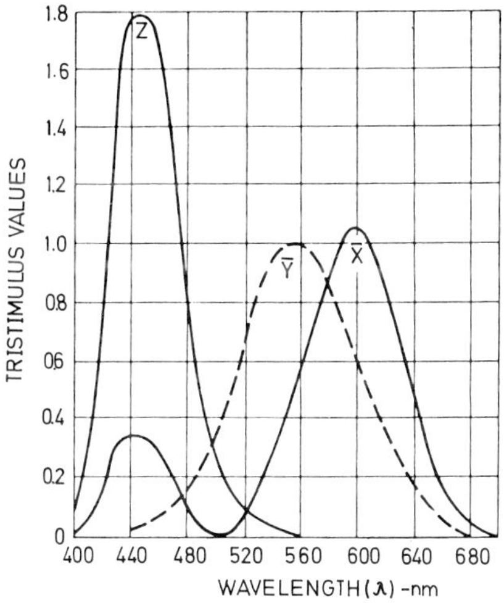

Fig. 1.3. CIE standard color-mixture curves

Fig. 1.4. CIE chromaticity diagram. The points of *ABCDE* were obtained with GaN LEDs, the spectra of which are reported in Fig. 2.15. The point Δ corresponds to the spectrum in Fig. 2.16

▼

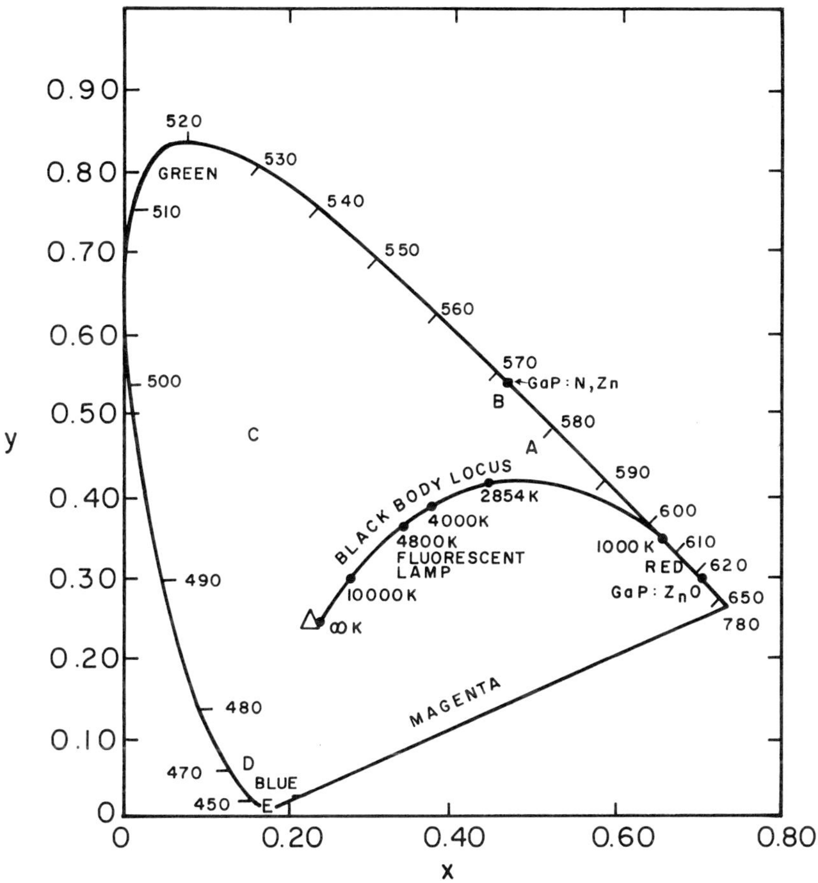

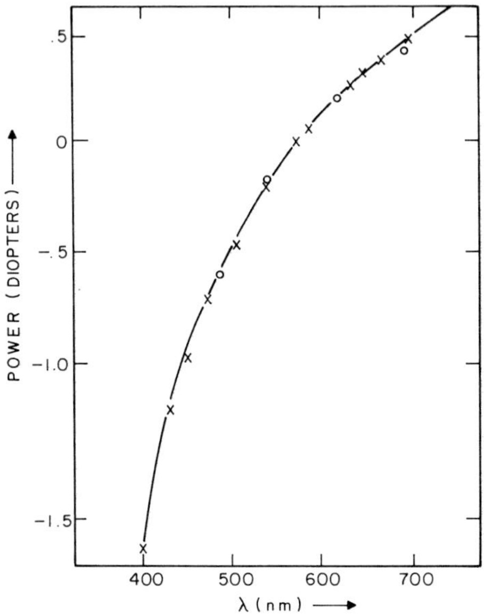

Fig. 1.5. Axial chromatic aberration of the human eye measured in diopters and normalized (0 diopter) at 578 nm. The crosses are the data of [1.8], the circles are data of [1.9]

Loebner [1.7] describes the many forms of color-perception abnormalities which affect about 5% of the male population and about 0.4% of the female population. Monochrome displays most suitable to color abnormals should be either yellow or white.

The eye is capable of adapting to a 10^7 range in light level. Most of the adaptation takes place within the retina. Changes in the size of the pupil account for only a 20:1 range in light-level adaptation. The purpose of increasing the f number of the eye is primarily to increase the sharpness of the image.

The chromatic aberration of the eye (Fig. 1.5) causes the shorter wavelengths of light to be focused much closer to the lens than the longer wavelengths [1.8]. In fact, blue light is never brought to a sharp focus; consequently, the resolving power in the blue is poor. Another consequence of the axial chromatic aberration is the difficulty in focusing on two objects in the same place when their colors are so different that an accomodation of more than 0.4 diopter is required. Thus, if the ambient illumination is mostly yellowgreen ($\sim$570 nm), it may be difficult to focus on either the red ($\lambda > 650$ nm) or the blue ($\lambda < 500$ nm) portions of a picture. Chromatic aberration may account for the eye fatigue experienced by viewers of red LED displays who wear bifocal or trifocal correcting lenses (i.e., those viewers who already have difficulty with focus accommodation). It should be pointed out that a special lens system has been designed to completely compensate for the chromatic aberration of the human eye [1.8].

In cases where contrast rather than resolution is important, e.g., visibility of a large display, another factor must be considered: The spectral responses of red- and green-sensitive cones have a considerable overlap, whereas the spectral overlap with the blue-sensitive cones is small. Hence, in artificial ambient illumination, which is already enriched with red light, the red LED display has a much lower contrast than a blue LED display of equal luminance. For comparable efficiency and cost, the most useful source of light is one capable of emitting white light.

Each active area of display is surrounded by a passive matrix the size of which cannot be reduced below limitations imposed by insulation between adjacent elements. The insulation must be large enough to minimize image washout. As the size of each active element is reduced, the proportion of active area and its perceptibility decrease. Furthermore, there is no need to increase the packing density beyond the resolution capability of the human eye. The picture element need not be smaller than the size needed to subtend 1.3 min of arc; this corresponds to 100 elements/in. (25 elements/cm) at a viewing distance of 28 in. (70 cm) [1.10].

In large area displays, a gradual nonuniformity of less than 2:1 brightness ratio is not noticed by the eye. However, the eye perceives slight discontinuities forming an extended line, such as the boundary between nonidentical display modules, or clusters of inoperative pixels.

A brightness of about 25 fL against a contrast-enhancing black background gives a satisfactory legibility for most ambients.

1.3 Technical Discussion of Display Devices

1.3.1 Light-Emitting Diodes – LED

The LED consists of a p–n junction in a semiconductor. When the junction is forward biased, electrons and holes recombine radiatively. Since the radiative transition of the electron occurs between the conduction and the valence band edges, or via intermediate states such as impurities, only materials with an energy gap greater than 1.8 eV are acceptable. This value corresponds to the red edge of the visible spectrum. To generate blue light an energy gap greater than ~ 2.8 eV is needed. LEDs operate at low voltages (<5 V) and are compatible with integrated-circuit devices.

LEDs have reached a very advanced stage of development. They have become an abundant low-cost commercial product. The fierce price-cutting they have experienced on the market place is the culmination of many man-years of research. Yet, there is still considerable room for improvement: only red light is emitted by the cheap LEDs. The technology for yellow-emitting diodes (operating in the color range where the eye is most sensitive) is still expensive. More effort is needed to develop efficient and inexpensive blue LEDs.

Another concern is the arraying of LEDs into large-area displays. Automated bonding is still too expensive a process for assembling numerous individual diodes. The largest crystal wafers from which LEDs are made are about 5 cm in diameter. Assembling these squared-up wafers into a mosaic presents a costly interconnect problem.

1.3.2 Plasma Panel

The basic mechanism of the plasma panel is the luminescence produced by the glow discharge of field-excited ions that collide with each other. At low pressures, a gas can be ionized with about 250 V. The discharge in the gas can be maintained with about 150 V. This bias is needed to activate the cathode because a regenerative feedback by photons and positive ions bombarding the cathode increases the electron emission at the cathode.

dc Plasma Display

In the dc plasma display, the electrodes are inside the plasma. The glow is most intense near the cathode. This principle was used to great advantage in the Numitron tube where an assortment of cathodes had the shape of numerals.

The most successful use of the dc plasma is the Burroughs "Self-Scan" display panel which has been thoroughly reviewed recently [1.11]. The unique feature of the Burroughs display system is the ability to propagate the glow from cell to cell along a line with a signal on the order of 25 V, i.e., without requiring the much higher voltage needed to ionize the gas. Once the plasma has been generated, it can be moved sequentially from one cell to the next. A brightness of about 20 fL can be achieved. Turn-on and turn-off times are about 10 μs.

In some embodiments, uv light emitted in the plasma excites photoluminescence in a phosphor. It is also possible to impact-excite a phosphor with electrons or ions from the plasma.

The display on a dc plasma panel can be stored if a sustaining voltage remains across each cell that has been turned on. A convenient means for establishing a low sustaining voltage is to place a high resistance in series with the cell. Advances in this technology have been made by incorporating the resistor into the cathode, i.e., by making the cathodes from a graphite film [1.12]. Under continuous operation, a high luminance can be achieved (870 fL for 25 μA in a cell 0.38 mm in diameter).

ac Plasma Display

In the ac plasma display, the electric field is capacitively coupled to the gas. The ac plasma display is described in great detail in Chap. 3.

Comparison of ac and dc Plasma Displays[1]

In dc plasma displays, the electrodes interface directly with the gas. ac plasma displays have thin dielectric films insulating the electrodes from the gas. Charge

1 This portion was kindly contributed by P. Pleshko.

stored on the capacitance formed by these films, provides the ac plasma display with memory. dc plasma displays with no integrated resistor per cell do not have memory. As such, there is a size limitation based on the maximum number of lines that can be scanned at a given luminance level. For typical luminance requirements, this seems to be of the order of 200 scanned lines. With refresh operation of dc plasma displays, gray scale can be more easily achieved than in ac plasma displays by duty-factor or amplitude modulation. For color panels using the uv generated by the glow discharge to excite phosphors, uv is more efficiently generated in the dc plasma devices as compared to ac plasma devices. Panel construction, however, is considerably more complex for dc plasma displays.

Color Control in Plasma Displays

Krupka et al. [1.13] have obtained white light from a dc Self-Scan structure by adding to the orange glow of the neon the green emission of a ZnO:Zn phosphor. This phosphor was excited by low-voltage electrons from the plasma. The output remained white independently of intensity. A spot intensity of 300 fL of white light was achieved at a current density of 0.33 A/cm^2. A penalty for broadening the spectrum is that the brightness is reduced by about a factor of two compared to that of a cell without phosphor. The resistance of this phosphor to sputtering still needs to be studied.

Fujino et al. [1.14] have demonstrated that the color of a plasma in an Hg–He discharge can be varied from red to blue by controlling the width, shape, or repetition rate of the driving pulses, or by changing the ambient temperature which controls the Hg vapor pressure. The controlled parameters change the discharge process via changes in the space charge or in electric field strength.

Rectangular pulses which produced red light at room temperature at low frequency produced blue light at higher frequencies. Also, increasing the Hg vapor pressure changes the plasma color from red to blue. With pulses having an asymmetric triangular shape, the color was blue for Λ-pulses and red for decaying pulses.

TV Plasma Panel

An experimental real-time color TV dc plasma panel was made by Japan Broadcasting Corporation utilizing the uv radiation from the negative glow of the plasma in a "self-scan" panel to excite an array of three different phosphor dots [1.15]. A contrast ratio of 30:1 with continuous gray scale was obtained. However, much work remains to improve the life of the panel beyond the 15 h limitation due to sputtering by the Hg ions used to enhance the vacuum-uv emission.

An He–Xe gas mixture was used at *Hitachi* [1.16] to generate the vacuum uv for a color TV panel using different phosphors to emit saturated colors, albeit at low luminance (5 fL). Increasing the intensity reduces the color saturation because of the enhanced visible glow of the plasma in the background.

1.3.3 Electroluminescence

There are essentially three types of electroluminescent, or EL displays: dc powder, ac powder, and thin film (also ac). Progress in all three types has been so full of surprises to the pessimists that one hesitates to call a "limitation" what may be only a present deficiency.

dc EL

EL was discovered by *Destriau* in ZnS powders doped with Cu in the mid-1930s [1.17]. The Destriau effect was a spectacular curiosity and an intriguiging phenomenon but without promising practicality because the luminescence efficiency degraded rapidly. But, thirty years later, *Lehman* succeeded in reducing the migration of Cu ions by suppressing sulfur vacancies in ZnS and thus greatly extended the operational life of dc EL devices [1.18].

The dc EL cell consists of a powdered phosphor coated with Cu in a binder deposited between a transparent conducting anode on a glass substrate and a backing aluminum cathode [1.19]. Forming is a critical part of the process. In fact, the forming procedure is an art which will determine the subsequent properties of the device. When the electric field is first applied, a large current flows; within a few minutes, the current drops markedly, then light emission begins and grows in intensity.

Although ZnS:Mn, Cu has received considerable attention from many researchers for already four decades, this phosphor may not be the ultimate material. Many other wide-gap compounds exhibit dc EL. It is possible that some of these, given the same effort as ZnS:Mn has received, would yield even better results.

Even the luminescence mechanism is not always well understood. The mechanism may be different with different impurities and in different hosts. The luminescence may result from energy transfer from one impurity to another which will be put in the excited state (e.g., from Cu to Mn). Or the luminescence may result from direct impact excitation of the luminescence center, e.g., Tb in ZnS:TbF_3 [1.20, 21]. Or the emission may result from donor-to-acceptor transition as in *Prener* and *Williams* and other workers [1.22].

ZnS:Mn, Cu emits yellow-orange light peaking at 580 nm. Other compounds emit other colors, but less efficiently. *Highton* et al. [1.23] proposed an attractively simple multicolored dc EL panel in which phosphors capable of emitting different colors are mixed together in appropriate proportions. The observed color change results from the different brightness-voltage characteristics of each phosphor: as the driving voltage is increased, first the phosphor having the lowest threshold voltage emits a narrow spectrum, then, as each higher-voltage phosphor is turned on, adding its narrow spectrum to the now slowly increasing brightness of the lower-voltage phosphor, the composite spectrum broadens, tending to white light.

The major problem of dc EL is the degradation of the cells operated at high brightness (100 fL). This degradation is sometimes associated with an increase in internal resistance. Microscopic examination of EL in $ZnS:TbF_3$ under two closely spaced point contacts [1.24] reveals the nucleation and growth of dark spots spreading away from the cathodic contact. When the polarity is reversed, the black spots gradually recede at the anode while new black spots appear at the cathode. This process is repeatedly reversible. Dark spots did not form under ac biasing. This observation suggests electric-field driven migration of impurities under the contact.

Attempts at prolonging the operational life of dc displays consist of pulsing the dc bias to reduce the duty cycle or of superimposing pulses from an integrated circuit driver on a sustained lower-voltage dc bias. Recently, *Abdalla* and *Thomas* [1.25a] succeeded in lowering the operating voltage of dc EL cells below 50 V by using an evaporated thin film of ZnS:Mn, Cu on In_2O_3-coated glass. Here also, a forming treatment at a high dc current is needed before luminescence is obtained. The brightness-voltage (B-V) characteristic is steep, rising about one order of magnitude in 3 V. Because of the steepness of the B-V characteristic, a gray scale is more easily obtained by pulse-width modulation rather than by amplitude modulation. A life in excess of 2500 h was achieved at 0.2% duty cycle and at 10 fL. At this rate, one would require a century to be able to compare it with the degradation of a cw-operated dc powder EL ($\sim 10^3$ h).

A thin layer of powdered phosphor particles can be embedded in a high-dielectric resin. This layer can then be driven by an ac voltage on the order of 50 V. An ingenious scheme for an inexpensive multicolor high-contrast ac EL panel has been proposed by *Fischer* [1.25b].

ac Thin-Film EL

Capacitive Coupling to Evaporated Phosphor. The ac thin-film EL display has reached a stage of very advanced development at *Sharp* [1.26]. A similar approach has been taken at *Rockwell* [1.27]. The Sharp display features high

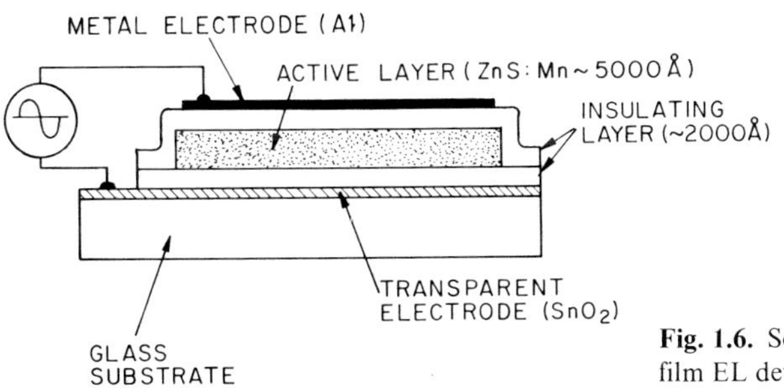

Fig. 1.6. Schematic structure of the thin-film EL device [1.26]

brightness, stability, long life, and inherent memory. *Sharp* has already demonstrated the feasibility of a flat-panel TV.

The active phosphor layer produced by evaporation is sandwiched between two insulating layers, one of which is transparent, which separate the active layer from the electrode (Fig. 1.6). All the layers in this device are made by vacuum evaporation on a glass substrate. The active layer is ZnS:Mn. An additional protective layer of Si_3N_4 (not shown in Fig. 1.6) is reactively sputtered onto the outer surface.

The emitted spectrum peaks at 580 nm. The luminescence increases rapidly above a threshold and tends to saturate at about 10^3 fL (Fig. 1.7). The brightness has been maintained absolutely flat for more than 2×10^4 h of constant operation.

This device is endowed with a memory effect attributed to polarization by charge storage at the ZnS–insulator interface. Because of this charging, if a train of pulses of the same polarity is used, there is negligible output after the first pulse. But when the polarity of a pulse is inverted, the repolarization of the cell emits a large pulse of light. Hence, maximum luminance is obtained by alternating the polarity of the driving pulses. Experiments are underway in which the two insulating layers are replaced by $BaTiO_3$ ferroelectric layers to maintain the electric field [1.28].

Note that the initial polarization can be induced by optical excitation (best with band-gap light at 3.6 eV) in the presence of a small electric field. A similar mechanism has been studied in ZnS:TbF_3 [1.29]. The photoelectrons accumulate at the ZnS–insulator interface. This optically written image is intensified by the subsequent pulses of alternating polarity. This procedure can lead to an image amplifier or to a display that can be written with a light pen [1.30a].

The brightness-voltage characteristic of the Sharp cell is endowed not only with saturation (Fig. 1.7) but also with hysteresis (Fig. 1.8). The hysteresis is believed due to the excitation and retrapping of electrons at deep centers in ZnS:Mn, the number of electrons depending on the voltage. At high enough voltage all the traps are emptied, causing the saturation of brightness. Recent observations at IBM indicate that the hysteresis is associated with the formation of localized bright regions less than 1 μm in diameter corresponding to the onset of filamentary regions of negative resistance [1.30b].

An *x*–*y*-addressed panel comprising an array of 90×120 lines has been demonstrated by researchers at *Sharp* [1.31]. Larger arrays have been built since then. The gray scale was realized by pulse-width modulation. A brightness of 60 fL and a maximum contrast of 50:1 has been achieved. No "crosstalk" was observed.

Instead of using pulse-width modulation, the brightness control, or gray scale, can be achived by exploiting the memory effect. In this case, the driving pulse alternates at a sustaining voltage V_s (Fig. 1.9). A synchronized excursion of the write pulse to V_{W_1} increases the brightness to a steady value B_1. When the pulse is dropped temporarily to the erase value V_E, the brightness drops to the low value B_O. The decrease from V_s to V_E is achieved by applying synchronous

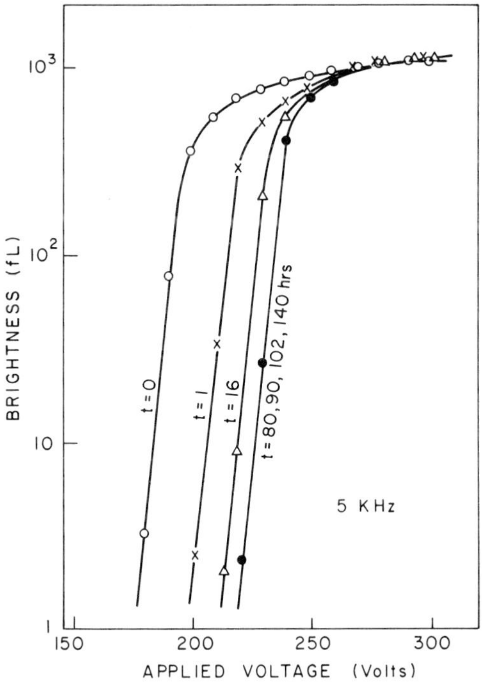

Fig. 1.7. Brightness vs applied voltage characteristics of the thin-film EL device [1.26]

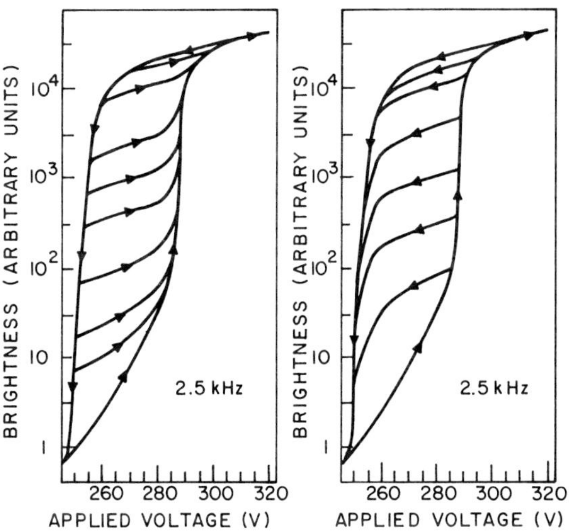

Fig. 1.8. Hysteresis behavior of the EL brightness vs voltage characteristics of the double insulating-layer structure thin-film EL device [1.26]

pulses of opposite polarity. Since any value of V_W can be used, a gray scale can be readily obtained.

The Rockwell group using a technique similar to Sharp's has realized a 100×100 matrix with a resolution of about 20 pixels/mm [1.27]. A gray scale of

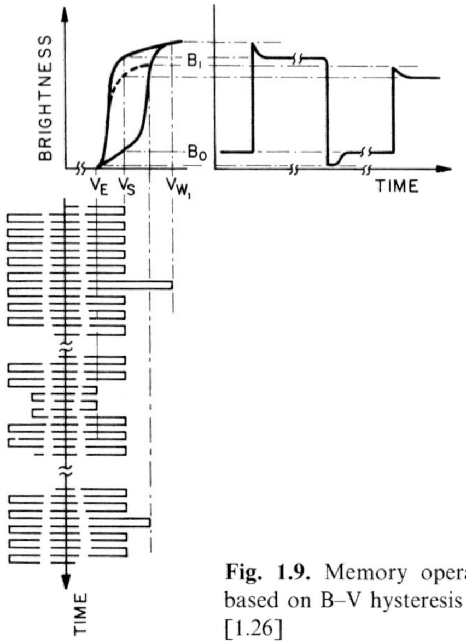

Fig. 1.9. Memory operation based on B–V hysteresis loop [1.26]

9 levels was obtained with a maximum brightness of 10 fL. No storage effect was mentioned.

Because of the inherent memory of the Sharp device, the image can be stored indefinitely in stop motion [1.32]. This was demonstrated with a 220×160 matrix having a resolution of 2 pixels/mm, a brightness of 30 fL, a contrast ratio of 15 : 1, a gray scale of 8 levels, and power consumption of less than 40 W.

Metal–Oxide–Metal Diode. A recent development in thin film EL is the radiative decay of plasmons excited by electrons that have tunnelled across a thin insulator [1.33]. The structure is a metal–oxide–metal diode having a roughened surface. The roughness is needed to couple the plasmons (collective oscillation of electrons in the metal) to the electromagnetic modes of the radiated photons. A bias of 2–4 V causes electrons to tunnel across the oxide layer with enough energy to emit visible light. The spectrum is tunable by the applied voltage which limits the maximum photon energy. Although efficiencies of only 10^{-5} have been obtained, theoretical calculations indicate that a value of 10^{-3} might be obtainable.

At this early stage, it is difficult to assess the future prospect of this new phenomenon. At room temperature and at low light levels ($\sim$2 fL) stable operation has been obtained for one month [1.34]. This device is attractive because of inherent low cost and compatibility with integrated circuit technology.

Fig. 1.10. IBM thin-film incandescent filament [1.35]

1.3.4 Incandescent Display

A miniaturized incandescent display has been demonstrated by *Hochberg* et al. at IBM [1.35]. The incandescent filament consists of a thin tungsten film shaped by photolithography into a meandering ribbon (Fig. 1.10). After the backing substrate has been etched off, the ribbon remains supported at the ends by posts so that the film is spaced from the ceramic base. The filament can be driven by an integrated circuit since only 23 mA at 2.6 V are required. The filament operates at 1200 °C producing a 0.5×0.5 [mm] spot with a luminance of 500 fL. Although its power efficiency is comparable to that of LEDs, the spectrum falls in a more visible region. The response times are: 45 ms rise and 15 ms decay. An operational life of over 10^4 h has been obtained. The main disadvantage of this device is that it must be vacuum sealed.

1.3.5 Cathodoluminescent Displays

Although cathode-ray tubes (CRT) have attained the stage of mature technology, they are still undergoing intense developmental work to increase brightness, contrast, resolution, phosphor efficiency at all colors, and local intensification of stored image, to reduce the aberations of electro-optic lenses, and even to improve the stress distribution in the glass [1.36]. As in electroluminescence, cathodoluminescence results from the impact excitation of a rare-earth atom in a host phosphor. However, in contrast to EL, the electron,

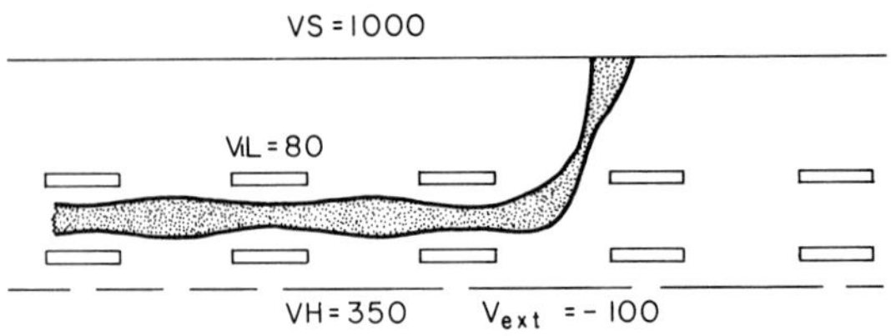

Fig. 1.11. Bar-guide periodic focusing constructed with metal grids. Beam extraction is illustrated [1.38]

being accelerated in vacuum, loses no kinetic energy before impact-exciting the material. The hot electron needs about three times the energy of the emitted light to excite the luminescence [1.37]. Hence, a maximum efficiency of 30% can be expected (and often obtained).

High voltage CL requires a long path in vacuum to accelerate and focus the electrons. In a more compact arrangement, electrons can be extracted from an ionized gas plasma at low voltage (~5 V) and drawn to a positively biased anode coated with a phosphor. The luminescent area is determined by the shape of the anode. A brightness of 60 fL has been obtained [1.38].

The greatest inovation which might lead to a cathodoluminescent flat-panel display is perhaps the use of periodic focusing of *Anderson* et al. [1.39]. This technique provides the electron beam with a long path behind the screen with convenient *x–y*-addressed extraction.

Periodic focusing prevents the electron beam from expanding due to space-charge repulsion between electrons. Alternating apertures held at high and low voltages form a series of lenses which periodically refocus the electron beam (Fig. 1.11). This stabilization is so effective that the beam can run at a low energy (~100 V) which reduces the switching voltage needed to extract the beam and direct it to the phosphor. In the example shown in Fig. 1.11, the segmented high-voltage electrode is biased negatively to repel the beam toward the screen. A similar structure transverse to the previous channel permits feeding the beam to the selected channel. This additional dimension allows *x–y*-addressing the extraction of the beam.

A flat-package cathodoluminescent display has been described by *Holton* et al. [1.40]. An array of filamentary oxide-coated tungsten cathodes assisted by focusing electrodes are located at the back of the panel. The cathodes provide a uniform flux of electrons at the rate of 1–5 mA/cm. Between the phosphor-coated anode on the front-face window and the cathodes, a switching-stack panel is inserted. The switching stack consists of a multilayer laminate of spaced aperture plates which control the addressing and intensity of the electron streams. With a potential of 18 kV at the anode, a brightness of 500 fL was obtained. The brightness was adjusted by pulse modulation.

A new type of cold cathode opens the feasibility of a large-area thin flat-panel TV. The principle of operation of the new cold cathode is based on the fact that under positive feedback conditions, the output of an electron multiplier saturates at a stable value determined by the space charge of the electron current [1.41]. Since the output of the multiplier does not depend on the quality of the initial electron source, unprecedented uniformity can be achieved. The feedback results from electrons emitted at the multiplier input either by ions from residual gas in the evacuated envelope or by photons from the excited phosphor.

These saturable electron multipliers are fabricated on the sides of glass stripes or vanes (Fig. 1.12a) that can be arrayed into a large-area display (Fig. 1.12b) [1.42]. A matrix array consists of 480 vertical multiplier vanes and 500 horizontal cathode-addressing electrodes. A clever multiplexing arrangement reduces the number of modulation circuits required. Electrons emitted from a suitably biased cathode traverse the multiplier section between two vanes and with the help of feedback produce a saturated current that can be subsequently modulated, accelerated, and deflected to a chosen color stripe. With such a system, a full-color well-resolved TV display could be made with 100 fL brightness. The special technologies needed to fabricate this display have been demonstrated on small-scale models [1.43].

1.3.6 Liquid-Crystal Displays – LCD

LCDs are now a widely used commercial product. Liquid crystals comprise an elastic continuum of elongated molecules which can reflect, scatter, or rotate the plane of polarization of light. The directions of alignment of these rodlike molecules are normally identically oriented either parallel or perpendicular to a window surface. The image is formed by introducing a local electric field which perturbs the orientation of the molecules between the electrodes.

In some nematic crystals of finite conductivity, a bias of about 20 V causes a turbulent flow which disturbs the local order and causes a dynamic scattering of light, whereas the unperturbed surrounding area remains totally transmitting.

The twisted structure of nematic crystals can rotate the plane of polarization of the light. Hence total extinction is possible with polarizing filters. A bias of about 2 V can rotate the orientation of the molecules perpendicular to the window, thus the polarization is lost and the medium becomes transparent. The surface properties of the window are of great importance since they control the spontaneous ordering of the crystals.

Liquid crystals can be used in bright sunlight without loss of contrast. The power consumption is very low since the display draws only a few $\mu A/cm^2$. The main drawback is that the viewing angle is somewhat limited and the contrast tends to be low (5:1).

An improvement in contrast has been achieved in LCD display by using a fluorescent backing plate larger than the display; the fluorescent plate converts the incident broad-spectrum ambient light into a narrower spectral band [1.44]. The fluorescence, which is trapped by total internal reflection in the back panel, can escape through beveled notches behind the display area. The liquid crystal then modulates the transmission of the fluorescence at the selected areas, making them appear self-luminous.

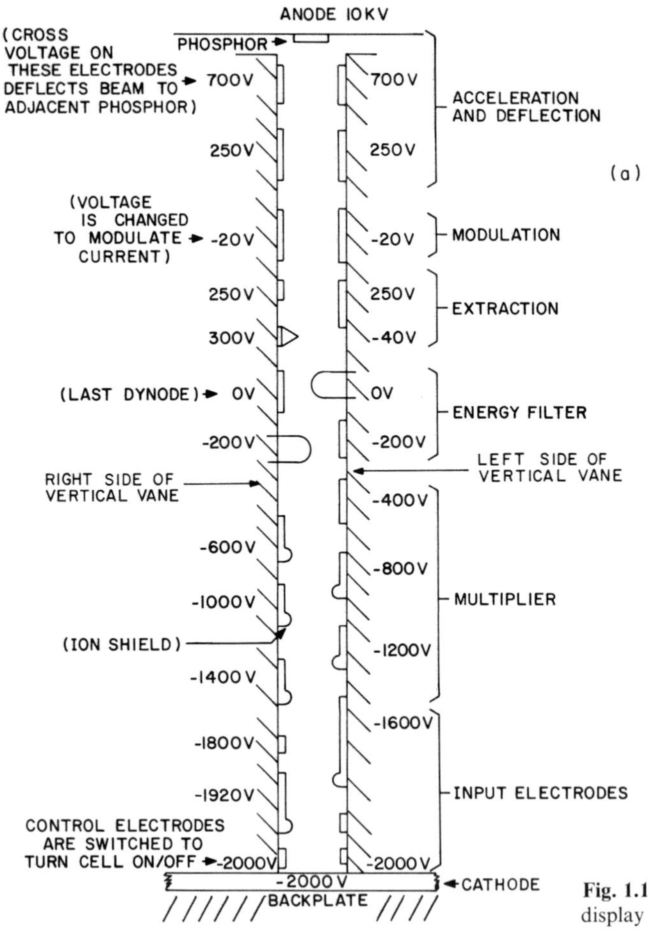

Fig. 1.12. (**a**) Cross section of display element [1.41]

The turn-on time of LCD may be as short as a few ms. When the field is turned off, the orientation of the surrounding medium provides a restoring force that realigns the molecules in a time of the order of 100 ms. Turn-off can be speeded up by a transverse field [1.45].

The driving voltage across the LCD can be controlled by a photoconductor in series with the cell. By projecting on the photoconductor an image, the resulting potential distribution over the LCD determines the displayed picture. Such a device can operate as a light amplifier since a very bright light can be used on the viewing side. A high-resolution image can be projected on a large screen in multiple colors with a brightness of 40 fL [1.46]. The device can also be used as an image converter if the photoconductor is sensitive to x-rays, uv or ir.

A more detailed discussion of LCD will be found in Chap. 4.

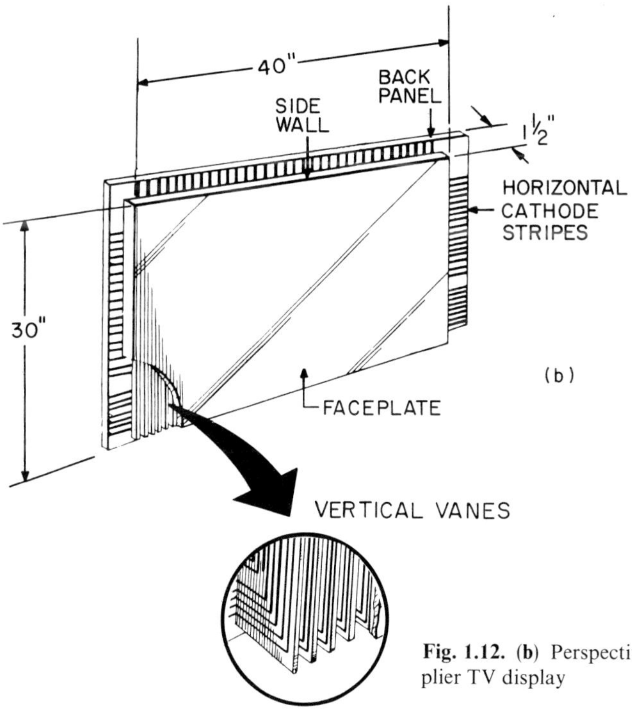

Fig. 1.12. (**b**) Perspective view of feedback multiplier TV display

1.3.7 Electrochromic Displays – ECD

An electric field can induce a color change by means of charge transfer. This charge transfer can occur at the boundary of a liquid cell. For example, electrons are transferred to viologen dye molecules at the cathode, whereupon the dye molecule turns dark blue and becomes insoluble. In time, more positively charged dyes are attracted to the cathode where the dark layer builds up. Electrochromism occurs also in the volume of a solid film such as WO_3. This will be discussed in greater detail in Chap. 5.

A contrast of 3:1 is obtained with a charge transfer of several mC/cm^2. An image can be produced in 0.1 s with tens of mA/cm^2; long storage times are obtained. The display can be erased by reversing the polarity of the bias. However, there is a conflict between contrast and device longevity. A high contrast requires the motion of a large charge, which tends to shorten the life. ECDs are best suited for infrequent operation, which also minimizes power consumption.

Depending on the material, electrons can be injected by any ionizing process: an electron beam causes cathodochromism (e.g., dark-trace kinescope); uv light induces photochromism; heat produces thermochromism.

A cathodochromic display has been made using sodalite glass which has a white area that can be illuminated to a brightness of 300 fL [1.47a]. The information is deposited by a low power electron beam producing a contrast of 6:1. The information is erased thermally (heating to 300 °C). Spot erasure with a more powerful electron beam requires about 1 ms. With a 10 W electron gun, the information can be erased at the rate of 0.1 s/cm^2. Therefore, only small area displays are practical. However, because of the inherent high resolution, a magnifying projection system is possible. Another advantage is the long-term storage of the image. The main disadvantage of the cathodochromic display is the slow erase which makes real-time TV impractical.

1.3.8 Electrophoretic Displays–EPID

The operation of this display is based on electrophoresis: the motion of charged particles in an electric field. The electrophoretic display consists of a suspension of charged pigmented particles in a dark liquid. This suspension forms a thin layer between two electrodes. A dc voltage across the layer draws the charged particles to one transparent electrode which then may look white while the surrounding area is dark. Reversing the polarity of the bias draws the white particles to the opposite electrode making the first electrode appear black. A variety of two-color combinations is possible. The image has a beautiful appearance and can be viewed from all angles.

A contrast ratio of 40:1 has been obtained at 75 V. The write and erase times are about 20 ms. The pigmented particle will cling to the electrode and provide image storage.

There is a problem with life due to the slow aglomeration of colloidal particles and their eventual migration outside the electrode area.

A more comprehensive discussion of EPID devices will be found in Chap. 6.

1.3.9 Electroplating Cell

Metal atoms can be deposited by plating onto a transparent electrode to generate a reflecting or an opaque surface depending on the smoothness of the deposit. A deposition of 10^{17} ions will control 99.9% of the ambient flux. Hence the current density required is 100 mA/cm^2 when the image is generated in 0.1 s. The plated image is stored indefinitely when the bias is removed. The image can be erased by electroetching when the bias polarity is reversed. The main drawback may be the danger of gas evolution.

Silver halide complexes have been electrodeposited onto transparent indium–tin oxide electrodes in less than 10 ms with a pulse of several volts [1.48]. The contrast of the cell is increased by adding an opacifier such as TiO_2 to the solution. The opacifier tends to slow down the silver deposition, but the speed can be recovered by the addition of fine particles of a superionic conductor. Contrasts of 5:1 have been demonstrated with current densities of 0.5 mA/cm^2 with 0.1 s write times [1.47b].

1.3.10 Light Valves

This category includes all the devices in which bodies much larger than the wavelength of light are physically moved. The size of the moved body can vary over a tremendous range. On one hand, there are very large panels using solenoid- or motor-operated flaps as seen in railroad stations, and on the other hand there are very small displays that need magnification by projecting the image on a screen.

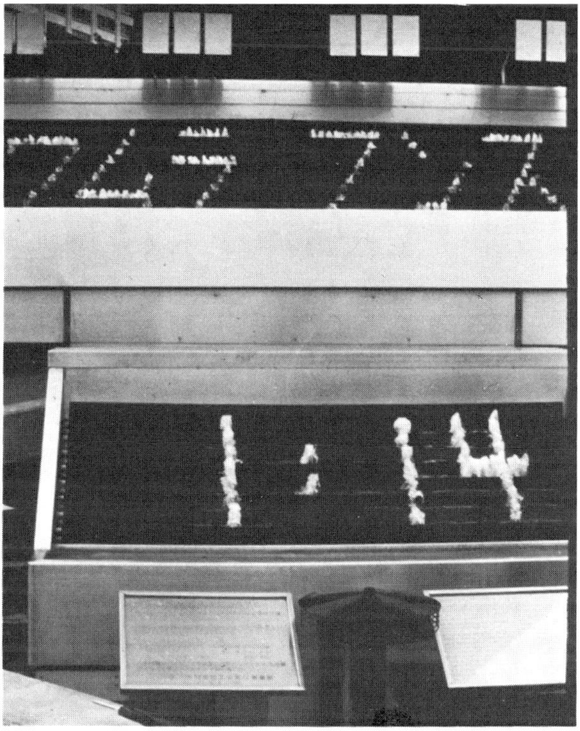

Fig. 1.13. Water-jet multilingual alphanumeric display in Tokyo

Water Jets

One of the most aesthetically appealing large displays is a fountain that can be seen in Tokyo (Fig. 1.13). This display panel consists of rows of water jets. Each individual jet is controlled by solenoid-operated valves providing fully on, half-height, and off conditions. The array is tilted so that each row is visible, the most distant row appearing on top of the panel. The advertisement is changed periodically in less than 1 s. The multiword and multiline message alternates between Japanese and English.

Rotating-Ball Displays

Intermediate-size displays can be made utilizing rotating magnetic or electrostatic balls that are painted on one hemisphere.

The magnetic balls described by *Lee* [1.49] are made of ferrite. The spherical particles have reflective and black areas. These balls are rotated by an external magnetic field produced by x–y-addressed pairs of conductors. The amount of rotation provides the gray scale. Once the field is turned off, the balls stay in the last orientation, providing indefinite storage. A switching time of 30 ms is claimed. The power dissipation is expected to be low.

The electrostatically polarized balls are electrets in a viscous fluid driven by x–y-addressed external fields. When the electric field is turned off, the display fades away very gradually as the particles drift into random orientations. However, fast erase is possible by the application of a transverse restoring field from auxiliary electrodes [1.50].

In the "Gyricon" rotating-ball display [1.51], small white glass balls coated with a black hemispherical layer are dispersed in an elastometer material. After heat curing, the elastometer is immersed in a liquid such as silicon oil or an organic solvent. This causes the elastometer to swell, leaving a space around each ball. This space is filled with the liquid, which acts as a dielectric lubricant. Because of different contact potentials over the painted and unpainted hemispheres with respect to the surrounding elastometer, a polarization develops that will react with an externally applied electric field and cause the ball to rotate. Black or white hemispheres can be selected by reversing the polarity of the applied voltage. The viscosity of the lubricant dampens the oscillations of the ball upon voltage reversal. Switching times of 0.1 s have been achieved with spheres less than 100 μm in diameter. A contrast ratio of 3:1 has been obtained. After switching, the ball remains in a set orientation for several days without applied bias. Power is consumed only during switching and this is due to the capacitive displacement current. With the Gyricon technique, one should be able to make very large-area, thin-sheet panels at low cost.

Moving Flaps

An opaque electret can swing between two electrodes in response to an applied voltage thus intercepting or transmitting the incident light [1.52]. The electret is a corona-charged polypropylene film. The electret film clings to one or the other electrode forming a bistable light switch with inherent memory of the last applied voltage polarity. Hence, there is no power consumed to sustain the display. However, in the early version of a 1 cm^2 valve, a switching voltage of 500 V was needed.

A very practical device is under development at RCA [1.53]. It consists of two pieces of a piezoelectric polymeric film assembled in opposition between two pairs of electrodes. The polymer is polyvinylidene fluoride (PVF_2). As shown in Fig. 1.14, this assembly forms a multilayer bimorph. A voltage applied across the bimorph causes one film to expand and the other to contract. The

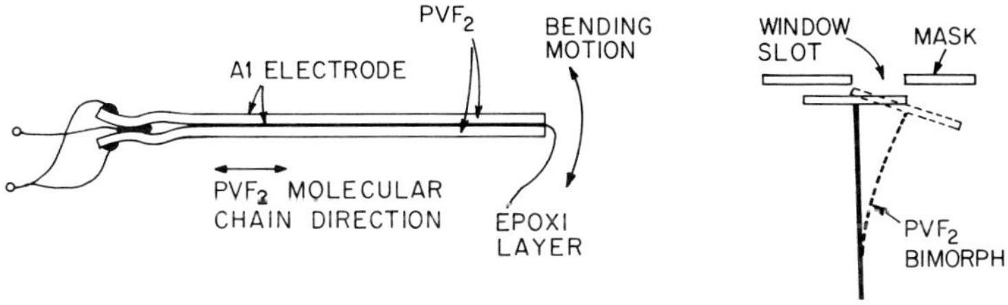

Fig. 1.14. Basic PVF_2 bimorph cantilever-device structure and display element

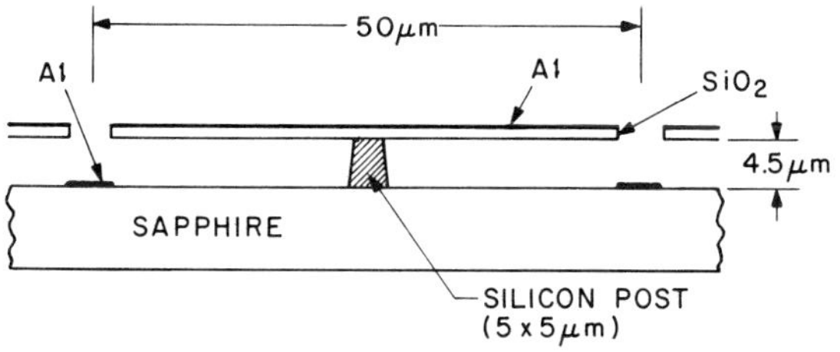

Fig. 1.15. Cross-sectional view of individual light valve in Westinghouse mirror-matrix tube [1.54]

net result is a flexing of the bimorph. With one end clamped, large swings of the cantilevered bimorph are possible with less than 50 V applied. A response time on the order of 25 ms is obtained. The power consumption for charging and discharging 1 cm^2 of bimorph is about 10^{-5} W for one switching per s. The bimorph can move a larger-area flag attached to the free end. This flag can be used in either the reflecting, scattering, or transmitting modes. A multicolor display is feasible.

The mirror-matrix tube of *Thomas* et al. [1.54] is a high-resolution projection display, wherein light is reflected from a dense array of individually controlled tiny mirrors. The mirrors consist of an aluminized SiO_2 membrane 0.3 μm thick supported centrally on a small Si post above a sapphire face-plate (Fig. 1.15). This structure, fabricated by standard silicon-on-sapphire technology, forms the target of an otherwise conventional Vidicon tube. The flat stress-free oxide membrane can be deflected electrostatically when addressed with the electron beam. In conjunction with Schlieren optics, the deformations of the reflective matrix are translated into a high resolution image projected onto a 1 m^2 screen with highlight brightness of 35 fL and 15:1 contrast ratio.

More recent work by *Peterson* [1.55] has shown that the SiO_2 movable membrane can be made on single-crystal Si. Thus, the light modulator is compatible with integrated circuits that could be made on the same chip to drive the mirrors. It is expected that 20 V would suffice to cause maximum deflection. The membranes have been flexed over 10^{10} times without degradation.

Suspended Particles

Waring [1.56] has proposed that the Christiansen effect could be used to generate a display of addressable and adjustable colors. The Christiansen effect

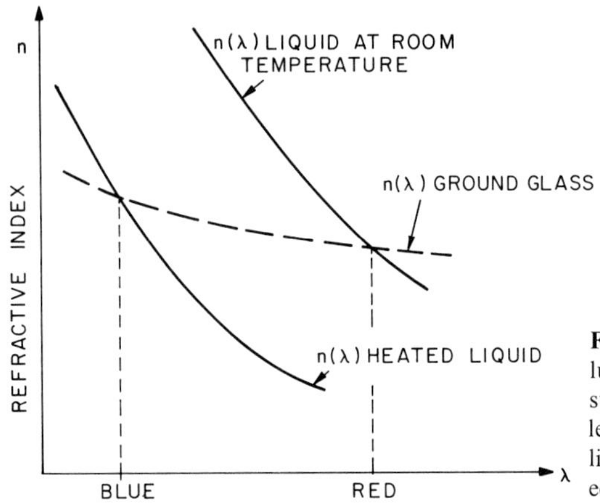

Fig. 1.16. Optical dispersion curves illustrating the Christiansen effect. The suspension is transparent at the wavelength where the refractive indices of liquid matrix and of suspended glass are equal [1.56]

is based on the different temperature dependences of the refractive indices of two components: a glass powder and its surrounding matrix. The two refractive indices match at one wavelength, λ_c, but scatter all other colors (Fig. 1.6). Hence, the material transmits only the color determined by λ_c. If λ_c is in the ir or uv, the cell appears opaque. By heating the material in the cell (glass powder in a liquid suspension) to 60 °C, λ_c could be changed from red to blue. The modulation is inherently slow, but interesting aesthetic effects could be achieved.

Latex is a water suspension of uniform polystyrene spheres. Their preparation is described in [1.57, 58]. When the ion concentration in the solution is properly adjusted, the polystyrene spheres align themselves in an orderly crystal-like array [1.59] with interesting optical properties [1.60]. This is due to bound negative charges at the surface of the spherical particles causing Coulomb repulsion between adjacent particles. Since the particles are spherically symmetric and of identical size, their position acquires periodicity. Bragg reflection of incident light can be observed as an irridescence and in some regions as an angle-dependent color (Fig. 1.17). The orientation and spacing of the latex particle array can be controlled by an applied electric field which transports the particles by electrophoresis. This can be used to produce a display in contrasting colors.

1.3.11 Electro-Optic Modulators

Imaging light valves modulate the light coming from an independent light source that can be of very high intensity. This allows the projection of TV pictures onto a large screen. The modulation can be derived from phenomena causing reflection, diffraction, scattering, or change in the polarization state of the light. Light absorption is not a suitable phenomenon to control intense light

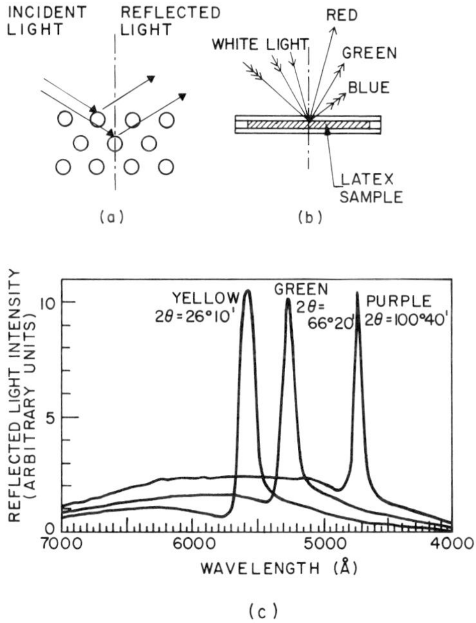

Fig. 1.17. (**a**) Bragg-reflection condition in a microparticle dispersion. (**b**) Macroscopic condition to obtain monochromatic light. (**c**) Bragg-reflection spectra of latex [1.57]

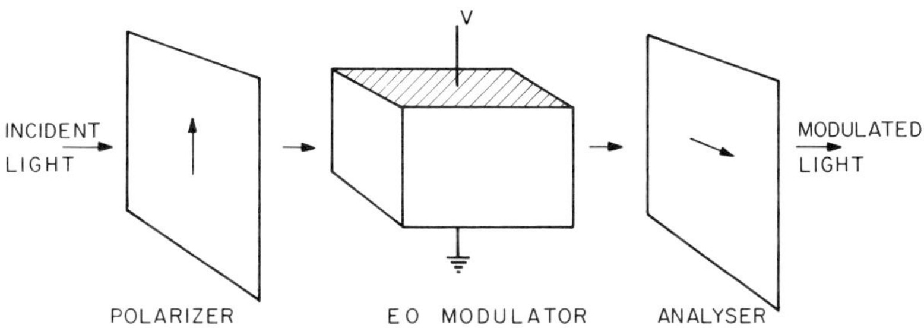

Fig. 1.18. Electro-optic modulation of light beam

because of the serious heat-dissipation problem. When high-intensity light is to be controlled at a fast rate, electro-optic polarization modulation is the most suitable mechanism.

For projection-type TV display using a laser beam [1.61] a fast light modulator is needed that will control the amplitude of the cw laser beam in real-time in response to the video signal while the beam is deflected in a raster.

The amplitude modulation of the polarized laser beam is achieved with an electro-optic modulator in which the electric-field-induced birefringence rotates the plane of polarization (Fig. 1.18). The field is applied transversely to the direction of propagation of the light. The analyzer transmits an amplitude proportional to the cosine of the angle of polarization with respect to the

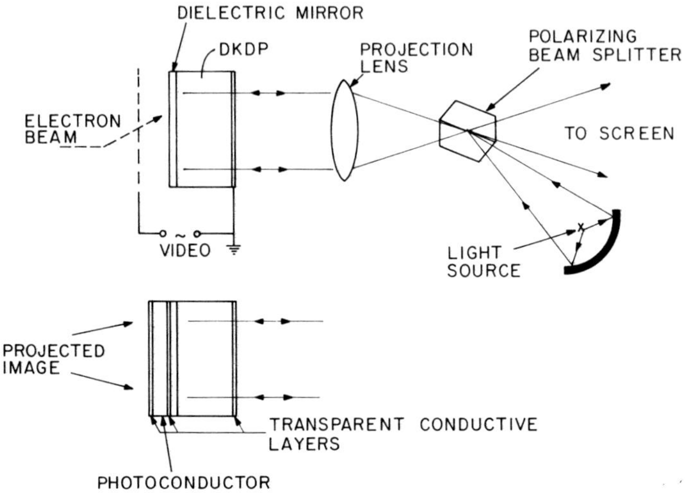

Fig. 1.19. Schematic diagram of Pockels-effect imaging device operating in the reflection mode. *Upper figure*: electron-beam-addressed Titus. *Lower figure*: Optically addressed Phototitus (a dc voltage is applied across the photoconductor during writing; the photoconductor is short-circuited during reading and erasing) [1.63]

orientation of the analyzer. The beam can be deflected by rotating mirrors or by acousto-optic (AO) deflectors [1.62]. In the AO deflector a sound wave produces a periodic change in refractive index. The incident light beam suffers a Bragg reflection at an angle that depends on the driving frequency.

The Pockels effect is the induced birefringence in some materials when the electric field is applied parallel to the direction of propagation of the light. This is a convenient relationship which allows imaging by depositing the right amount of electrical charge at the right place on the image-producing surface. The charge can be produced by a voltage applied to electrodes, by an electron beam, or optically by controlling the local charge leakage with a photoconductor.

Some birefringent materials are endowed with photoconductive properties – then a separate photoconductor is not required. The image can be optically addressed using uv light to control lower energy light in the visible spectrum to which the photoconductor does not respond.

The Pockels effect has been successfully applied in the "Titus" device by *Marie* and co-workers [1.62]. In the Titus device, schematically illustrated in Fig. 1.19, an electron beam scans a KD_2PO_4 (DKDP) crystalline plate ($3 \times 4\,cm^2$) through a video-modulating grid and through a dielectric mirror. The DKDP plate is thermoelectrically cooled just above the Curie temperature ($-50\,°C$). Under such thermal conditions, DKDP is ferroelectric and, therefore, can store the polarization for a long time. The Titus device operates in the reflective mode. A resolution of 750 elements/line is possible (50 µm elements).

A beam current density of 1.5 A/cm^2 is adequate, providing a contrast ratio of 40:1.

A similar device using a selenium photoconductor instead of the electron beam is the Phototitus [1.63]. The photoconductor is decoupled from the light to be reflected by an intermediate dielectric mirror. This device (that I have seen demonstrated) has a fantastic resolution of 1500 elements/line with a contrast ratio of 70:1. By optimizing the design, especially reducing the thickness of the photoconductor, a resolution of more than 5000 elements/line is possible. Phototitus can be operated very rapidly (1000 frames/s), or the picture can be stored for more than 1 h in the dark.

Phototitus is adaptable to image processing; for example, images from different pictures of the same subject can be superposed (integration) to improve the signal-to-noise ratio. Pictures can be subtracted by inverting the bias polarity during the second picture. Thus, differentiation can be achieved to suppress the unchanged portions of similar pictures. And contour enhancement is also possible. Phototitus is directly adaptable to image conversion, for example converting an x-ray image into a visible picture in real-time.

Lead lanthanum zirconium titanate (PLZT) ceramics are highly transparent and can be used for display purposes. Like the LCD, the PLZT ceramics have the property of scattering light or of rotating its plane of polarization. The latter mode is due to field-induced birefringence (Pockels effect). A half-wave phase retardation at 633 nm can be obtained with 100 V across a 75 μm plate of PLZT ceramic [1.64].

The main advantage of the PLZT ceramics is their storage property because this material is ferroelectric. Once the material is polarized, the polarization remains. Since PLZT ceramics are piezoelectric, the biasing field can be strain induced.

In one mode of operation, the writing field is applied across the ceramic plate at the various image points in the desired intensity. The image is erased by a field along the plane of the PLZT. The erase field can be induced by electrodes along two opposed edges or by a grid of interleaved electrodes on one surface. (The image can also be erased thermally.)

The image can be written optically using a photoconducting film in series with the PLZT ceramic, as shown at the bottom of Fig. 1.19. Once the image has been stored, it can be viewed by transmission or it can be projected on a screen. A resolution of 30 lines/mm could be obtained [1.64]. When the image was projected on a 30 cm square screen 150 cm away, a luminance of 150 fL was achieved, the contrast ratio depends on the aperture used. By incorporating an array of very small mirrors, the display can be viewed in the reflection mode.

The power requirement is very low. A seven-bar numeric display (3×1 mm^2 bars) with a backing mirror requires only ~ 13 μW of driving power at 15 V for a 0.1 s change every minute.

1.4 Applications

The major consumer market in display devices is found in personal electronic systems that can be carried on the wrist or in the pocket – mostly watches and calculators. Portable mini-TV receivers that will fit in the pocket or the purse will probably appear within the next five years.

The next largest user of display devices is instrumentation: digital displays for cash registers, weighing scales, ranging equipment, cockpits of aircraft, etc. These applications require larger size alphanumeric digits.

For business use where a larger amount of data must be displayed simultaneously, flat-panel displays are desirable. These are found in brokerage offices, for seat reservations in travel and theatre agencies, and on some secretarial desks for letter editing. The CRT display is dominant in these applications.

The very large display panels are found in railroad and bus stations and at air terminals. When a large amount of information is to be presented to many viewers, image projection is more suitable. It can be used to project maps, data from computers, graphs, advertisements, instructional material. If high speed is not required, the image can be transmitted over telephone lines and projected as still pictures to one or more groups of people.

References

1.1 C.R.Carlson: Photogr. Sci. Eng. **22**, 69 (1978)
R.W.Cohen: Photogr. Sci. Eng. **22**, 56 (1978)
1.2 P.K.Weimer: Proc. IRE **50**, 1462 (1962)
1.3 S.R.Hofstein, F.P.Heiman: Proc. IEEE **51**, 1190 (1963)
1.4 T.P.Brody, F.C.Luo, Z.P.Szepesi, D.H.Davies: IEEE Trans. ED-**22**, 739 (1975)
1.5 C.H.Graham (ed.): *Vision and Visual Perception* (Wiley, New York 1965) p. 154
1.6 J.I.Pankove: J. Lumin. **7**, 114 (1973)
1.7 E.E.Loebner: Proc. IEEE **61**, 837 (1973)
1.8 R.E.Bedford, G.Wyszecki: J. Opt. Soc. Am. **47**, 564 (1957)
1.9 G.Wald, D.R.Griffin: J. Opt. Soc. Am. **37**, 321 (1947)
1.10 W.L.Martin: SID Intern. Symp. Dig. **7**, 46 (1976)
1.11 R.Cola, J.Gaur, G.Holz, J.Ogle, J.Siegel, A.Somlyody: In *Advances in Imaging and Display*, Vol. 3, ed. by B.Kazan (Academic Press, New York 1977) p. 83
1.12 J.Smith: IEEE Trans. ED-**22**, 642 (1975)
1.13 D.C.Krupka, Y.S.Chen, H.Fukui: Proc. IEEE **61**, 1025 (1973)
1.14 T.Fujino, T.Kubo, R.Itatani: Presented at Phys. Soc. Japan Meeting, Sendai (April 1978)
1.15 I.Ohishi, T.Kojima, H.Ikeda, R.Toyonaga, H.Murakami, J.Koike, T.Tajima: IEEE Trans. ED-**22**, 650 (1975)
1.16 M.Fukushima, S.Murayama, T.Kaji, S.Mokoshiba: IEEE Trans. ED-**22**, 657 (1975)
1.17 G.Destriau: J. Chim. Phys. **33**, 587 (1936); **34**, 117 (1937)
1.18 W.Lehman: J. Electrochem. Soc. **113**, 40 (1966)
1.19 A.Vecht: J. Lumin. **7**, 213 (1973)
1.20 D.Kahng: Appl. Phys. Lett. **13**, 210 (1968)

1.21 J.I.Pankove, M.A.Lampert, J.J.Hanak, J.E.Berkeyheiser: J. Lumin. **17**, 349 (1977)
1.22 J.S.Prener, F.E.Williams: Phys. Rev. **101**, 1427 (1956)
H.Reiss, C.S.Fuller, F.J.Morin: Bell Syst. Tech. J. **35**, 535 (1956)
J.J.Hopfield, D.G.Thomas, M.Gershenzon: Phys. Rev. Lett **10**, 162 (1963)
F.Williams: J. Lumin. **7**, 35 (1976)
1.23 M.Highton, A.Vecht, J. Mayo: SID Intern. Symp. Dig. **9**, 136 (1978)
1.24 J.I.Pankove: Unpublished observations
1.25a M.I.Abdalla, J.A.Thomas: SID Intern. Symp. Dig. **9**, 130 (1978)
1.25b A.G.Fischer: Appl. Phys. **9**, 277 (1976)
1.26 T.Inoguchi, S.Mito: In *Electroluminescence*, ed. by J.I.Pankove, Topics in Applied Physics, Vol. 17 (Springer Berlin, Heidelberg, New York 1977) p. 197
1.27 R.D.Ketchpel, I.S.Santha, L.G.Hale, T.C.Lim: SID Intern. Symp. Dig **9**, 138 (1978)
1.28 N.Yamamura, K.Funabiki, M.Koide: Collected Abstracts 39th Meeting of the Jpn. Soc. Appl. Phys. Osaka, No. 3a-R-10 (1978) (in Japanese) p. 332
1.29 J.I.Pankove, M.A.Lampert, J.J.Hanak, J.E.Berkeyheiser: RCA Rev. **38**, 443 (1977)
1.30a C.Suzuki, M.Ise, K.Inazaki, K.Machino: SID Intern. Symp. D. **9**, 134 (1978)
1.30b V.Marrello, W.Ruhle, A.Onton: Appl. Phys. Lett. **31**, 452 (1977)
1.31 S.Mito, C.Suzuki, Y.Kanatani, M.Ise: SID Intern. Symp. Dig. **5**, 86 (1974)
1.32 N.Kako, Y.Yamane, C.Suzuki: SID Intern. Symp. Dig. **9**, 132 (1978)
1.33 R.K.Jain, S.Wagner, D.H.Olson: Appl. Phys. Lett. **32**, 62 (1978)
1.34 S.L.McCarthy, J.Lambe: Tech. Dig. Intern. Electron Devices Meeting (IEEE, New York 1978) p. 261
1.35 F.Hochberg, H.K.Seitz, A.V.Brown: IEEE Trans. ED-**20**, 1002 (1973)
1.36 R.E.Enstrom, R.S.Stepelman, J.R.Appert: RCA Rev. **39**, 665 (1978)
1.37 R.C.Alig, S.Bloom: Phys. Rev. Lett. **35**, 1522 (1975)
1.38 B.Kazan, W.B.Pennebaker: Proc. IEEE **59**, 1130 (1971)
1.39 C.Anderson, A.Pelios, T.Credelle, W.Siekanowicz, F.Vaccaro: Tech. Dig. Intern. Electron Devices Meeting (IEEE, New York 1978) p. 264
1.40 W.C.Holton, W.C.Scott, W.G.Manns, D.F.Wierauch, M.R.Namordi, F.H.Doerbeck, R.Pitts, K.H.Surtani, R.W.Gooch, J.E.Gunther: Tech. Dig. Intern. Electron Devices Meeting (IEEE, New York 1977) p. 78
1.41 C.A.Catanese, J.G.Endriz: J. Appl. Phys. **50**, 731 (1979)
1.42 J.G.Endriz, S.A.Keneman, C.A.Catanese, L.B.Johnston: IEEE Trans. ED-**26**, 1324 (1979)
1.43 L.S.Cosentino, V.Christiano, J.G.Endriz, J.Dresner, G.F.Stockdale, J.L.Cooper, J.N.Hewitt, J.J.Harrison, Jr.: SID Intern. Symp. Dig. **9**, 126 (1978)
1.44 G.Bauer, W.Greubel: SID Intern. Symp. Dig. **8**, 99 (1977)
1.45 D.J.Channin, D.E.Carlson: Appl. Phys. Lett. **28**, 300 (1976)
1.46 J.Grinberg, W.P.Bleha, A.D.Jacobson, A.M.Lackner, G.D.Myer, L.J.Miller, J.D.Margerum, L.M.Fraas, D.D.Boswell: IEEE Trans. ED-**22**, 775 (1975)
1.47a L.T.Todd, C.J.Starkey: Tech. Dig. Intern. Electron Devices Meeting (IEEE, New York 1977) p. 80A
1.47b J.Duchene, R.Meyer, G.Delapierre: Joint IEEE-SID-AGED Conf. Record **78 CH 1323 SED**, 34 (1978)
1.48 I.Camlibel, S.Singh, H.J.Stocker, L.G.VanUitert, G.J.Zydzik: Appl. Phys. Lett. **33**, 793 (1978)
1.49 L.L.Lee: IEEE Trans. ED-**22**, 758 (1975)
1.50 J.I.Pankove: RCA Tech. Note Nr. 535 (March 1962)
1.51 N.K.Sheridon, M.A.Berkovitz: SID Intern. Symp. Dig. **8**, 114 (1977)
1.52 J.L.Bruneel, F.Micheron: Appl. Phys. Lett. **30**, 382 (1977)
1.53 M.Toda, S.Osaka: SID Intern. Symp. Dig. **9**, 18 (1978)
1.54 R.N.Thomas, J.Guldberg, H.C.Nathanson, P.R.Malmberg: IEEE Trans. ED-**22**, 765 (1975)
1.55 K.E.Petersen: Appl. Phys. Lett. **31**, 521 (1977)
1.56 R.K.Waring, Jr.: Proc. IEEE **65**, 1074 (1977)
1.57 H.Fujita, K.Ametani, M.Inoue: RCA Rev. **36**, 108 (1975)
1.58 K.Ametani, H.Fujita: Jpn. J. Appl. Phys. **17**, 17 (1978)

1.59 P.A.Hiltner, I.M.Krieger: J. Phys. Chem. **73**, 2386 (1969)
1.60 H.Fujita, K.Ametani: Jpn. J. Appl. Phys. **16**, 1907 (1977)
1.61 M.Yamamoto, T.Taneda: In *Advances in Image Pickup and Display*, Vol. 1, ed. by B.Kazan (Academic Press, New York 1975) p. 1
1.62 I.Gorog, J.D.Knox, P.V.Goedertier: RCA Rev. **33**, 623 (1972)
1.64 G.Marie, J.Donjon, J.P.Hazan: In *Advances in Image Pickup and Display*, Vol. 1, ed. by B.Kazan (Academic Press, New York 1974) p. 241
1.64 J.R.Maldonado, D.B.Fraser, A.H.Meitzler: In *Advances in Image Pickup and Display*, Vol. 2, ed. by B.Kazan (Academic Press, New York 1975) p. 65

2. Light-Emitting Diodes – LEDs

C. J. Nuese and J. I. Pankove

With 41 Figures

The light-emitting diode (LED) is a solid-state light source with several attractive properties for display applications. First, it is a low-voltage device, and is compatible with battery operation and conventional silicon logic circuits. Compared to the cathode-ray tube, the LED is about one order of magnitude less efficient, but is many orders of magnitude more compact. In fact, its small size provides thousands of individual elements from a single wafer, thereby enhancing its cost effectiveness; the large-scale production of visible LEDs for the watch and calculator industries attests to their commercial viability. LEDs also can be directly modulated simply by controlling their input current, with switching speeds often less than 50 ns. Finally, a variety of semiconductor materials have been developed that provide emission from red to blue for multiple-function displays. LEDs can be mass-produced as single picture elements of various shapes (dots or bars), as monolithic arrays of dots that are individually controlled or, most commonly, as monolithic numeric or alphanumeric displays.

Historically, injection electroluminescence was discovered by *Losev* in 1923 in SiC containing accidentally produced p–n junctions [2.1]. A fuller understanding of injection luminescence was developed by *Haynes* using p–n junctions in Ge [2.2]. However, the practical potential of injection luminescence became apparent only with the demonstration of efficient luminescence from direct-gap semiconductors such as GaAs [2.3]. The spectral range was soon extended into the visible by *Holonyak* and *Bevaqua*, who chose the wider-band-gap $GaAs_{1-x}P_x$ alloy, and produced red-light-emitting diodes [2.4]. Prior to about 1964, a direct-band-gap semiconductor was assumed to be a prerequisite for efficient LEDs. However, the indirect-band-gap material GaP also can provide reasonably efficient electroluminescence, as first illustrated in 1964 by *Grimmeiss* and *Scholz* [2.5]. Efficient LEDs emitting orange, yellow, and eventually, green light awaited the discovery and understanding of new luminescence centers in GaP [2.6]. Blue-emitting LEDs are the most recent arrivals on the diode display scene, and are fabricated by forming Schottky barriers to large-band-gap n-type or semi-insulating GaN [2.7], ZnS [2.8], or ZnSe [2.9].

2.1 Carrier Excitation

Luminescence in a semiconductor is the result of a radiative transition between atomic energy states. The radiative recombination rate is determined by the product of the electron density in the upper energy states, the density of empty lower energy states, and the transition probability between the two sets of states. Since the separation of neighboring atoms in a semiconductor is small (~ 5 Å), the wave functions of their orbital electrons interact, thereby providing *bands* of allowed states, rather than discrete energy levels (as in an isolated atom). The luminescence that arises from interband transitions is, therefore, not monochromatic, but is broadened over a wavelength interval typically several hundred angstroms wide.

Electrical conductivity is imparted to a semiconductor by doping to provide a majority carrier, either donors to produce mobile electrons in the conduction band or acceptors to provide holes in the valence band. In general, although the excess carriers consist of equal numbers of electrons and holes (needed for local charge neutrality), for luminescence purposes, only the minority carriers are important.

In a semiconductor, luminescence occurs via the recombination of electrons or holes generated in concentrations greater than those at thermal equilibrium. The recombination process is a strong characteristic of the physical and electrical properties of the material, but is nearly independent of the several techniques that can be used to generate the excess carriers. Producing minority carriers efficiently always requires an expenditure of energy, and often requires experimental ingenuity. Three distinctly different techniques are commonly used to create excess carriers, each of which is described briefly in the subsections below.

2.1.1 Photoluminescence

Photoluminescence is the process in which the minority carriers are generated by the absorption of incident light whose energy is larger than the semiconductor energy band gap. This technique is primarily used as a research tool to conveniently probe the fundamental radiative characteristics of a *material.* Since the radiative recombination is appreciably altered by the type and concentration of impurities, photoluminescence can be used to monitor the doping concentration and impurity ionization energy, as well as the material band-gap energy.

2.1.2 Cathodoluminescence

In cathodoluminescence, the excess carriers are generated by the ionization of electron–hole pairs by high-energy electrons. The number of electron–hole pairs generated per incident electron is very large (typically 10^3 for a 10 keV electron). Cathodoluminescence often is induced in a scanning electron microscope (SEM), where the radiative characteristics of a device or material can be *microscopically* examined. This has been found to be valuable for determining the presence of localized nonradiative defect centers in an otherwise radiative material.

A projection TV system has been made using a semiconductor laser excited by an electron beam [2.10].

2.1.3 Injection Electroluminescence

Electroluminescence is the process by which the minority carriers are generated as a result of an applied electric field. The most useful technique involves the application of forward bias to a p–n junction; in this case, minority carriers are "injected" across the junction, as in the base-emitter of a bipolar transistor.

In a semiconductor, the efficiency of converting the input electrical energy (for electroluminescence and cathodoluminescence) or optical energy (for photoluminescence) to the desired radiative energy depends on three factors: (i) excitation of excess minority carriers, (ii) radiative recombination of these carriers, and (iii) extraction of the generated light from within the crystal. The power conversion efficiency for this process can be expressed in terms of these factors by an expression of the form

$$\eta_{\text{power}} \equiv \eta_g \eta_{\text{int}} \eta_x , \tag{2.1}$$

where η_g is the power efficiency for generating the excess minority carriers, η_{int} is the internal quantum efficiency for minority-carrier radiative recombination, and η_x is the efficiency for extracting a photon to the external ambient. In the literature, the product of the terms η_{int} and η_x is usually called the external quantum efficiency, η_{ext}, so that

$$\eta_{\text{ext}} \equiv \eta_{\text{int}} \eta_x \tag{2.2}$$

and

$$\eta_{\text{power}} \equiv \eta_g \eta_{\text{ext}} . \tag{2.3}$$

To conclude this section, we describe briefly the power efficiency, η_g, for generating excess minority carriers by the common techniques listed above. In Table 2.1, we provide simple expressions for η_g for injection luminescence, photoluminescence, and cathodoluminescence. Immediately apparent is the large loss that occurs in cathodoluminescence, where only about 30% of the incident beam energy is effectively used to produce electron–hole pairs. The remainder of the energy is converted to heat.

Of the techniques introduced above, only injection luminescence is practical for LEDs. A simple p–n junction requires 2–3 V for bias, which is consistent

Table 2.1. Minority-carrier generation efficiency for common excitation techniques[a]

Excitation	Effective distance	Efficiency, η_g, per incident electron or photon	
		Relation	Max. value
p–n junction injection luminescence	L	$\frac{h\nu_r}{qV_j}$	~100%
Photoluminescence	$1/\alpha_i + L$	$\frac{h\nu_r}{h\nu_i}(1-R)$	~ 70%
Cathodoluminescence	$d+L$	$\frac{h\nu_r}{E_0}$	~ 30%

R: reflection coefficient, air-semiconductor (~0.3)
V_j: voltage across the p–n junction
$h\nu_r$: energy of generated radiation recombination ($\sim E_g$)
$h\nu_i$: energy of incident photon ($h\nu_i > E_g$)
E_0: energy for pair generation in an electron beam ($\sim 3\ E_g$)
L: minority-carrier diffusion length
α_i: absorption coefficient
d: electron penetration depth, $d \propto E_i^{3/2}$
E_i: energy of incident electron beam

[a] Note: A more accurate description of η_g is the efficiency to place an excess electron–hole pair in the proper energy relationship to emit a photon of energy $h\nu_r$.

with inexpensive battery operation. The power efficiency of such devices is sufficiently high to require only moderately low currents, which is essential for long battery life. In the remainder of this chapter, most of the treatment will be devoted to injection luminescence, in particular the manner in which it is optimized in the fabrication of efficient LEDs.

p–n Junction

A forward bias across a p–n junction raises the potential energy of electrons in the n-type region, reducing the height of the barrier that prevents their spilling into the p-type region (Fig. 2.1). Similarly, the barrier that blocks the flow of holes from p-type to n-type regions is reduced by the forward bias. As soon as an overlap of electrons and holes occurs, their recombination becomes possible via processes to be described in subsequent sections.

Once across the junction, the excess carriers significantly increase the minority-carrier concentrations (A p–n junction passing 10 A/cm^2 injects minority carriers at a rate of about 10^{24} carriers/cm^3 s within a narrow region, typically 1–2 µm wide, adjoining the junction. Assuming a minority-carrier lifetime of 10^{-9} s, this generation rate yields a net excess carrier concentration of $10^{24} \times 10^{-9} = 10^{15}$ carriers/cm^3, which is much in excess of typical equilibrium minority-carrier concentrations at room temperature) (Fig. 2.1c). The "excess" carriers then recombine with some of the many oppositely charged

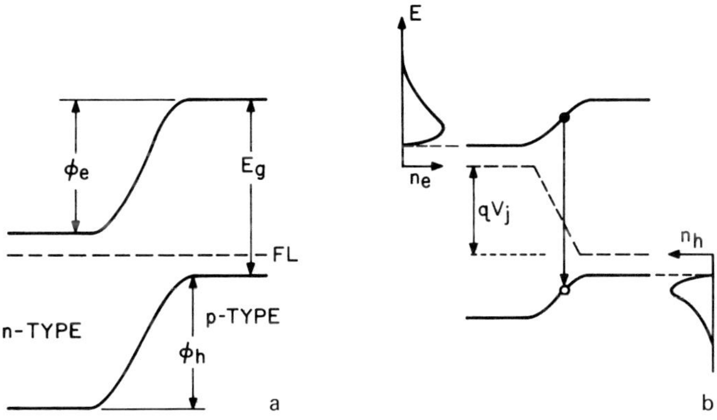

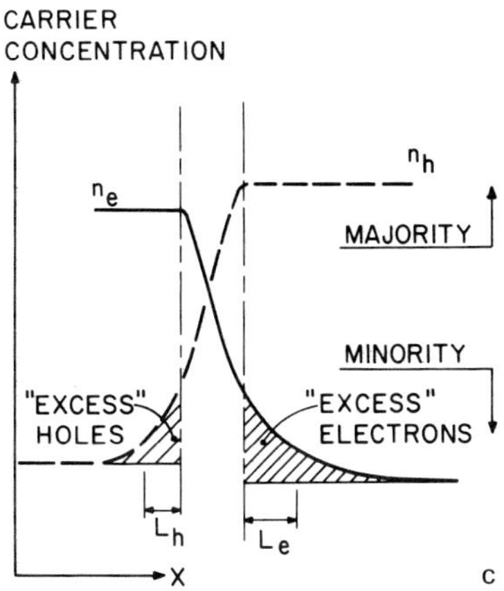

Fig. 2.1a–c. Diagram of a p–n junction. (**a**) Barriers ϕ seen by electrons and holes at equilibrium. (**b**) Barrier reduction by a forward-bias V_j; the electron and hole distributions are labeled n_e and n_h respectively; the arrow illustrates a band-to-band recombination. (**c**) Spatial distribution of majority- and minority-carrier concentrations on the n- and p-sides of a forward-biased p–n junction

majority carriers, thereby tending to return the minority-carrier concentrations to their equilibrium values. The recombination of the excess minority carriers within a diffusion length L of the junction is the mechanism by which light is generated.

Note that the tail of the carrier's Boltzmann distribution extends to energies greater than the barrier height. Therefore, some current will flow in narrow-gap materials even at a low forward bias. However, since the number of free carriers increases rapidly toward the Fermi level, the current grows exponentially with applied forward bias. The current obeys the diode equation [2.11]

$$I = I_0[\exp(qV_j/\beta kT) - 1], \tag{2.4}$$

where V_j is the voltage appearing at the junction. V_j may differ from the applied voltage V_a by the ohmic drop across the internal resistance R

$$V_j = V_a - IR. \tag{2.5}$$

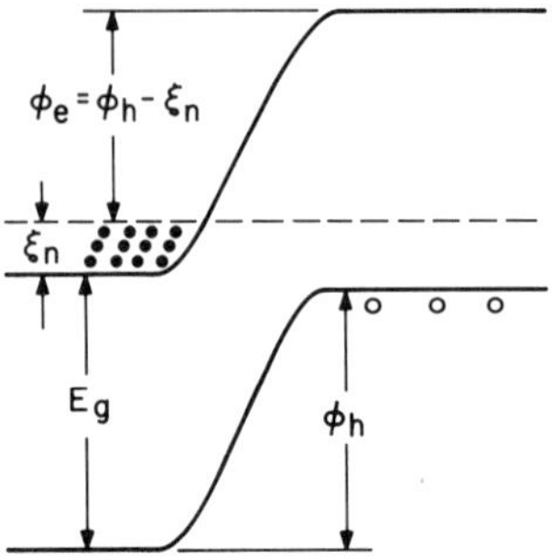

Fig. 2.2. Diagram of a p–n junction between a degenerately doped n-type and a lightly doped p-type semiconductor

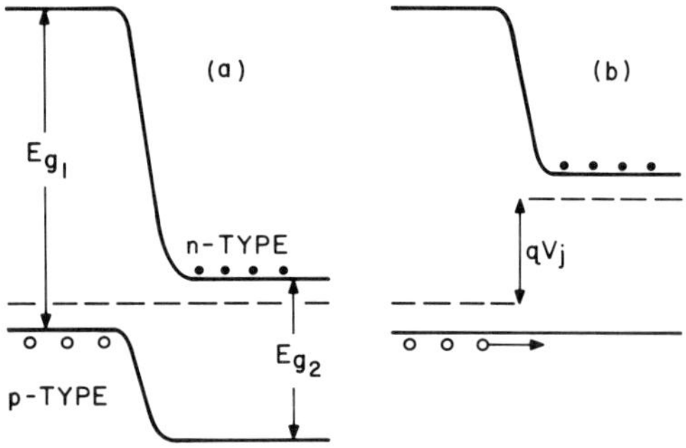

Fig. 2.3a, b. Diagram of a heterojunction. (**a**) At equilibrium. (**b**) With a forward bias V_j

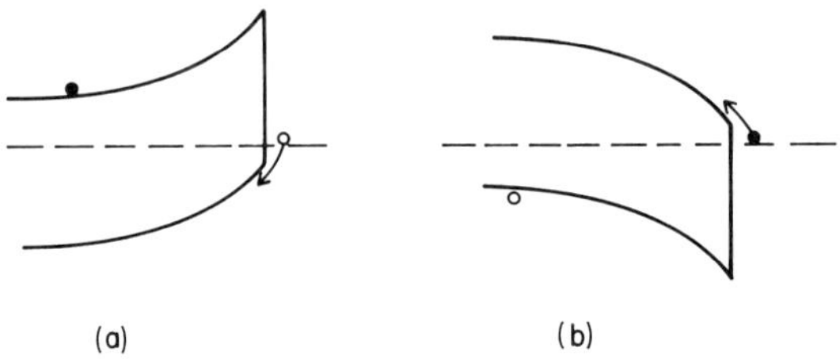

Fig. 2.4a, b. Diagram of Schottky barriers. (**a**) To n-type. (**b**) To p-type material. The arrows show minority-carrier injection during forward bias

I_0 in (2.4) is the so-called "saturation current", the current obtained with a reverse bias, i.e., with V_j negative. Usually, the current does not actually saturate because it is dominated by the presence of generation–recombination centers [2.12] inside the p–n junction and at the junction intersection with the surface [2.13]. This last point will be discussed further in Sect. 2.2.3 on nonradiative recombination. In the ideal p–n junction, the saturation current is controlled by the number of minority carriers that are able to diffuse to the junction; in this ideal case, I_0 depends on the density of states in the band, the energy gap, the position of the quasi-Fermi levels, and the minority-carrier diffusion lengths. The coefficient β in (2.4) then has a value of 1.0. When I_0 is dominated by generation–recombination centers at midgap, $\beta=2$. In the realistic case when both mechanisms are operative, or when the centers are not at midgap, β can take values between 1 and 2.

Light-emitting diodes often have one side of the junction more heavily doped than the other. In fact, on the heavily doped side, the material is often degenerate, i.e., the Fermi level is inside the band by an amount ξ. As can be seen in the illustration of Fig. 2.2, the carriers in the degenerately doped band have the advantage of a reduction of barrier height by ξ compared to those in the more lightly doped side. A second benefit of heavy doping in the n^+ region of Fig. 2.2 is that band-filling raises the threshold energy for absorption by ξ, so that the emitted light can traverse the n^+ region without much absorption. When both sides are degenerately doped, an additional phenomenon must be considered: the relative shrinkage of the energy gap on either side due to band-perturbation effects [2.14].

Heterojunction

In order to control which carrier is injected into a luminescent material, a heterojunction can be used (Fig. 2.3). The source of injected carriers is a material having a wider gap than that of the luminescent region. Then, the asymmetry of the barriers seen by electrons and holes assures a high injection efficiency from the wider-gap to the narrower-gap layer [2.15, 16]. Although heterojunctions can be made between very different materials, the most successful devices are those made from materials having similar lattice constants at the fabrication temperature – a precaution which minimizes dislocations and interfacial stress. A particularly attractive heterojunction structure to minimize lattice mismatch is $Al_{1-x}Ga_xAs$ with a different value of x on each side of the junction [2.17].

Schottky Barrier

A Schottky barrier usually occurs at the surface of a semiconductor in contact with a metal. Such a barrier is similar to a p–n junction; whenever surface states induce an inversion layer, the surface has the opposite conductivity type than the bulk. In effect, this is equivalent to a p–n junction immediately below the surface. Figure 2.4 illustrates inversion in n- and p-type materials, respectively. A forward bias tends to flatten the bands, allowing the injection of minority carriers into the bulk. The minority carriers can then recombine radiatively, just as they would in a p–n junction.

One concern with a forward-biased Schottky barrier is the relative ease with which majority carriers (electrons in Fig. 2.4a) can flow to the metal electrode, thus reducing the injection efficiency and, therefore, the luminescence efficiency. The injection efficiency depends on the relative population of majority carriers at the surface and in the bulk. Strong inversion on a lightly doped crystal is the most favorable combination for high injection efficiency. This can be readily visualized by examining Fig. 2.4. For example, if the n-type layer in Fig. 2.4a is lightly doped, the Fermi level is farther from the conduction band in the bulk than it is from the valence band at the surface. In other words, the barrier to electrons will be higher than the barrier to holes. There will be fewer electrons flowing from the semiconductor to the metal than holes flowing in the opposite direction.

Schottky barriers are easy to make: a point contact or an evaporated layer on the surface forms the Schottky barrier. A thin metal layer can be simultaneously conducting and partially transparent (with a flat spectral transmission). Although such a structure facilitates the emission of light generated near the surface, it is not as practical as a p–n junction or a heterojunction because of the limited transmission of the metal electrode and

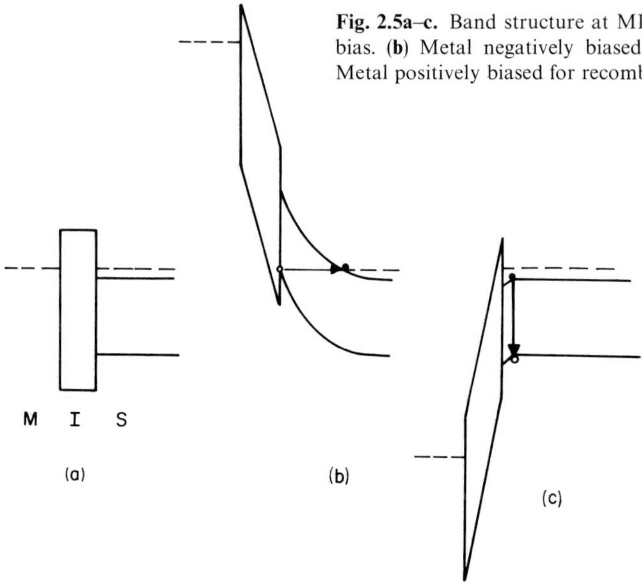

Fig. 2.5a–c. Band structure at MIS transition. (**a**) Without bias. (**b**) Metal negatively biased for hole generation. (**c**) Metal positively biased for recombination

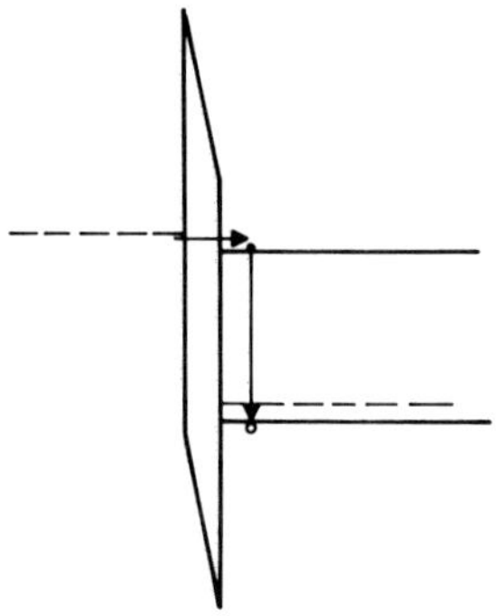

Fig. 2.6. Electron injection by tunneling through a thin insulator

because of the relatively low minority-carrier injection efficiency. On the other hand, it is a convenient technique for a fast experiment with materials in which good control of the p–n junction technology has not been achieved.

MIS Structure

When surface charges are insufficient to induce the desired inversion layer, one can induce surface charges by a capacitive coupling to the surface through an insulator. In a metal–insulator–semiconductor (MIS) structure, the surface charge on the semiconductor is determined by the potential applied to the metal [2.18]. For the present discussion, we shall call ZnS and other luminescent insulators "semiconductors" to distinguish them from the wider-gap insulator placed on their surface. An immediate benefit of the MIS structure is that the band bending can be controlled by the applied voltage. Thus, an inversion layer can be induced by one polarity, then the carriers accumulated at the surface can be injected by reversing the polarity of the metal electrode (Fig. 2.5). This method has been used to obtain electroluminescence in GaAs without a p–n junction [2.19]. Another example of this mode of excitation is the uv electroluminescence of GaN [2.20].

If the insulator is so thin ($\lesssim 100$ Å) that the exponentially decaying probability of finding the electron extends from both sides of the insulator without too much attenuation, the electron can readily reappear on the other side of the insulator without loss of energy [2.21]. This tunneling property of thin layers can be used to inject electrons into the conduction band of a p-type semiconductor (Fig. 2.6). However, with rare exceptions [2.22], it is not suitable for hole injection (electron extraction by tunneling from the valence band) because, with n-type material, there is a high probability for electrons to simultaneously tunnel out of the conduction band.

2.2 Recombination Processes

Once electron–hole pairs have been produced by one of the above-mentioned excitation processes, the electron will return to its lower equilibrium energy and recombine with the hole. This recombination can occur via many pathways; some are radiative, but others are not. Radiative transitions can occur from the conduction band to the valence band or via intermediate impurity levels. Nonradiative transitions dissipate the energy into mechanical vibrations (heat).

2.2.1 Band-to-Band Radiative Recombination

When an excess electron density Δn is injected into the p side of an electroluminescent junction, the probability of an excess electron in the conduction band recombining with one of the many available holes, $p+\Delta p$ in the valence band, is proportional to the product $(\Delta n)(p+\Delta p)$. For low or moderate injection levels, $p \gg \Delta p$, and the radiative-recombination rate r for such a process is

$$r = R(\Delta n)p\,. \tag{2.6}$$

Here, the proportionality factor R is called the radiative-recombination coefficient, which can be determined from experimental absorption data and available theory. Both energy and momentum must be conserved in order for an electron and hole to recombine and emit a photon. Although a photon can have considerable energy, its momentum ($h\nu/c$) is very small. Therefore, the most simple (and most probable) recombination process will be that in which the electron and hole have the same value of momentum. Such a situation actually occurs in many III–V and all II–VI compounds, where the conduction-

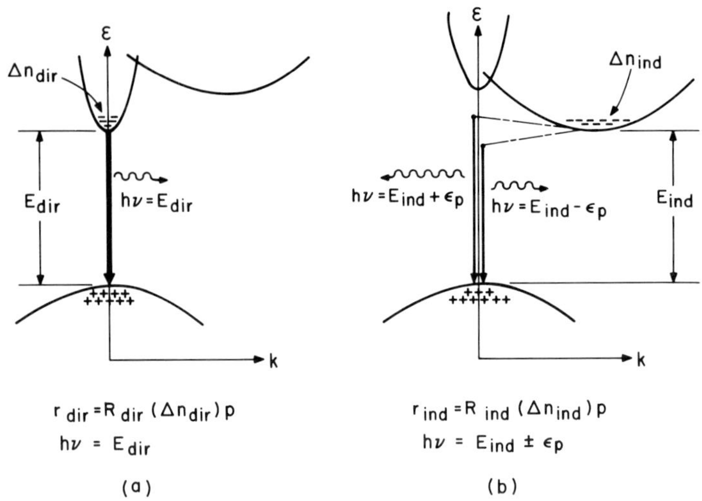

Fig. 2.7a, b. Electron energy E vs wave number k for energy states near the valence-band maximum and the conduction-band minima. (**a**) For direct. (**b**) For indirect semiconductors. k is directly proportional to carrier momentum. ε_p is the phonon energy

Table 2.2. Radiative recombination coefficient for representative direct- and indirect-band-gap semiconductors [2.23a][a]

Material	E_g [eV]	Energy-band-gap type	Recombination coefficient R [cm^3/s]
Si	1.1	indirect	1.79×10^{-15}
Ge	0.67	indirect	5.25×10^{-14}
GaP	2.25	indirect	5.37×10^{-14}
GaAs	1.4	direct	7.21×10^{-10}
GaSb	0.71	direct	2.39×10^{-10}
InAs	0.31	direct	8.5×10^{-11}
InSb	0.18	direct	4.58×10^{-11}

[a] More recent values have been obtained for several of these materials; however, the values above are from a single study for the purpose of comparison.

band minimum and the valence-band maximum both lie at the zero momentum position, as shown in Fig. 2.7a. A semiconductor energy-band structure such as this is said to be "direct".

Figure 2.7b illustrates an "indirect" band structure, where the conduction-band minimum and the valence-band maximum occur at different values of momentum; here, the band-to-band recombination necessarily involves a third particle to conserve momentum. Phonons (i.e., lattice vibrations) serve this purpose; however, the probability of electron–hole recombination in this three-particle process is drastically reduced relative to the simpler two-particle direct recombination process. This difference is clearly reflected in significantly larger values of R (by more than three orders of magnitude) for direct- than for indirect-band-gap semiconductors, as illustrated in Table 2.2 [2.23a] for several III–V compounds and for Si and Ge.

2.2.2 Radiative Recombination via Impurities

Donor and acceptor impurities determine not only the conductivity type and resistivity of the material, but also the dominant radiative-recombination processes when the impurity is a luminescence center. Impurities substituting for host atoms within a crystal are located somewhat randomly, and do not form a truly periodic array. Therefore, the impurity energy level is spread out in momentum space and, in particular, appears at $k=0$. This explains why transitions involving impurities can be so efficient. The radiative part of the recombination may occur from the conduction band to the acceptor or from the donor to the valence band. More frequently, transitions occur between the donor and acceptor states since they individually provide a lower energy state for the electron and hole, respectively. The choice of a particular impurity in III–V compounds is usually made from the group VI donors, Te, Si, and S, and from the group II acceptors Zn, Cd, and Mg. In some cases, "amphoteric" dopants such as Ge, Si, and Sn from group IV of the periodic table can be used as either donors or acceptors, depending on which of the two available types of lattice sites (group III or V) they occupy.

For each particular dopant and semiconductor combination, an optimum impurity concentration is usually empirically determined for efficient luminescence. In general, the following limits can be placed on such concentrations: They cannot be too *small* because (i) the recombination probability for an injected minority carrier is directly proportional to the concentration of majority carriers with which it will recombine (or to the concentration of luminescence centers where it recombines), and (ii) the higher series resistance that results from low carrier concentrations can cause excessive heating and large voltage drops at high currents. On the other hand, the concentration cannot be too *large* because of "concentration quenching" when metallurgical imperfections such as precipitates and metal complexes are introduced at concentrations near the solubility limit of the impurity. Such imperfections are known to introduce competing nonradiative-recombination centers which drastically reduce the luminescence efficiency. For these reasons, impurity concentrations in the range of 10^{17}–$10^{18}\,\mathrm{cm}^{-3}$ for donors and 10^{18}–$5\times10^{19}\,\mathrm{cm}^{-3}$ for acceptors generally have been found to be best for high luminescence efficiency. In Fig. 2.8, we illustrate the dependence of luminescence intensity on doping concentration for p-type GaAs excited by an electron or photon beam. Note that the maximum intensity occurs at a concentration of $\sim 3\times10^{19}\,\mathrm{cm}^{-3}$, as indicated above.

Concerning the relative efficiency of the radiative recombination from n- and p-type material, it was originally thought that the radiative processes in

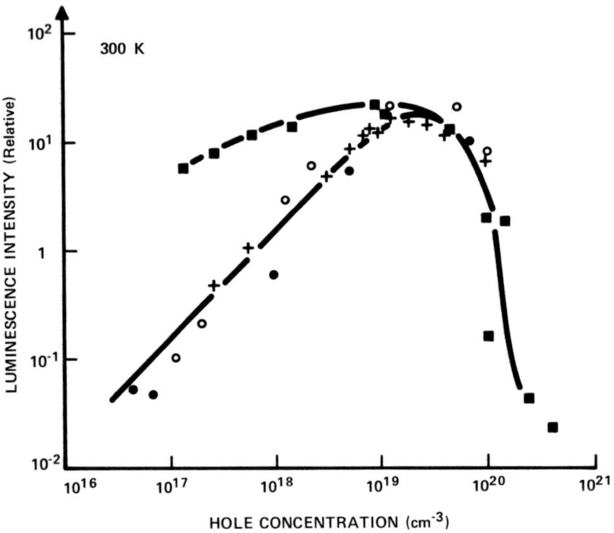

Fig. 2.8. Luminescence intensity vs hole concentration for p-type GaAs. Data is from cathodoluminescence (○) [2.23b] and photoluminescence (+ ■ ●) [2.23c, 2.23d] at room temperature

p-type material were the more efficient, perhaps due to the involvement of acceptor states. This idea was based largely on the fact that the dominant emission from most LEDs was found to originate from the p side of the electroluminescent junction. Detailed cathodoluminescence measurements revealed, however, that the luminescence intensities for n- and p-type GaAs at optimum doping concentrations are almost identical. The fact that the observed emission from many LEDs originates on the p-side of the junction stems from several impurity effects. First, the injection of electrons into the p-region is usually favored over the injection of holes into the n-region by the large ratio of electron-to-hole mobility for most III–V compounds. Furthermore, since the emission energy is usually slightly higher in n-type than in p-type material, any radiation which does originate from the n-side of the junction is very strongly absorbed in the p-type material, where strong band tailing has effectively shrunk the energy gap. Conversely, the radiation originating in a heavily doped p-type layer will pass through n^+ material with reduced absorption losses due to the reduced emission energy for p-type material. This consideration is usually used to good advantage in designing efficient LEDs.

In indirect-gap materials, impurities provide the only pathway for efficient luminescence since their spread in momentum space facilitates momentum conservation. Thus in GaP, Zn and O are used to obtain efficient red emission [2.24, 25]. Not only donors and acceptors have the property of extended momentum, but also deeper neutral centers where the electron is so localized in real space that its momentum is very spread out. Thus, nitrogen in GaP was found to greatly enhance the efficiency of the near-gap green emission [2.6]. Nitrogen produces only a short-range, non-Coulombic attractive potential

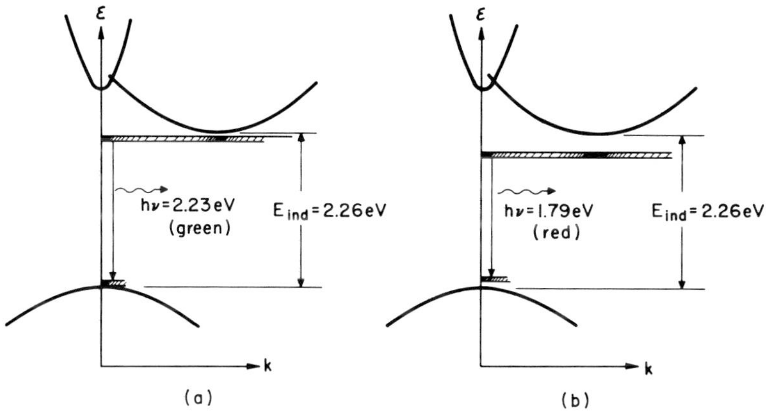

Fig. 2.9a, b. Relaxation of momentum conservation requirement for indirect-band-gap materials. (**a**) Nitrogen doping in GaP, resulting in 2.23-eV green emission. (**b**) Zn, O doping in GaP, resulting in 1.79-eV red emission

with an electron binding energy of about 8 MeV. After the capture of an electron, the center becomes negatively charged and the long-range Coulombic potential then allows it to capture a free hole, thus forming a bound exciton. The annihilation of this bound exciton by radiative recombination gives rise to green luminescence at room temperature (Fig. 2.9). Unfortunately, the exciton may dissociate by thermal activation prior to radiative recombination, allowing nonradiative-recombination processes to dominate. As a result, the internal quantum efficiency for green luminescence in GaP is on the order of 1 %, which is much lower than that for direct-band-gap radiative-recombination processes.

2.2.3 Nonradiative Recombination

The *internal* quantum efficiency is highly dependent on the perfection of the material in the vicinity of the p–n junction, where the radiative recombination occurs. Defects (and contaminants such as copper) give rise to deep recombination centers with consequent infrared emission or nonradiative recombination. Relatively little is known about the details of the nonradiative-recombination processes via defect centers. Surface recombination is known to be nonradiative, possibly because a continuum (or quasi-continuum) of states may join the conduction band to the valence band [2.26, 27]. The recombination at surface states dissipates the excess energy by phonon emission. Crystal defects such as pores, grain boundaries, and dislocations may provide regions where a localized continuum of states bridges the energy gap and similarly allows nonradiative recombination. In the limit, a cluster of vacancies or a precipitate of impurities (e.g., a metallic phase) also could form such a nonradiative center.

Although the continuum-of-states type of recombination center is very localized, its effect extends, on the average, over a carrier diffusion length. As we have seen in Sect. 2.1.3, the relationship between injection current and the junction voltage in the ideal diode is

$$I \approx I_0 \exp(qV_j/kT), \tag{2.7}$$

where V_j is the junction voltage and kT is the thermal energy.

At low forward bias, a nonradiative-recombination current shunts the ideal injection current responsible for light emission. These processes result from recombination via deep traps in the space-charge region, as first described by *Sah* et al. [2.28a]. The current due to nonradiative space-charge recombination follows the expression

$$I_{sc} = I_1 \exp(qV_j/\beta kT), \tag{2.8}$$

where β generally takes on a value near 2. In Fig. 2.10, a vapor-grown p–n junction of GaAs in forward bias is illustrated and is found to show clearly the two current regimes described above; the transition between the two regimes occurs at a current density of 10^{-2} A/cm^2. In the same figure, the light is found to vary as $\exp(qV_j/kT)$, just as the component of current due to minority injection over the junction barrier.

Recently, *Henry* et al. [2.13] have shown that the "$2kT$" current component also can arise through nonradiative surface recombination at locations where the p–n junction intersects the semiconductor surface. Thus, while it is not readily apparent for a particular diode whether this current component is due primarily to space-charge or surface recombination, it is clearly entirely nonradiative and undesirable. In practice, the $2kT$ component due to surface recombination can be reduced by avoiding surface damage or by treating the surface with various etchants. However, trapping states that contribute to internal space-charge recombination have not yet been identified, and cannot be readily eliminated.

The effect of the $2kT$ current component on the electroluminescence efficiency can be illustrated from the I–V characteristics of Fig. 2.10. Since the $2kT$ component is entirely nonradiative while the $1kT$ component is entirely responsible for the luminescence, the internal quantum efficiency can be simply expressed by the equation

$$\frac{\eta_{int}}{\eta_{int_0}} = \frac{I_{1kT}}{I_{1kT} + I_{2kT}} = \frac{I_{1kT}}{I_j}, \tag{2.9}$$

where η_{int_0} is the value of the efficiency at moderately large current values, where the $2kT$ component becomes negligible. In Fig. 2.11, we plot the relative electroluminescence efficiency from the L–V curve in Fig. 2.10. Also shown

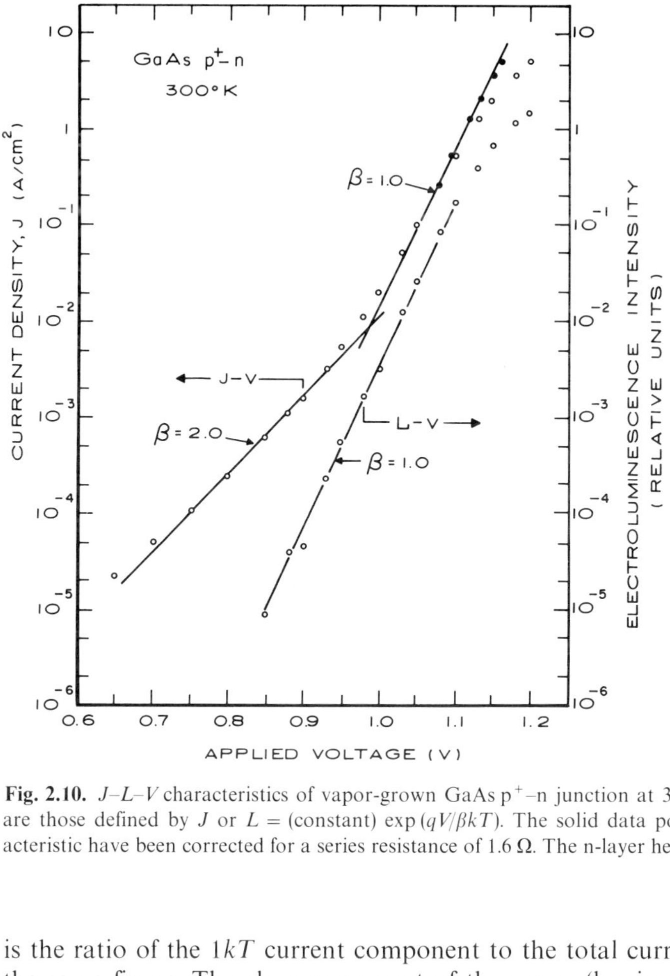

Fig. 2.10. *J–L–V* characteristics of vapor-grown GaAs p^+–n junction at 300 K. The values of β are those defined by J or L = (constant) exp $(qV/\beta kT)$. The solid data points in the *J–V* characteristic have been corrected for a series resistance of 1.6 Ω. The n-layer here is Se-doped [2.28b]

is the ratio of the $1kT$ current component to the total current I_j taken from the same figure. The close agreement of the upper (luminescence) curve and the lower (I_{kT}/I_j) curve clearly illustrates that the injection-luminescence process is effectively shunted by the $2kT$ leakage current component at currents less than about 2×10^{-2} A/cm^2. For a typical-sized commercial LED $(\sim 0.25 \times 0.25$ mm$^2)$, this corresponds to a current of 10 μA. Fortunately, the operating current of most LEDs is on the order of milliamperes, where $1kT$ currents dominate. However, for diodes with nonoptimized structures, the $2kT$ component can be much larger than that of Fig. 2.10, and luminescence efficiencies are reduced at appreciably higher current levels.

Not all defects are recombination centers. In some instances, the local strain induced by the defect may either widen the energy gap, thus repelling both carriers, or create a local field which separates the electron–hole pairs. In either case, although the carriers are conserved, they do not recombine radiatively at such a site [2.29].

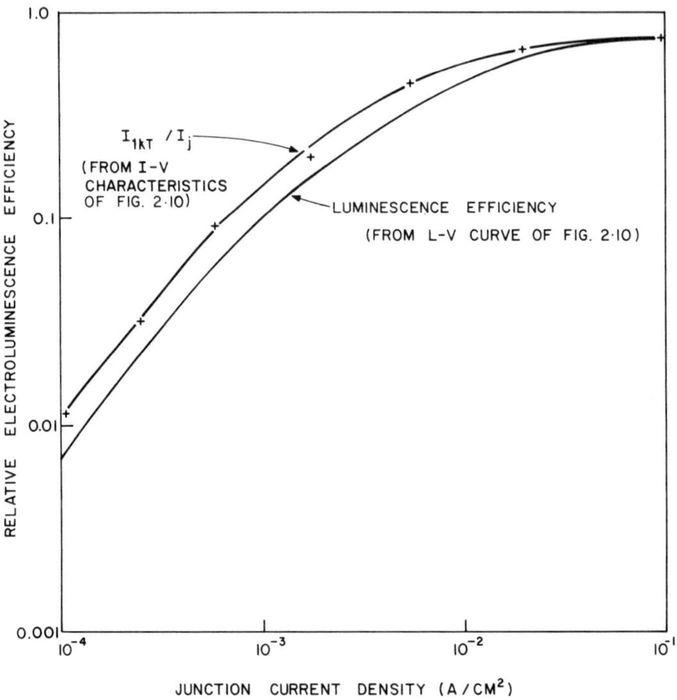

Fig. 2.11. Variation of luminescence efficiency with current for the GaAs diode of Fig. 2.10

Increasingly, the adverse effect of Ga vacancies on the radiative efficiency of GaAs [2.30] and GaP compounds and their alloys is being documented. In the case of GaP, for example, photoluminescence studies have clearly shown that growth of the material under conditions of increased Ga vacancy formation results in reduced radiative efficiency for green emission. Effects of this type possibly explain the lower efficiency values typically obtained by the use of vapor-phase epitaxy for GaP:N diodes compared to liquid-phase epitaxy, where the formation of Ga vacancies is lower.

In the Auger effect, a recombining electron transfers to another electron the energy it would have radiated. The second electron is then raised to a higher energy, from which it returns by multiple phonon emission. For example, a conduction-band electron could be excited to a higher energy within the conduction band and then it would be thermalized within the band by phonon emission. In this process, recombination occurs, but no radiation is emitted.

Sometimes is is possible to identify the Auger process by its kinetics, since three particles are involved: two electrons and one minority hole, or a minority electron and two holes. The probability, P_A, for an Auger event increases as either n^2p or np^2, where either the electron concentration n or the hole concentration p (or both) increase with the excitation current [2.31]. The Auger effect, in addition to decreasing the luminescence efficiency, reduces the carrier lifetime $\tau_A(\tau_A = 1/P_A)$. Of all the nonradiative processes, the most tractable one experimentally is the Auger effect, because in this process, a hot carrier is generated at about twice the energy needed to excite the luminescence. The hot carrier can signal its existence by overcoming a barrier either inside the device [2.32] or at its surface [2.33].

2.3 Materials for LEDs

2.3.1 Brightness Considerations

The response of the human eye is limited to a wavelength range between about 400 nm (violet–ultraviolet) and 720 nm (red–infrared), as illustrated in Fig. 1.1. Because of the immense variation in the visual sensitivity over this spectral range, the performance of a visible-light-emitting diode must be appraised not only by the absolute intensity of the emitted radiation, but also by the relative response (i.e., the luminous efficiency) of the eye at the wavelength of interest. Hence, two important figures of merit have evolved for LEDs. The first is the *quantum efficiency*, which is simply the ratio of the number of photons produced to the number of electrons passing through the diode, i.e., the conversion efficiency of the device independent of the eye's response to it. The *internal* quantum efficiency η_{int} is evaluated at the p–n junction, while the *external* quantum efficiency η_{ext} is evaluated at the exterior of the diode. The external quantum efficiency is always less than the internal quantum efficiency owing to optical losses that occur in extracting the emitted radiation from the semiconductor. Representative values of η_{ext} for visible LEDs fall between 0.1 and 7% at room temperature, while values of η_{int} can exceed 50% under optimum conditions.

The second figure of merit for an LED is the *brightness*, which is a measure of the *visual* response to the radiation emitted from the diode surface. The brightness B, in foot-Lamberts (fL), is proportional to the external quantum efficiency of the diode and to the sensitivity of the eye, and can be calculated from the equation

$$B = 1150\frac{\eta_{\mathrm{ext}} L J A_{\mathrm{j}}}{\lambda A_{\mathrm{s}}}. \tag{2.10}$$

Here, λ is the emission wavelength (in μm), J is the current density through the junction (in A/cm^2), and L is the luminous efficiency of the eye (see Fig. 1.1), which has a maximum value of 680 lm/W for green emission at 555 nm. A_{j} and A_{s} are the area of the p–n junction and the observed emitting surface, respectively. The ratio $(A_{\mathrm{j}}/A_{\mathrm{s}})$ is unity for the simplest case of a flat, unencapsulated, surface-emitting LED, but can be much less than unity in some cases (e.g., dome-shaped diodes, diodes with encapsulated plastic lens, etc.).

Since the brightness of a diode can be varied by changing its size or the junction current, B is often normalized with respect to the current density J so as to more fairly compare the relative performance of different electroluminescent materials. Values of brightness in excess of 1000 fL are readily available in commercial LEDs at normal current densities of 10 A/cm^2. By comparison, the brightness of the frosted surface of a 40 W incandescent bulb is about 7000 fL.

From the considerations just discussed, it is apparent that efficient LEDs emitting near the peak of the eye sensitivity are required. Since the upper limit for the emission energy is approximately equal to the semiconductor energy gap, values of E_g greater than 1.72 eV (720 nm) and as close as possible to 2.23 eV (550 nm) are needed.

2.3.2 Binary Compounds

In all of the elemental and practical III–V materials, and many of the II–VIs, the atoms are tetrahedrally bonded in a zinc-blende structure. The covalent component of bonding, which is dominant in most of these materials, arises from the sharing of the outer four electrons of each atom with a single electron from each of its four nearest neighbors. In the case of III–V and II–VI materials, there also is an ionic component of bonding, which increases in materials having atoms located progressively closer to the outermost columns of the periodic table, (i.e., the II–VIs are more ionic than the III–Vs, which are more ionic than the elemental semiconductors). The increased binding energy for the III–Vs and II–VIs is illustrated by progressively larger energy band gaps, as shown in Fig. 2.12. It also should be noted that the binding energy of the orbital electrons shrinks as we form progressively heavier materials by descending the columns of the periodic table. The reduced binding energy for the heavier materials is reflected by a shrinkage in the band-gap energy.

Table 2.3 lists the binary compounds of interest for luminescence.

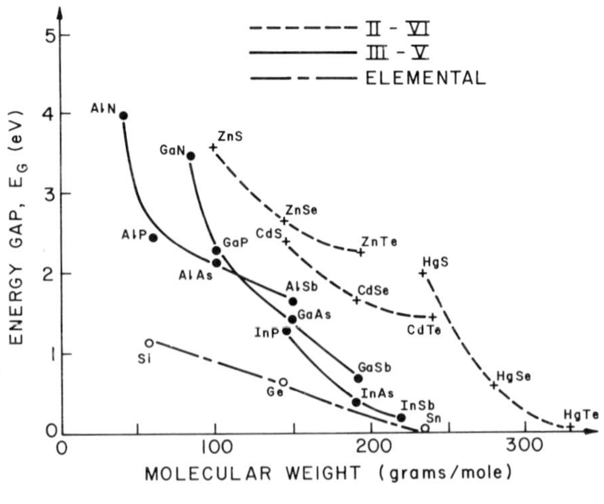

Fig. 2.12. Dependences of the energy gap at room temperature on molecular weight for elemental, III–V, and II–VI semiconductors

Table 2.3. Binary compound semiconductors

		E_g[eV]	Direct or indirect	a/c or a_0 [Å]
II–VI	ZnO	3.2	direct	3.249/5.206
	ZnS α	3.8	direct	3.814/6.257
	ZnS β	3.6	direct	5.406
	ZnSe	2.58	direct	5.667
	ZnTe	2.28	direct	6.101
	CdS	2.53	direct	4.136/6.713
	CdSe	1.74	direct	4.299/7.010
	CdTe	1.50	direct	6.477
III–V	AlP	2.43	indirect	5.463
	AlAs	2.16	indirect	5.661
	AlSb	1.6	indirect	6.138
	GaN	3.4	direct	3.18/5.16
	GaP	2.25	indirect	5.449
	GaAs	1.43	direct	5.653
	InN	1.7	direct	3.54/5.705
	InP	1.28	direct	5.868
	InAs	0.36	direct	6.058
IV–IV	SiC α	2.8–3.2[a]	indirect	3.082/15.112
	SiC β	2.2	indirect	4.359

α hexagonal; β cubic.
[a] depending on polytype.

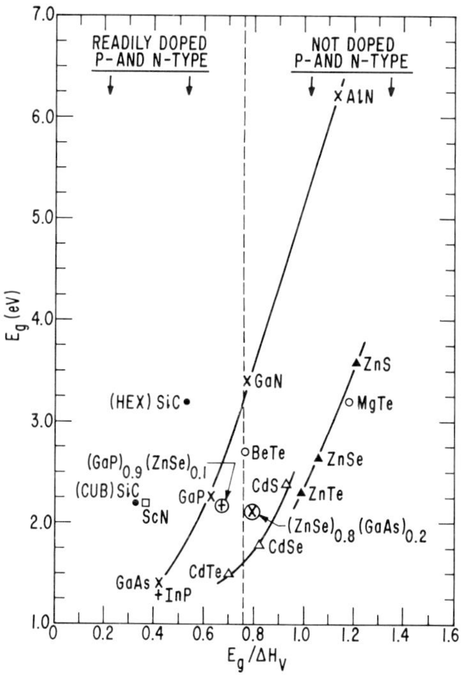

Fig. 2.13. Dependence of energy gap, E_g, upon $E_g/\Delta H_v$ for a number of semiconductors. ΔH_v is the enthalpy for vacancy formation in the compound [2.34]

Another consideration is the dopability of compounds, i.e., the ability to make p–n junctions. Compounds with small lattice constant and molecular weight tend to have a large enthalpy for vacancy formation, ΔH_v. In these compounds, the introduction of an impurity is readily compensated by a vacancy. In Fig. 2.13 *Dismukes* et al. have mapped various compounds as a function of the energy gap, E_g, and $E_g/\Delta H_v$ (the ease of vacancy formation) [2.34]. The vertical dashed line separates those compounds which can be readily doped n- or p-type from those for which amphoteric doping has not been achieved. The amphoteric dopability of compounds close to the line is uncertain.

III–V Compounds

Referring to Table 2.3, of the III–V compounds with an energy gap large enough to provide visible luminescence, only GaN and InN have a direct gap. These materials, which have not yet reached the commercial stage of development, will be discussed later.

GaP. The only binary III–V compound used in the mass production of LEDs is GaP, in spite of its indirect energy gap. The state-of-the-art performance of commercially available LEDs is summarized in Table 2.4. Red-emitting diodes of GaP are commercially available with external quantum efficiencies as high as 3%. Laboratory efficiency values of 7–15% have been reported [2.36, 37].

Table 2.4. State-of-the-art performance of p–n junction LEDs

Material	Color	Peak wavelength [nm]	Lum. eff. [lm/W]	η_{ext} [a] [%]	B/J [b] [fL/A cm^{-2}]	Ref.
GaP:Zn, O	red	690	20[c]	3–15[d]	350[e]	[2.36, 37]
$GaAs_{0.6}P_{0.4}$	red	660	42	0.5	145	[2.42]
$GaAs_{0.35}P_{0.65}$:N	orange	632	190	0.2		[2.87b]
$GaAs_{0.25}P_{0.75}$:N	amber	610	342	0.04	40–100[f]	[2.35]
SiC	yellow	590	515	0.003	10	[2.43]
$GaAs_{0.15}P_{0.85}$:N	yellow	589	450	0.05		[2.87b]
GaP:N	yellow	590	450	0.05		[2.87b]
GaP:N	green	550	677	0.05–0.7[d]	470[g]	[2.38]

[a] Except where noted, efficiencies for diodes with plastic encapsulants.
[b] Except where noted, B/J calculated from (2.10) using the efficiency of *unencapsulated* diode with $(A_j/A_s)=1$. Diode efficiency assumed to be 2.5 times less without encapsulation.
[c] Mean value for nonmonochromatic emission spectrum.
[d] Range between commercially practical and best laboratory results.
[e] Assumed 3% unencapsulated diode efficiency; (A_j/A_s) assumed to be 1/3 to compensate for significant edge emission.
[f] *Typical* values of B/J reported as 40–60 fL/A cm^{-2} in [2.35]. Value of 100 fL/A cm^{-2} calculated from (2.10) using efficiency value found in [2.35].
[g] Calculated for representative *dc* efficiency of 0.1% for unencapsulated diode. (A_j/A_s) assumed to be 1/3.

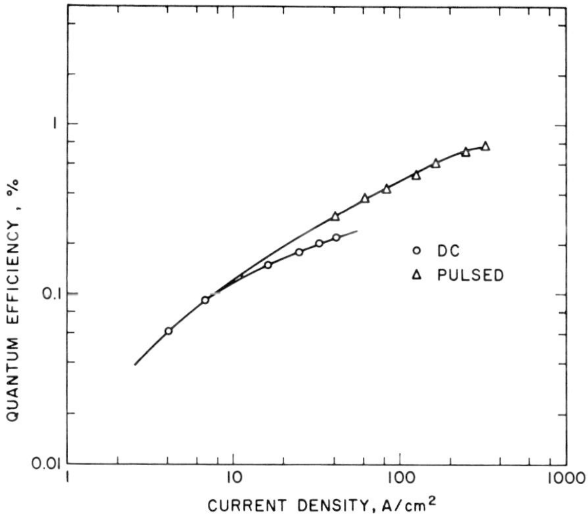

Fig. 2.14. Quantum efficiency as a function of current density of a high-efficiency green-emitting GaP diode consisting of two epitaxial layers deposited by the multiple-bin technique. The lower dc efficiency is due to heating of the diodes [2.38]

However, these diodes were prepared with low Zn doping concentrations, which enhance the external efficiency by reducing the internal absorption coefficient, but lead to an efficiency saturation at current densities as low as 1 A/cm^2. By comparison, GaP LEDs with 3% efficiency values saturate at a current density of about 10 A/cm^2. The brightness value of 350 fL/A cm^{-2} entered in Table 2.4 was obtained by multiplying (2.10) by the ratio of the primary surface area to the total (primary surface plus edge) area, since a large portion of the red emission escapes from the edges of GaP diodes. This brightness value appears to be more than adequate for standard display applications. At the moment, it does not appear likely that useful efficiency increases will be made for the GaP red emission, but we can expect refinements in the manufacturing technology that will further reduce the cost of these devices.

High-quality green-emitting diodes of GaP also are commercially available. The efficiency of routinely fabricated diodes by liquid-phase epitaxy (see Sect. 2.5.2) is about 0.1% at 10 A/cm^2, while the best laboratory results have reached 0.7% when operated pulsed at ~200 A/cm^2, and 0.2% at 33 A/cm^2 (dc) [2.38]. The latter results were obtained from GaP made by a multiple-bin process of growth, involving two epitaxial layers. Values of 0.1% are readily achieved by the vapor-transport method of Zn doping described in Sect. 2.5.1.

It is noteworthy that the efficiency of green-emitting GaP diodes prepared by LPE increases nearly linearly with current density, typically up to 100–200 A/cm^2 (see Fig. 2.14); in contrast, red LEDs of GaP show significant efficiency reductions at current densities above ~5 A/cm^2. This effect is thought to be due to a saturation of the radiative-recombination centers (Zn and O) at high injection levels.

It is possible to obtain orange- or yellow-appearing emission from GaP diodes by introducing N and O into the *n* and *p* sides of the junction, respectively. Green emission arises predominantly from the *n* side, and red emission from the *p* side of the junction. The same diode can therefore simultaneously emit *both* colors, whose integration by the eye gives an orange or yellow appearance to the device. Because of the saturation of the red emission at relatively low current densities, the hue of such diodes shifts toward green with increasing current [2.39].

It is also possible to obtain yellow emission from vapor-grown GaP LEDs by adding N in exceptionally large concentrations ($>10^{19}\,cm^{-3}$) to the crystal during growth [2.40]. At these concentrations, the dominant room-temperature radiation shifts from 2.225 to 2.10 eV (yellow light). p–n junction structures are formed by diffusing Zn into the N-doped (via NH_3) vapor-grown layers, and provide external quantum efficiencies as high as 0.1% for encapsulated diodes [2.40]. Such devices are now available commercially. Finally, acceptor dopants other than Zn are being explored for orange and yellow emission. However, so far, these offer no advantage for room-temperature emission [2.38, 41].

GaN. One of the most promising materials for blue and green emission is the wide-band-gap material GaN ($E_g=3.4\,eV$) [2.44]. Low-resistivity GaN is obtainable only in n-type material. Although p-type doping has been reported for poly-crystalline GaN [2.45], this has not been obtained reproducibly.

GaN has the Wurtzite crystal structure, with lattice constants $a=3.18$ Å and $c=5.18$ Å [2.46]. A cubic phase is sometimes found, with a lattice constant of 4.51 Å [2.47]. The energy gap of both structures is direct. Photoluminescence measurements have shown that the near-gap emission occurs at 3.47 eV with approximately 100% internal quantum efficiency [2.48]. When such an efficiently radiative direct-gap material is optically pumped, laser action can be demonstrated at 2 K [2.49] and more recently at 78 K [2.50]. The various radiative transitions have been identified and reviewed in [2.51].

Undoped GaN is always *n*-type and highly conducting due to a high concentration of N vacancies. Carrier concentrations on the order of 10^{18} electrons/cm^3 and mobilities of about 130 cm^2/Vs can be readily obtained. However, an electron concentration of $10^{17}\,cm^{-3}$ and a mobility of 400 cm^2/Vs have been reported [2.52] in undoped material. *Rode* [2.53] has calculated the influence of ionized impurity scattering on drift mobility, and has found that a mobility of 900 cm^2/Vs should be attainable in uncompensated GaN containing 10^{16} electrons/cm^3.

The electroluminescence of the GaN M–I–n diode occurs in the visible portion of the spectrum. As shown by the solid lines of Fig. 2.15, blue [2.7], green [2.54], yellow [2.55], or red emission can be obtained. Even a broad spectrum "white light" has been observed, as shown in Fig. 2.16. The color of

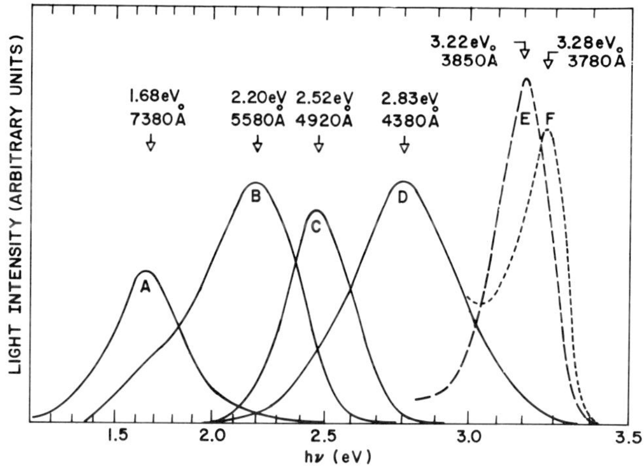

Fig. 2.15. Emission spectra of various GaN LEDs. Solid lines are for i–n structures using Zn-doped material. Dashed and dotted lines are for diodes made on undoped GaN, as discussed in text

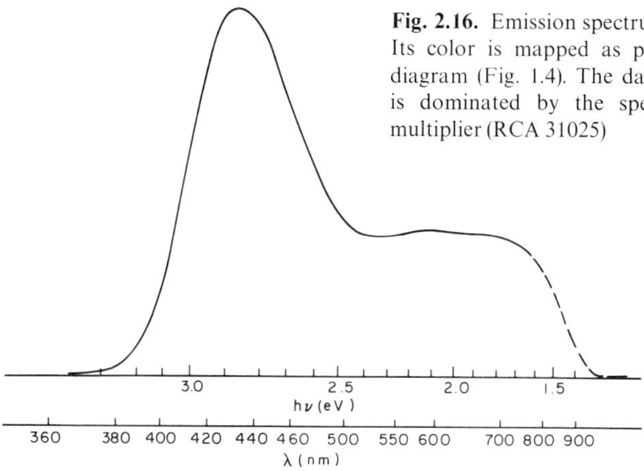

Fig. 2.16. Emission spectrum of a Zn-doped GaN LED. Its color is mapped as point Δ on the chromaticity diagram (Fig. 1.4). The dashed portion in the infrared is dominated by the spectral cutoff of the photomultiplier (RCA 31025)

the light is determined by growth conditions such as the partial pressure of Zn during growth, the duration of the growth, and the thickness of the layers. The color of the emission also can depend on the polarity of the bias. Thus, blue and red have always been obtained with a negative bias on the cathode, whereas green and yellow occur with either polarity.

The electroluminescence process is believed to consist of impact excitation of electrons from deep centers with a subsequent radiative transition of a conduction-band electron to the empty center [2.56]. A careful observation of electroluminescence in thick Zn-doped layers, where the depth of focus can be resolved, shows that the luminescence is not distributed throughout the insulating layer, but rather occurs at the cathode, either near the metal-insulator or near the I–n transition, depending on the polarity of the bias.

When Mg-doped insulating GaN is used in the I–n diode structure, violet dc electroluminescence is obtained at 2.9 eV at room temperature [2.57]. Usually, the intensity of the light is proportional to the electrical-power input until heating sets in. External power efficiencies of 1% in the green and 0.1% in the blue have been obtained uniformly and reproducibly [2.58]. The diodes operate continuously at room temperature. At the highest power level, where heating occurs, the efficiency can be improved by pulsing to reduce the duty cycle.

A spot brightness of 850 fL has been measured [2.59]. Because of the low current requirements of GaN LEDs, their application to watch displays was considered, leading to the favorable figure of merit of 2 lm/A [2.60].

Electroluminescence in the ultraviolet has been obtained in undoped GaN (curve *E* of Fig. 2.15). This diode consisted of a surface barrier under colloidal carbon "Aquadag". When the carbon is biased positively with respect to the *n*-type GaN, an emission peak in the uv at 3.22 eV is obtained together with a broad spectrum in the visible. With a negative bias on the carbon, only the broad visible emission is obtained. The uv component is evidence for hole injection, since this luminescence corresponds to a near-band-gap transition.

UV electroluminescence also has been generated in undoped GaN using an MIS structure, with Si_3N_4 as the insulator [2.61]. Aluminum dots are evaporated on the insulator to form the "field electrode", and ohmic contact is made to the n-type crystal. A negative pulse is applied across the insulator to produce an inversion layer at the I–S interface; this negative pulse is followed immediately by a positive pulse, which drives the holes from the I–S interface into the n-type bulk, where radiative recombination produces ultraviolet radiation (curve *F* of Fig. 2.15). Light is emitted only with the $-\ +$ sequence of polarization. The emission is independent of the duration of either the negative or positive portions of the applied voltage.

The potential of GaN has not been realized commercially, mostly because of market conditions. In order for a new LED to become an acceptable product, it should compete at least price-wise with existing LEDs. Since GaP and GaAsP LEDs had already benefited from several orders of magnitude more developmental effort than GaN, a comparison in real-time appears to have doomed the GaN LED to a laboratory curiosity.

Silicon Carbide

SiC, being a wide-gap semiconductor, offers the attraction of allowing radiative transitions throughout the visible region of the spectrum, where the material is transparent. Since SiC can be doped n-type or p-type at will, p–n junctions can be fabricated. The choice of polytype allows a 0.8 eV change in energy gap from 2.4 eV for cubic SiC to 3.2 eV for hexagonal 4 H SiC. Many other polytypes are possible with intermediate band gaps [2.62]. This material is stable at elevated temperatues and is extremely hard. Both these properties endow SiC devices with longevity. However, because of the unavailability of large-area single crystals and because of the high processing temperatures required, SiC LEDs have not remained competitive with recent commercial devices.

As an impurity, N makes SiC n-type, while Al, a shallow acceptor, renders SiC p-type and conducting. Be, B, Ga, and Sc are somewhat deeper acceptors, and are responsible for luminescent transitions.

The spectral ranges readily available from SiC LEDs are yellow with B, blue with Al, green with Sc, and red with Be. Boron is an efficient luminescence center, but is too deep to make the material p-type conducting. Therefore, a combination of B and Al is more suitable for LEDs [2.63]. Yellow-emitting LEDs of $2\,cm^2$ area have been fabricated with 1% external power efficiency [2.64]. One drawback of SiC yellow LEDs is a brightness saturation with increasing bias due to internal resistance. Blue-emitting p–n junctions can be made by diffusing Al at 2200 °C from an Al-doped p^+ layer containing $5\times10^{19}\,Al/cm^3$ [2.62]. Blue LEDs have a low external efficiency (10^{-5}). SiC LEDs can operate at temperatures in excess of 50 °C without a significant drop in output.

II–VI Compounds

Several good reviews describe II–VI compounds [2.65–67]. All the II–VI compounds have a direct energy gap in which radiative recombination is very efficient. Unfortunately, most of these materials are rather resistive, and can be made with only one type of conductivity. The difficulty in producing amphoteric doping is attributed to compensation by native defects during crystal growth, diffusion, or annealing (after implantation). p-n junctions have been made only in the small-band-gap compound CdTe. Although p–n junctions have been claimed for higher-gap materials such as CdS or ZnSe, there is no definitive evidence that type conversion was achieved. All the claims for a p-type layer in II–VI compounds were made for very thin layers, where it is difficult to distinguish between an inversion layer and an equally thin p-type layer. Moreover, a channeling effect could join Schottky barrier contacts used in Hall-effect and conductivity measurements, and give a p-type reading. Hence, although the distinction between p-type and inversion layer may be academic in the case of thin regions, this possibility for confusion must be indicated. From a practical point of view, if one can inject minority carriers after some "magic" surface treatment, it does not matter if the inversion is only skin deep; such a surface is useful for efficient light generation.

Robinson and *Kun* [2.9] have obtained injection luminescence in forward-biased ZnSe diodes treated with Iodine and Ga. *Park* and *Shin* [2.68] have used P implantation into ZnSe to make LEDs (P substitutes for Se to form an acceptor).

Of the wider-gap II–VI compounds, only ZnTe can be made conducting p-type. In principle, this compound can, therefore, be used to form p–n heterojunctions to inject holes into other II–VI compounds having a smaller energy gap ($\lesssim 2.2\,eV$); however, because of high internal resistance, the output

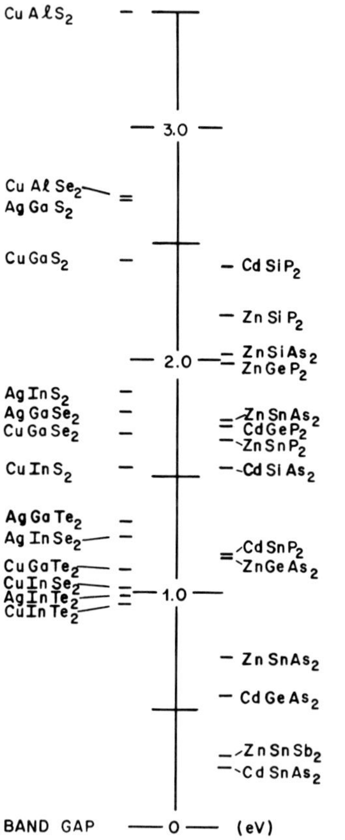

Fig. 2.17. Magnitude of the lowest energy gap of I–III–VI$_2$ and II–IV–V$_2$ compounds

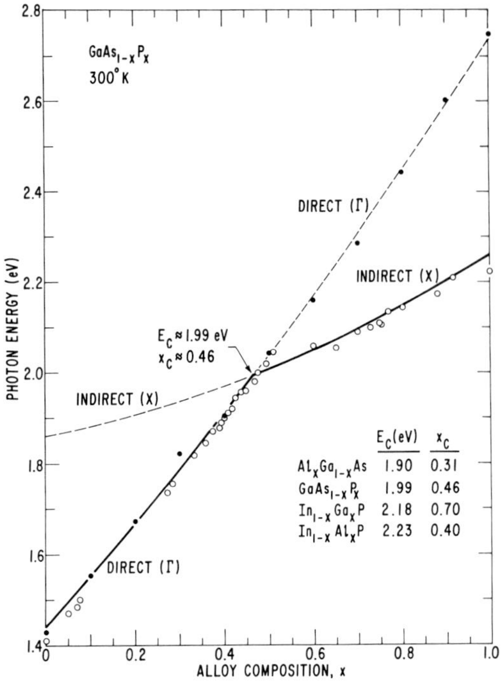

Fig. 2.18. Direct (Γ) and indirect (X) conduction-band minima for $GaAs_{1-x}P_x$ as a function of alloy composition x. The closed data points are from electroreflectance measurements [2.70], while the open points are from electroluminescence spectra [2.35, 71]. The values of E_c and x_c for $Al_xGa_{1-x}As$ and $In_{1-x}Al_xP$ are from [2.72]

of LEDs from these materials tends to saturate with increasing current. Moreover, the brightness of the prototype II–VI LEDs is not yet comparable to that of commercial III–V LEDs.

Chalcopyrites

Chalcopyrites are II–IV–V$_2$ or I–III–VI$_2$ compounds with III–V-like or II–VI-like properties; all have direct allowed transitions, and some can be made p-type and conducting [2.69]. Hence, they can be used to make p–n heterojunctions. The energy band gaps of these compounds are shown in Fig. 2.17. Of greatest interest is the heterojunction between n-type CdS and p-type $CuGaS_2$ [2.70]. Unfortunately, early results with low emission efficiency are not encouraging.

2.3.3 Ternary Alloys

A widely used approach to obtain materials with predetermined direct energy gaps is to form ternary alloys between two binary compounds having a common element, but one compound having too small a gap and the other one too large a gap. Both binary compounds need not have a direct gap. Thus, one can form a ternary alloy between a binary compound with a small direct energy gap and a second binary compound with a larger indirect energy gap (for example, $GaAs_{1-x}P_x$ from GaAs and GaP). In this way, the direct band structure of the smaller-energy-gap material is increased monotonically in the alloy as the content of the second compound is increased, as shown in Fig. 2.18 for a representative ternary alloy, $GaAs_{1-x}P_x$. For this alloy, as well as for the other materials listed in Fig. 2.18, the direct (Γ) conduction-band minimum lies below the indirect (X) conduction-band minima over a significant fraction of the alloy system. For values of x somewhat less than the crossover value x_c, each of the semiconductors in Fig. 2.18 is direct and of sufficiently high energy to provide visible emission. However, at alloy compositions close to x_c, some of the electrons that would otherwise occupy the direct conduction-band minimum at $k=0$ transfer to the indirect minima ($k \neq 0$) as a result of their small but finite thermal energy. This effect is particularly dramatic since there are typically 50–100 times more available energy states in the heavy-mass indirect minima than there are in the direct minimum [2.74]. It has been found experimentally for $Al_xGa_{1-x}As$ [2.75], $GaAs_{1-x}P_x$ [2.74], and $In_{1-x}Ga_xP$ [2.76], that such a transfer results in a severe reduction in the luminescence intensity, because the radiative recombination coefficient for electrons in the indirect valley is much smaller than for those in the direct valley (recall Table 2.2). Therefore, most electrons in the indirect valley do not radiate, while only the small porportion of electrons in the direct valley emit light.

The position of the conduction-band minima can be used to calculate the dependence of the emission spectrum, luminescence efficiency, and brightness on alloy composition [2.73, 77]. For these calculations, equal radiative (direct) and nonradiative lifetimes are assumed, which is equivalent to the assumption of a 50% internal quantum efficiency for the direct-band-gap materials. The calculated curves for the external quantum efficiency are shown in Fig. 2.19. Also superimposed on the efficiency curves are data points obtained for the respective alloys by electroluminescence or by cathodoluminescence. Figure 2.19 shows that $GaAs_{1-x}P_x$ has potential for higher efficiencies than $Al_xGa_{1-x}As$ for red-light emission, but that $In_{1-x}Ga_xP$ and $In_{1-x}Al_xP$ alloys have clearly higher potential than either $GaAs_{1-x}P_x$ or $Al_xGa_{1-x}As$ for red, yellow, or orange emission.

The potential brightness (per unit current density) for these alloys can be calculated from Fig. 2.19a and from (2.10). From these calculations, plotted in Fig. 2.19b, it is clear that the potential brightness of $In_{1-x}Ga_xP$ and $In_{1-x}Al_xP$ is about 20 times higher than that for $GaAs_{1-x}P_x$ and $Al_xGa_{1-x}As$. Equally important, the brightness maximum for $In_{1-x}Ga_xP$ occurs at about 2.03 eV,

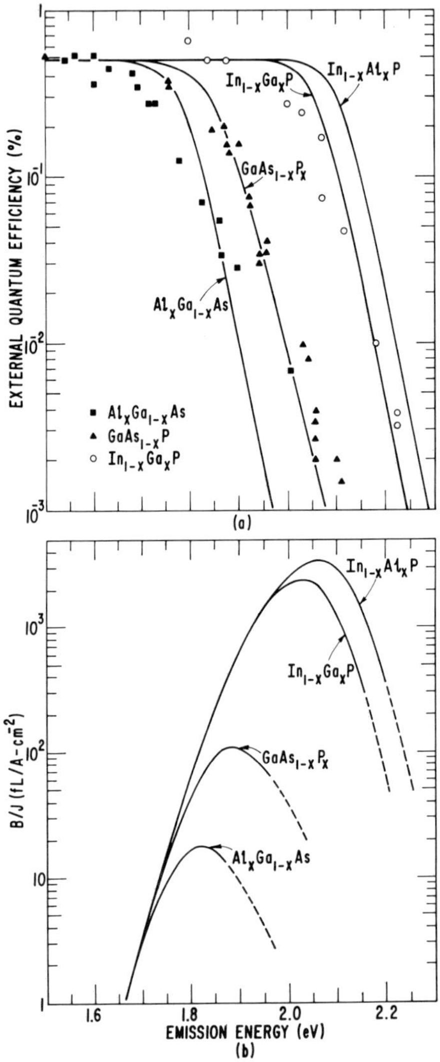

Fig. 2.19. (**a**) Calculated external quantum efficiency [2.72], for $Al_xGa_{1-x}As$, $GaAs_{1-x}P_x$, $In_{1-x}Ga_xP_x$, and $In_{1-x}Al_xP$ LEDs as a function of photon emission energy. The data points for $Al_xGa_{1-x}As$ [2.73, 77] and for $GaAs_{1-x}P_x$ [2.71] are relative electroluminescence efficiencies. The $In_{1-x}Ga_xP$ data [2.75] are relative photoluminescence efficiency values. (**b**) Calculated [2.72] brightness values (per unit current density) for the same alloys in (a)

which is in the orange portion of the optical spectrum. The curves in Fig. 2.19 are based solely on the energy-band structure of these alloys and the assumption of equal radiative and nonradiative lifetimes. These calculations do *not* take into consideration such important problems as the stability of the compound, the extent of lattice mismatch for the alloy, or the relative ease of crystal preparation. Such considerations are treated further in the following sections.

At values of x where the indirect valley robs electrons from the direct valley, nitrogen-doping enhances the luminescence efficiency. Just as in GaP, N in $GaAs_{1-x}P_x$ is an isoelectronic center, which relieves the momentum con-

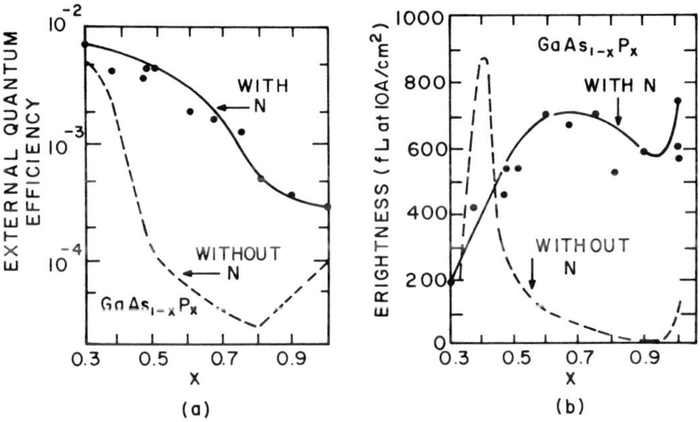

Fig. 2.20a, b. Electroluminescence comparison [2.79a] of GaAsP LEDs with and without nitrogen. (**a**) External quantum efficiency. (**b**) Brightness

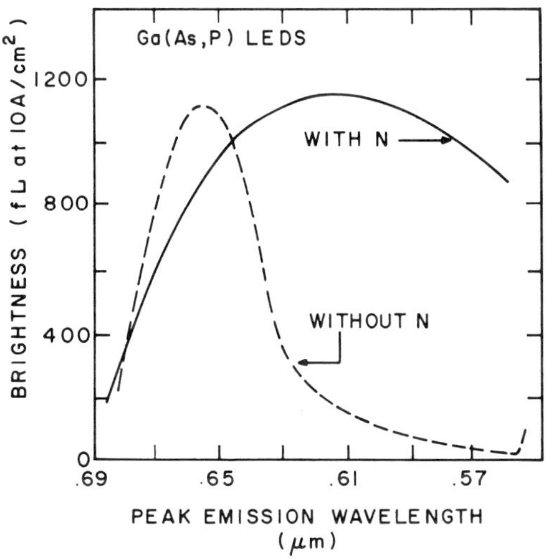

Fig. 2.21. Brightness vs peak emission wavelength for nitrogen-doped GaAsP LEDs [2.79a]

servation dilema. Hence, N-doping is widely used in commercial $GaAs_{1-x}P_x$ LEDs designed to emit yellow and orange light. The extent to which the diode efficiency and brightness are enhanced in GaAsP LEDs is illustrated in Figs. 2.20a, b, respectively [2.79a]. In these figures, only the portion of the GaAsP alloy system influenced by the indirect (X) conduction-band minima is shown, since the direct (Γ) recombination is not affected appreciably by the presence of N. In Fig. 2.21, we plot the brightness versus *emission wavelength* for N-doped GaAsP LEDs. Note that brightness values in excess of 800 fL can be obtained (at a current density of 10 A/cm^2) for all wavelengths between 650 nm (red) and 570 nm (yellow-green).

Alloys of some II–VI compounds can be made in all proportions, allowing a continuous variation of the energy gap over a range corresponding to the visible spectrum. Some of these are shown in Fig. 2.22. The alloy ZnS–ZnSe is

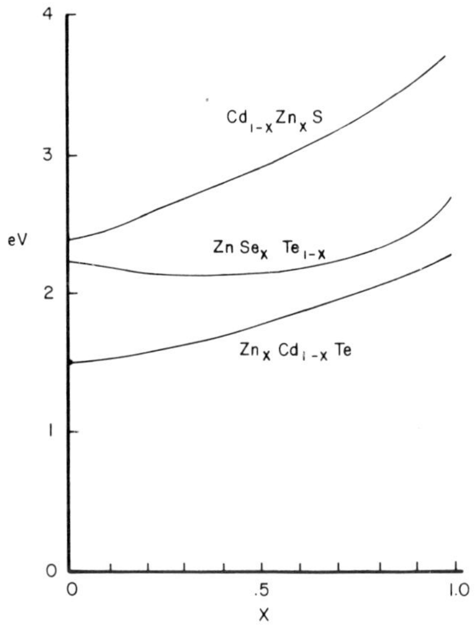

Fig. 2.22. Variation of band gap with composition in mixed crystals of II–VI compounds [2.67]

of particular interest because it can be made conducting n-type and is amenable to the formation of Schottky barriers. The system ZnTe–CdTe also is miscible in all proportions, and has the distinction of forming the only p-type conducting wide-gap II–VI material ($1.50\,\mathrm{eV} \leqq \mathrm{Eg} \leqq 2.28\,\mathrm{eV}$).

2.3.4 Quaternary Alloys

Quaternary alloys have one more degree of freedom than ternary alloys. This allows the independent control of energy gap and lattice constant over rather wide ranges. By way of example, the three-dimensional drawing of Fig. 2.23 [2.79b] illustrates the dependence of the energy band gap on alloy composition for the quaternary alloy InGaAsP [2.80a]. This system is bounded on four sides by the ternary alloys InGaP, InGaAs, GaAsP, and InAsP. The solid curves running approximately horizontally across the alloy system are constant-band-gap curves, and are seen to vary between 0.36 and 2.2 eV over the direct-band-gap portion of this alloy. The small GaP-rich portion of the alloy at the top of the illustration has an indirect energy-band structure. In principle, the InGaAsP system can provide wavelengths between 0.55 and 3.4 μm at room temperature.

The iso-lattice-constant curves, which are dashed in Fig. 2.23, are of particular importance. Any single iso-lattice-constant curve is seen to cross several iso-band-gap curves, illustrating that quaternary alloys with appreciably different energy-band-gap values (or refractive-index values) can be obtained without altering the lattice constant of these materials.

To date, quaternary alloys have generally been incorporated into multilayer heterojunction structures for the fabrication of laser diodes. Much of this

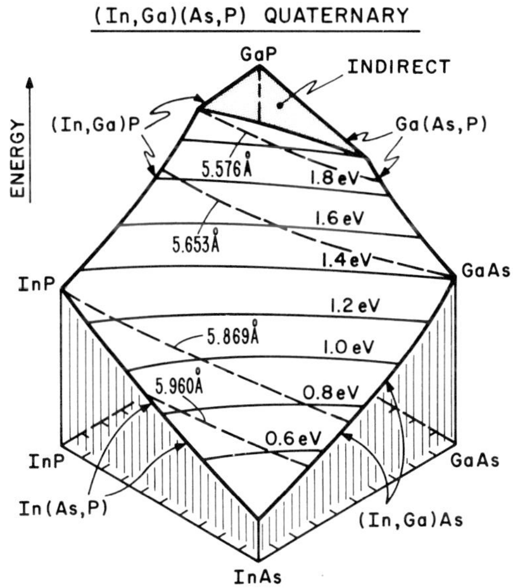

Fig. 2.23. Energy band gap vs composition for InGaAsP quaternary alloys. Nearly horizontal solid curves are iso-band-gap. Dashed curves are iso-lattice-constant [2.79b]

research involves the deposition of InGaAsP layers on InP substrates for long-wavelength (1.0–1.6 μm) lasers and LEDs for fiber optics. However, some research has been directed toward the deposition of InGaAsP lattice-matched alloys onto GaAsP substrates for visible-emitting diodes. This work has been pioneered by *Holonyak* and co-workers [2.80b–d], who have demonstrated low-temperature (77 K) laser operation at wavelengths between 670 nm (red) [2.80b] and 585 nm (yellow) [2.80c]. Although the emphasis of their work was not on room-temperature LEDs, the same structures should ultimately be capable of efficient visible (red, orange, or yellow) emission at room temperature.

2.4 Metallurgical Considerations

At this point, it may be relevant to point out potential problems that can result from neglect of subtle material properties. These problems manifest themselves as losses, and are listed in Table 2.5.

2.4.1 Crystal Defects

Over the past several years, it has become very clear that metallurgical defects, introduced into semiconductor crystals during growth or subsequent device processing, are very efficient *nonradiative* recombination centers. In particular, point defects such as precipitates and vacancies, and line defects such as

Table 2.5. Losses in semiconductor LEDs

Losses in excitation	Nonradiative recombination	Optical losses
1) Series resistance: a) at contacts b) within semiconductor	1) Surface recombination due to work damage; surface contamination during device fabrication. 2) Point imperfections: vacancies due to deviation from stoichiometry during growth, metal complexes and precipitates due to impurity concentrations approaching solubility in the material, and dislocations due to lattice mismatch between substrate and epitaxial layer. 3) Auger recombination due to high carrier concentrations. 4) Deep-level recombination due to contamination (e.g., Cu, O).	1) Absorption in bulk semiconductor 2) Shadowing by metal contacts 3) Reflection 4) Refraction

dislocations have been found to reduce minority carrier lifetimes and diode luminescence efficiencies by nonradiatively shunting the recombination process.

The fact that dislocations are localized nonradiative recombination centers is clearly illustrated in Fig. 2.24. Here, the intersection of dislocations, with the surface of a GaP crystal is delineated by "etch pits" (Fig. 2.24a), which are produced by preferential chemical etching. In Fig. 2.24b the cathodoluminescence emission from the same portion of the crystal is illustrated, and the vicinity of each etch pit is seen to be nonluminescent. The one-to-one

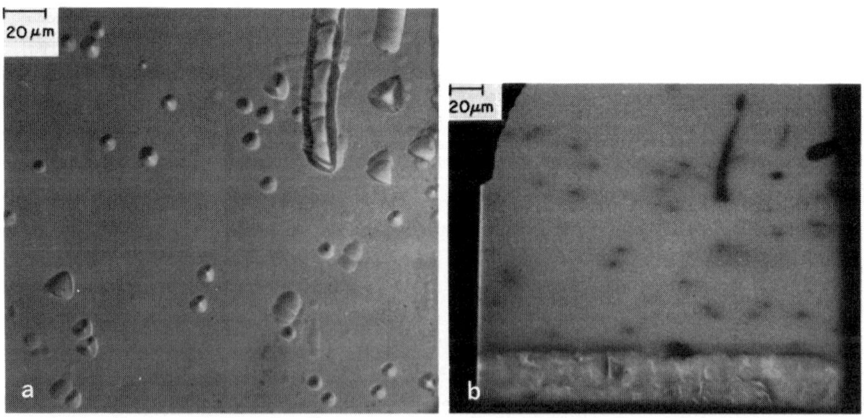

Fig. 2.24a, b. Topographical maps of dislocations in GaP. (**a**) Revealed by chemical etching. (**b**) Revealed by cathodoluminescence [2.82]

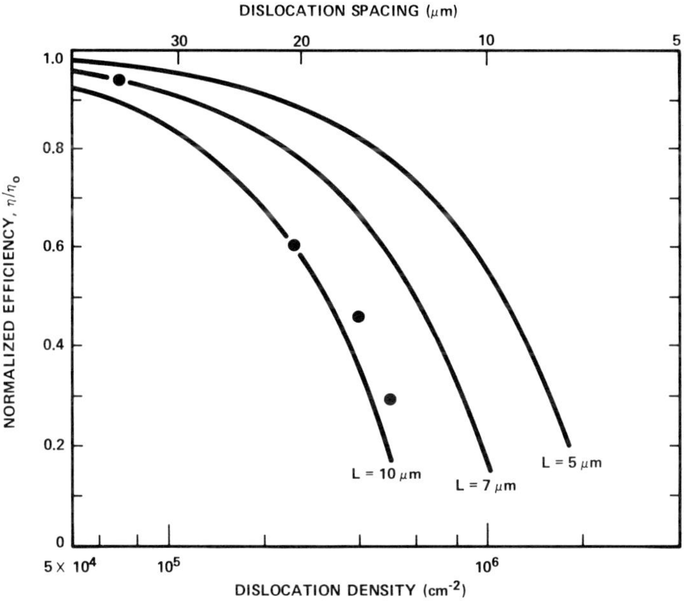

Fig. 2.25. Dependence of luminescence intensity on dislocation density and spacing in GaP. Solid curves are theoretical fits for various minority-carrier diffusion lengths [2.82]

correspondence between dislocations and nonradiative centers shown in Fig. 2.24, also has been observed in GaAsP [2.81]. In fact, it is generally believed that the formation of precipitates is largely responsible for the luminescence falloff that occurs at high Zn impurity concentrations ($\gtrsim 3 \times 10^{19}\,\mathrm{cm}^{-3}$), as indicated previously in Fig. 2.8.

Since dislocations exist in most semiconductor substrates in large concentrations (typically 10^2–10^4 dislocations/cm^2), and usually propagate into epitaxial layers during deposition onto the substrate, it is important to determine the net effect of localized nonradiative recombination at dislocations on LED performance. The impact of point defects on performance has been found to be determined by the defect spacing relative to the minority-carrier diffusion length. In general, luminescence degradation occurs when defect densities become sufficiently large that their spacing is only one or two times the diffusion length, as illustrated in Fig. 2.25 for dislocations in GaP [2.82]. Here, the points are efficiency values of LEDs fabricated on GaP chips having different dislocation densities. Note that the luminescence has degraded by 50% when the dislocation spacing (20 μm) is about twice as large as the calculated diffusion length (10 μm). Similar observations have been made on other luminescent semiconductors, especially GaAs [2.83a].

The deleterious effect of defects such as dislocations, precipitates, and stacking faults has become apparent simply because the size of such imperfections allows their ready observation by electron microscopy. Point defects such

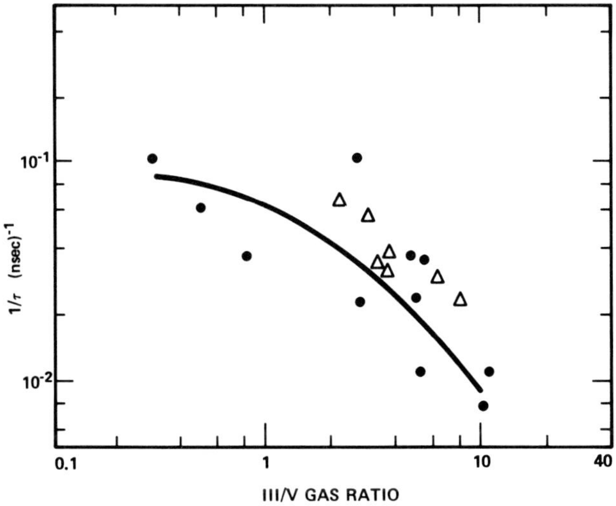

Fig. 2.26. Minority-carrier lifetime vs gas-phase stoichiometry (III/V gas ratio) in vapor-grown GaP [2.83b]

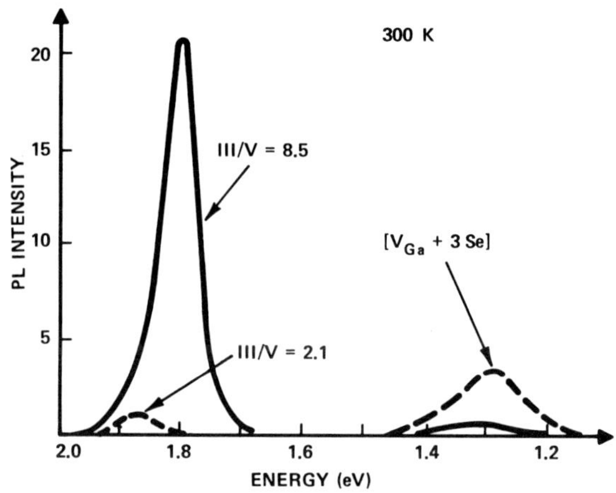

Fig. 2.27. Emission spectrum vs gas-phase stoichiometry in vapor-grown GaAsP. The low energy peak at 1.3 eV is thought to be due to Ga vacancy complexes [2.85]

as vacancies (and complexes of vacancies with impurities) are not readily observable, but have come to be recognized as dominant nonradiative centers. For example, a variety of indirect observations in III–V materials indicate that group III vacancies are particularly harmful. This is thought to be the major reason why the liquid-phase epitaxy of III–Vs such as GaP (from Ga-rich melts) has provided higher luminescence efficiency than has vapor-phase epitaxy (where growth conditions are not as rich in Ga). An indication of such an effect is shown in Fig. 2.26, where the minority-carrier lifetime of vapor-grown GaP layers is found to be strongly dependent on the ratio of Ga atoms to P atoms in the vapor ambient used during growth [2.83b]. The longer lifetimes produced under Ga-rich vapor conditions were found to provide higher brightness values

Table 2.6. III–V compound substrates

	E_g [eV]	T_g [°C]	a [Å]	Technique	Size [cm^2]	Disl.[a] Dens. [cm^{-2}]	Cost. [$/g]
GaP	2.25	1480	5.450	LE	10–20	10^4	15–20
GaAs	1.43	1240	5.653	CZ or GF	10–20	10^2–10^4	5–10
GaSb	0.69	712	6.095	CZ or GF	3–5	10^3	20
InP	1.28	1070	5.869	LE	5–15	5×10^3–10^5[b]	50
InAs	0.36	943	6.058	LE	2–7	10^3	30–35
InSb	0.17	523	6.479	CZ or BR	3–5	10^2–10^3	10–25

LE: Liquid Encapsulation; GF: Gradient Freeze; CZ: Czochralski; BR: Bridgeman.

[a] For a given compound, lower dislocation densities are usually obtained by Bridgeman or gradient-freeze techniques.

[b] Recently, zero-dislocation substrates of InP have been developed by Varian Associates, with a considerably higher cost.

in the LEDs of this experiment, as expected. A similar relationship between lifetime and vapor-phase stoichiometry has been reported in GaAs [2.84].

Another indication of the importance of crystal stoichiometry is presented in Fig. 2.27. Here, the photoluminescence spectra from vapor-grown GaAsP layers is shown for samples prepared with two different III/V gas-phase ratios [2.85]. Consistent with our expectations, the condition with the higher group III concentration (III/V = 8.5) significantly enhanced the near-band-gap photoluminescence signal at 1.8–1.9 eV. Interestingly, the low-energy peak near 1.3 eV is thought to be due to Ga complexes with the doping impurity (Se), and as expected, is smaller for the sample grown under Ga-rich vapor conditions.

2.4.2 Lattice Mismatch

Imperfections also can be introduced into a material during epitaxial growth if the lattice constants for the substrate and the epitaxial film differ. This problem can be especially important since only six binary compounds comprise the entire selection of commercially available substrates appropriate for epitaxial deposition, as described in Table 2.6. Since the density of lattice-misfit dislocations is directly related to the compositional gradient, one technique that is frequently used in ternary alloys to reduce these dislocations is to grade the composition from that at the substrate to that desired at the p–n junction. In this way, epitaxial layers with acceptably low dislocation densities can be grown routinely. Figure 2.28 illustrates the dependence of dislocation density on a (linear) compositional gradient for $GaAs_{1-x}P_x$ epitaxial layers deposited on GaAs substrates [2.86]. Recently, *Olsen* et al. [2.87a] have found that a series of *abrupt* grading steps can provide close to dislocation-free material in

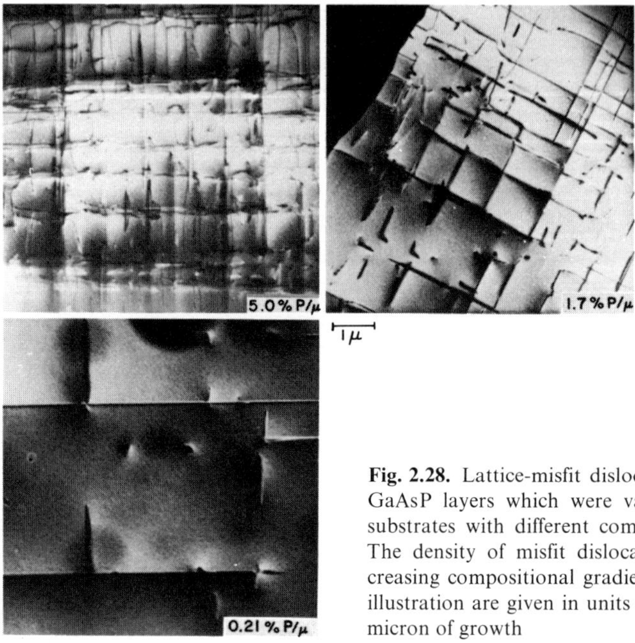

Fig. 2.28. Lattice-misfit dislocation networks [2.84] in GaAsP layers which were vapor deposited on GaAs substrates with different compositional gradient rates. The density of misfit dislocations decreases with decreasing compositional gradient. The gradients in each illustration are given in units of mole fraction GaP per micron of growth

the epitaxial layers deposited subsequent to the grade. Finally, we should add that even with "perfect" epitaxial layers, imperfections may be introduced during subsequent diffusion processes used to form the p-n junction.

2.4.3 Optical Losses

Another cause for losses in LEDs is the absorption of the light on the way out of the crystal. Hence, it is desirable to choose an appropriate substrate that is transparent to the emitted light. The use of GaP substrates on which narrower band-gap GaAsP is deposited has greatly improved the performance of these LEDs, as illustrated in Fig. 2.29. Other wide-gap semiconductors such as SiC, ZnS, and GaN are extremely transparent to visible light. So is sapphire, which can be used as a substrate for the deposition of many different semi-conductors.

The *extraction* of the emitted radiation from the semiconductor crystal also is a major problem. Only light approaching a flat crystal–air boundary within 16° of normal incidence is able to escape. This value is derived from Snell's law with a refractive index of 3.5 that is typical of III–V semiconductors. A major fraction of the light (i.e., all rays striking the surface at angles deviating from the normal by more than the critical angle of 16°) suffers total internal reflection. The total optical path inside the diode therefore exceeds the diode dimension, with the result that severe internal reabsorption of the emitted light occurs.

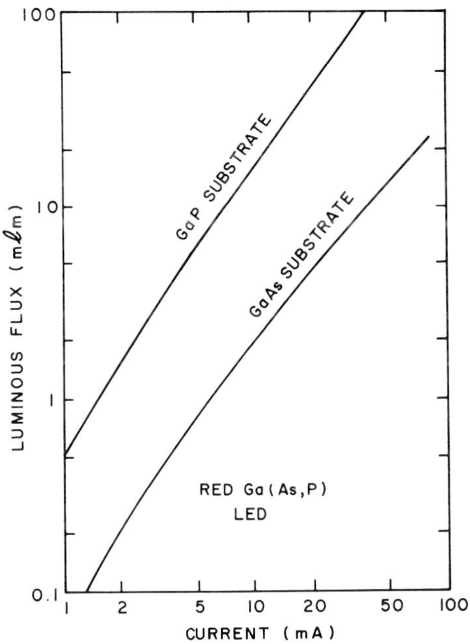

Fig. 2.29. Luminous flux vs applied current for red-light-emitting GaAsP LEDs prepared on GaAs and GaP substrates

Furthermore, even the small fraction of light approaching the surface within the critical angle does not totally escape because of partial reflection at the surface. A simple expression relating the external quantum efficiency η_{ext} and the internal quantum efficiency η_{int} is [2.88]

$$\eta_{ext} = \frac{\eta_{int}}{1 + \alpha \underline{V} / T_{av} A} . \tag{2.11}$$

Here, α is the average absorption coefficient in the diode, $\underline{V}$ is the diode volume, A is the area of the emitting surface, and T_{av} is the average surface transmissivity. For near-band-gap emission from flat direct-band-gap LEDs, the external quantum efficiency is typically 50–100 times less than the internal efficiency due to the reflection and absorption effects just described.

To improve the external efficiency, a number of techniques can be used to decrease the absorption coefficient α or to increase the transmissivity of the surface, as listed in Table 2.7. Dome shaping is effective [2.89], but also is the most expensive of these techniques, since it involves machining the semiconductor and uses a significant volume of material. Hemispherical domes and lenses cast from epoxy or acrylic–polyester resin (with a refractive index of about 1.5) simplify device construction, and are quite effective in increasing the external efficiency by a factor of typically 2–3, depending on the material.

The average internal absorption coefficient is appreciably higher in diodes that rely on near-band-gap emission than in diodes where the emission is much

Table 2.7. Methods for increasing external diode efficiency

Objective	Techniques
Minimize absorption coefficient α	1) Keep absorbing layers thin 2) Provide optical "window" with higher-energy-gap material. 3) Reduce carrier concentrations 4) Generate below-band-gap radiation.
Maximize surface transmissivity T	1) Shape semiconductor for normal incidence 2) Encapsulate diode with transparent material with high refractive index and shaped for normal incidence. 3) Apply antireflection coating.

below the band-gap energy. For example, for GaP:Zn, O diodes, the emission is far below the band-gap energy and α is very small ($<10\,\mathrm{cm}^{-1}$); however, since the *internal* quantum efficiency is only 10–20% [2.90, 91], the external quantum efficiency is still limited to a few percent.

2.4.4 Series Resistance

In an ideal LED, the external voltage is applied entirely across the p–n junction; however, in practice, a series resistance occurs at the contacts to the diode and across the bulk semiconductor layers comprising the diode structure. The total series resistance is on the order of 1–3 Ω for direct-band-gap III–V LEDs, but substantially higher (~10 Ω) in indirect-band-gap III–V LEDs due to reduced carrier mobilities and poor ohmic contacts. The voltage drop across the series resistance does *not* affect the *quantum* efficiency of the diode (defined as the photons generated per electron injected), but does reduce the *power efficiency* of the diode, as shown below. The power efficiency is

$$\eta_\mathrm{p}=\left(\frac{hv}{qV_\mathrm{a}}\right)\eta_\mathrm{int}\eta_x\,. \tag{2.12}$$

But if the applied voltage, V_a is dropped partly across a series resistance, R_s,

$$V_\mathrm{a}=V_\mathrm{j}+I_\mathrm{j}R_\mathrm{s}\,, \tag{2.13}$$

The power efficiency then becomes

$$\eta_\mathrm{p}=\frac{\left(\dfrac{hv}{qV_\mathrm{j}}\right)\eta_\mathrm{int}\eta_x}{1+\dfrac{I_\mathrm{j}R_\mathrm{s}}{V_\mathrm{j}}}=\frac{\eta_\mathrm{po}}{1+\dfrac{I_\mathrm{j}R_\mathrm{s}}{V_\mathrm{j}}}, \tag{2.14}$$

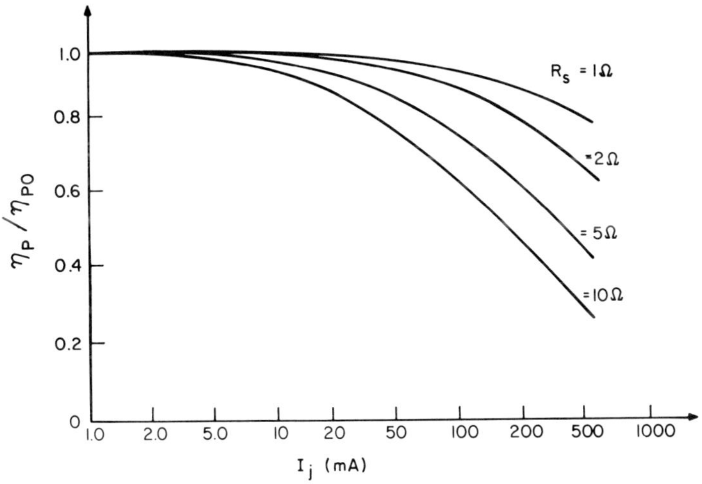

Fig. 2.30. Effect of LED series resistance R_s on power efficiency

where η_{po} is the power efficiency of an ideal junction diode (without series resistance). The reduction in power efficiency caused by the series resistance is illustrated in Fig. 2.30 as a function of applied current. It is clear that a series resistance of only 10 Ω can reduce the power efficiency by a factor of 2 at high current levels, but that resistance values of 1–2 Ω have little effect.

The effect is even more severe in several II–VI semiconductors such as ZnS, in which self-compensation prevents the attainment of high conductivities. In such materials, applied voltages on the order of 5–10 V are not uncommon, and power efficiencies can be one or two orders of magnitude smaller than quantum efficiencies. Hence, great care must be given to the doping of the layers adjacent to the electrodes to minimize the contact resistance (even though such optimized doping may be undesirable for efficient luminescence). It may even be advantageous to change the composition of the semiconductor (to reduce its band-gap) near the contact. The thickness of all the other layers also must be made small to minimize series resistance and maximize heat flow.

2.4.5 Degradation

The quantum efficiency and brightness of LEDs may decrease without evidence of mechanical damage in the course of forward-bias operation [2.92]. Several mechanisms are factors in such degradation: (i) increased surface leakage, (ii) in-diffusion of contaminants such as Cu [2.93], and (iii) formation of nonradiative recombination centers in the junction vicinity [2.94, 95].

Surface leakage and contamination problems are the simplest to eliminate with proper passivation, encapsulation, and cleaning techniques. The formation of nonradiative centers, however, is a far more difficult problem, and the causes are still not well understood. It is known that the degradation rate depends on the current density, and that a high concentration of imperfections such as dislocations increases the degradation rate for a given current density [2.94].

As in injection lasers, one mode of LED degradation during operation, especially in high-radiance GaAlAs LEDs, is by the generation of dark-line defects (DLD). These defects usually propagate along the ⟨100⟩ direction, and sometimes along the ⟨110⟩ direction. The growth of DLD requires current injection; it does not happen from moderate heating alone [2.96]. Not all diodes degrade by DLD. Many diodes fade uniformly and much more slowly than those exhibiting the DLD. The dicing of GaP chips induces a strain of $\sim 3 \times 10^7$ dyn/cm^2, which is responsible for the degradation of green LEDs. Etching away more than 2 μm of the n-type crystal surface removes the strain and extends the operating life beyond 10^3 h [2.97]. It is therefore essential that the defect density in LEDs be as low as possible.

Although data on LED reliability are limited, it has been established that properly fabricated diodes of GaP, $GaAs_{1-x}P_x$, and $Al_xGa_{1-x}As$ can be operated for many thousands of hours at moderate current densities (a few tens of A/cm^2). In fact, accelerated life tests on GaP [2.98] indicate an extrapolated half-life of many years. Clearly a long operating life is one of the features of LEDs, as compared to miniature incandescent sources. It is also clear, however, that careful life tests should be included in the evaluation of new materials and alloys where high defect densities are likely, before reaching a conclusion on the utility of such materials.

A green-emitting GaN LED has operated continuously for two years at a constant input of 32 mW without degradation.

2.5 LED Technology

Light-emitting diodes are generally fabricated from epitaxial material deposited on single-crystal GaAs or GaP substrates cut from melt-grown ingots. Both GaAs and GaP substrates are commercially available, with ingot cross sections of 2–4 cm. Ge substrates also have been used in recent years for the deposition of GaAsP epitaxial layers. Ge has nearly the same lattice constant as GaAs, and is somewhat less expensive. The choice of substrate depends on the lattice constant of the desired epitaxial layer and the thermal coefficient of expansion.

Both vapor-phase epitaxy (VPE) and liquid-phase epitaxy (LPE) are used to prepare materials for LEDs. Each technique has unique advantages in some cases. Vapor-phase epitaxy is used exclusively for GaAsP synthesis, while LPE yields the best GaP and AlGaAs devices at this time. In virtually all III–V LEDs, the epitaxial material is doped n-type, while the p–n junction is formed *after growth* by localized acceptor diffusion through masks formed by photolithographically-etched patterns in SiO_2. Zinc is the customary choice for the p-type dopant due to its large diffusion coefficient (short diffusion time) and high surface concentrations.

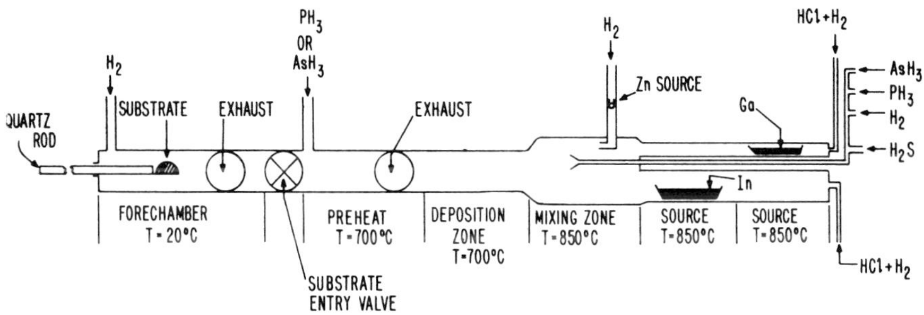

Fig. 2.31. Schematic view of the RCA vapor-phase epitaxy system

2.5.1 Vapor-Phase Epitaxy

Vapor-phase epitaxy has been applied to the deposition of a wide variety of electroluminescent III–V compounds and alloys [2.99]. This method not only permits materials to be prepared in a state of high purity and homogeneity, but also allows precise and independent control of layer thickness, alloy composition, and both n- and p-type doping. As a result, multilayer structures can be incorporated into a crystal entirely during the growth process.

A laboratory VPE apparatus used to prepare binary, ternary, or quaternary alloys of the elements In, Ga, As, and P is shown schematically in Fig. 2.31. The use of separate source zones for In and Ga allows independent control of the temperature and the HCl concentration at each metal source. The HCl reacts with the hot metals to form their (gaseous) monochlorides. These are carried in hydrogen into the mixing zone, where they combine but do not react with gaseous arsine or phosphine, the sources of arsenic and phosphorus. Mass-flow control valves are used on all input lines of the reactant species to ensure stable flow rates. The gaseous components react in the deposition zone, forming an epitaxial layer on the substrate which, during growth, is located in the deposition zone.

A typical growth process begins by insertion of the substrate into the forechamber while the flows of HCl are established over the metals and the Group V hydride gas(es) is (are) introduced. Composition is controlled by adjusting one or more of the alloying species (e.g., AsH_3 and PH_3 flows for GaAsP). When all flows have been established, the substrate is briefly preheated and then inserted into the deposition zone, the coolest region in the growth system. The reduced temperature allows the reaction between the chlorides, arsenic, phosphorus, and hydrogen to deposit the desired ternary or quaternary alloy as an epitaxial film and to form HCl as a gaseous product. Growth is terminated by withdrawing the wafer to the forechamber, where it cools to room temperature in a hydrogen atmosphere. A similar technique is employed to make the transition between adjoining layers of a heterojunction LED structure. Growth rates are typically 20–40 μm/h, and epitaxial layers as thin as

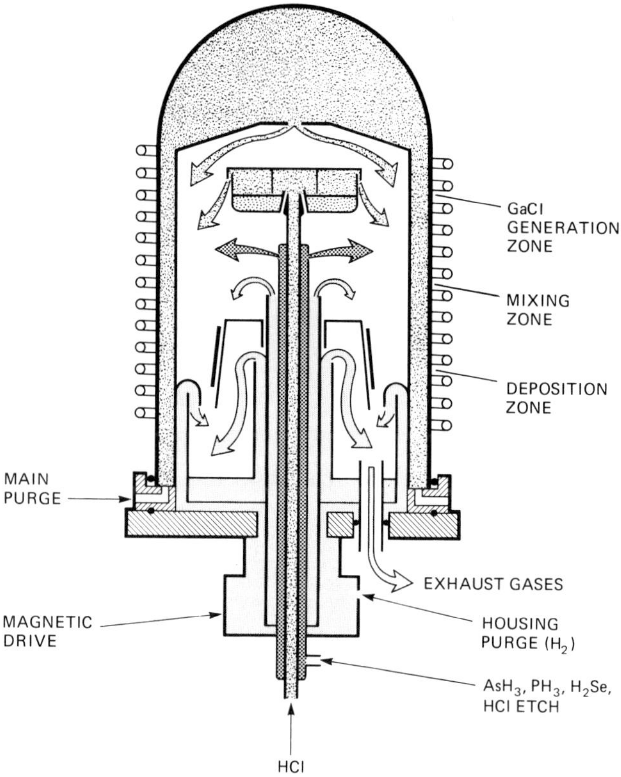

Fig. 2.32. Large commercial vertical reactor used for the vapor deposition of GaAsP material for electroluminescence applications. Schematic was graciously supplied by [2.102]

200 Å can be deposited [2.100] with present VPE techniques. Recent modifications to this system provide substrate rotation (for enhanced layer uniformity) and microprocessor control of all growth steps (for high reproducibility).

In vapor-grown epitaxial layers, the p–n junction can be incorporated during growth by the sequential introduction of gaseous donor and acceptor impurities such as H_2S (for sulfur) and Zn; however, the monolithic technology commonly used to form seven-segment numerical displays is more consistent with a postgrowth Zn diffusion through a photolithographically defined mask. Highest efficiencies to date also have been prepared with diffused p–n junctions.

Horizontal vapor-growth systems such as that shown in Fig. 2.31 were the first type to be developed for the preparation of III–V epitaxial layers [2.101], and are still widely used in several research and small-scale production facilities. However, for large-scale production, much larger, vertical reactors have been developed that can hold a large number of substrates simultaneously, thereby reducing the manufacturing cost per wafer (Fig. 2.32). The

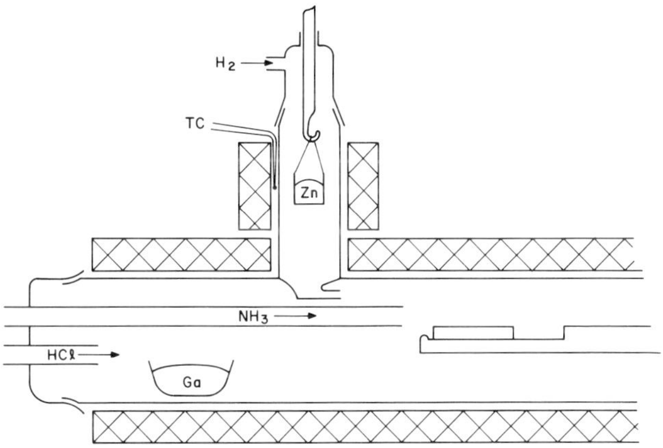

Fig. 2.33.
Furnace for growing GaN

primary advantage of the vertical system is the ease with which wafers can be supported in a concentric pattern about the growth axis with a single, large substrate holder. In present systems, the substrate holder is usually rotated slowly about the growth axis to compensate for irregularities in the vapor-flow patterns or the temperature profiles, thereby enhancing the uniformity of the growth rate and layer thicknesses. In some production systems, rf heating is used rather than conventional resistance heating, in order to hasten the heating–cooling cycles. Such large-scale reactors are now commercially available for the vapor deposition of n-type GaAsP layers for subsequent LED fabrication.

There are several reasons why GaAsP LEDs have received widespread commercial attention. The most important is the availability of growth equipment with multiwafer capability, as just described. Second, the surfaces of the epitaxial layers following growth are usually smooth and free of surface imperfections, eliminating the need for additional lapping or polishing steps. Finally, the ready availability of high quality GaAs and GaP substrates adds to the commercial viability of this process.

Historically, AlGaAs has *not* been prepared by vapor-phase epitaxy due to strong chemical reactions between AlCl and quartz. However, recently, high-quality AlGaAs heterojunction structures have been prepared (primarily for *laser* diodes) by an organometallic VPE process [2.103, 104].

Although many techniques have been used to prepare GaN, growth by vapor transport has enjoyed the greatest favor among researchers because of the availability of many volatile compounds of Ga: gallium chloride [2.44, 105], methylated gallium [2.106], and gallium amide [2.107]. Thus far, the best results have been obtained by the method of *Maruska* and *Tietjen* [2.44] using GaCl generated by passing HCl over Ga. This technique is illustrated in Fig. 2.33. The most commonly used substrate is sapphire. Here, it is convenient to introduce the impurity at a controlled vapor pressure by evaporating the element in a separately heated side-arm of the furnace (see

Fig. 2.33). Zn is the preferred acceptor, since it was found to compensate the native donors and make the material insulating [2.108]. The photoluminescence of Zn-doped GaN peaks at 2.85 eV, indicating that Zn forms a deep level about 0.6 eV above the valence band. Magnesium also compensates GaN, making it insulating with a luminescence peak at 2.92–2.98 eV; hence Mg forms a level about 0.5 eV above the valence band [2.109, 57].

SiC can be grown from the vapor phase by the thermal decomposition [2.110] of silicon–organic compounds such as methyltrichlorosilane, or by sublimation of powdered SiC – the Lely method [2.111]. Doping of SiC can be obtained during growth by providing the appropriate gaseous ambient (e.g., N_2).

2.5.2 Liquid-Phase Epitaxy

Beginning with the preparation of Ge for tunnel diodes [2.112], the deposition of semiconductor films by liquid-phase epitaxy (LPE) has rapidly been developed into a generally useful technique for the preparation of many III–V compounds. Basically, LPE involves the precipitation of material from a cooling solution onto an underlying substrate. The solution and the substrate are kept apart in the growth apparatus, and the solution is saturated with the growth material until the desired growth temperature is reached. The solution is then brought into contact with the substrate surface and allowed to cool at a rate and a time interval that are appropriate for the generation of the desired layer. When the substrate is single crystalline and the lattice constant of the precipitating material is the same or nearly the same as that of the substrate, the precipitating material forms an epitaxial layer on the substrate.

In the original "tipping" LPE technique (Fig. 2.34), a material such as GaP was grown from a Ga melt consisting of GaP (polycrystalline) and a dopant such as Zn (for p-type) or Te (for n-type). The melt was positioned at one end of a graphite boat, while a polished substrate was placed at the other end. The graphite boat was heated in a pure hydrogen atmosphere in a quartz furnace tube. The furnace, mounted on a hinge at its center, was tipped at the desired temperature ($\sim 1060\,^{\circ}$C), causing the melt to cover the substrate. Some dissolution of the substrate could be made to occur on contact, depending on the initial GaP content of the melt. As the furnace was slowly cooled, GaP precipitated from the melt in the form of an epitaxial deposition on the substrate.

A second LPE, called "dipping", also was used for early electroluminescent-diode preparation [2.113]. In this technique, the substrate was positioned in a holder just above the solution while the furnace temperature was increased; at the desired temperature, the substrate was immersed in the solution. Growth was terminated by withdrawal of the substrate from the solution at an appropriate temperature. The dipping technique was useful for cases when an oxide film forms on the solution during heating because the substrate was

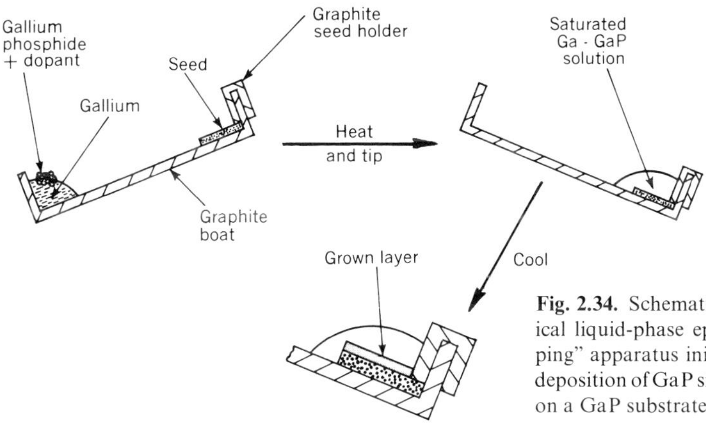

Fig. 2.34. Schematic diagram of typical liquid-phase epitaxy (LPE) "tipping" apparatus initially used for the deposition of GaP single-crystal layers on a GaP substrate

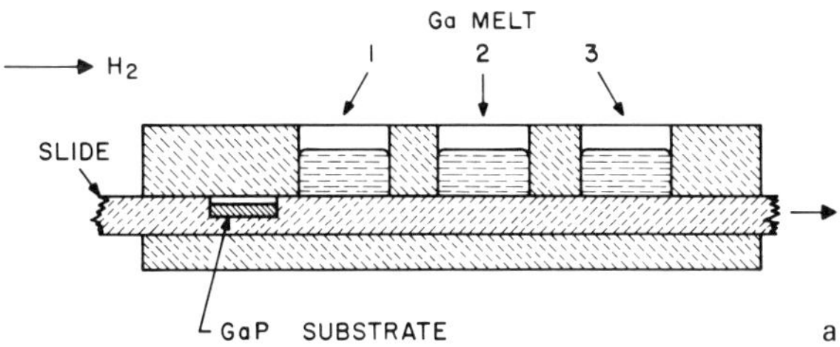

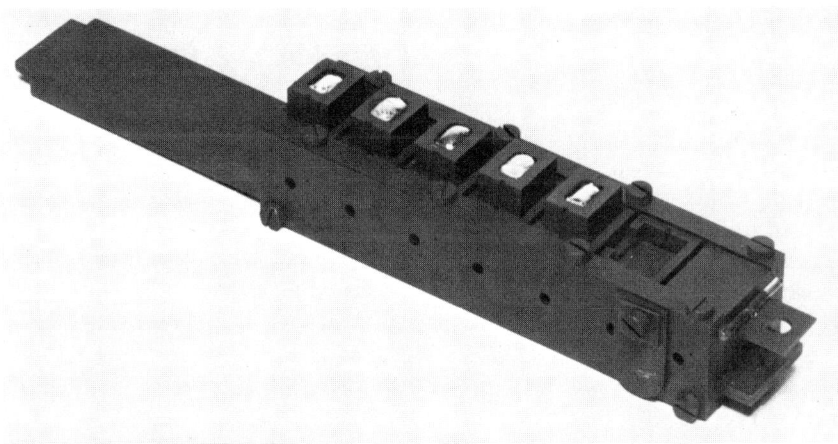

Fig. 2.35a, b. Modern multiple-bin boat used for the successive deposition of epitaxial layers on one substrate. (**a**) Schematic representation. (**b**) Actual photograph

inserted through the film and brought into contact with clean liquid. The vertical design also allowed doping of the solution during the growth cycle to form a p–n junction within the grown film by inserting a second dopant through a port in the substrate holder.

A third basic growth system uses a multibin boat [2.114]; this is an important innovation in LPE technology because it allows sequential deposition of several semiconductor layers during one growth cycle. The boat for this, illustrated in Fig. 2.35, is made of graphite, and is provided with several reservoirs and a moveable slide, the upper surface of which constitutes the bottom of the reservoirs. The substrate is placed in a recessed area of the slide and prior to growth is outside the reservoirs. The boat, with its bins filled with appropriate solutions, is inserted into a quartz furnace tube. A long rod fitted into a sleeve at the end of the furnace is used to slide the substrate sequentially into the bottom of the various reservoirs. In this manner, and with a choice of appropriate solutions and temperature schedules, different types of films can be deposited sequentially onto the substrate surface. Nitrogen-doped GaP LEDs usually are formed by passing NH_3 gas through the furnace at appropriate growth cycles. Multibin boats also have proven to be useful in the growth of AlGaAs alloys, where oxidation is a problem. More refined versions of the multiple bin have been developed [2.115, 116] that allow the controlled growth of layers as thin as 300 Å.

In addition to multilayer growth, the multiple-bin boat can be used to process several substrates at one time. Thus, economies due to multiple LPE wafer processing are possible, and in fact have been used for commercial production. For the most part, however, the LPE technology is used for AlGaAs laser diodes and infrared-emitting LEDs, where high performance is more essential than economy.

SiC has been grown from a solution of carbon in silicon at about 1800 °C [2.62]. A p–n junction in SiC can be obtained by alloying Si in a nitrogen ambient to p-type SiC.

2.5.3 Impurity Diffusion

In contrast to the routine diffusion techniques used to make silicon devices, diffusion into compound semiconductors must be tailored to the individual properties of each material, since many compounds tend to dissociate at the diffusion temperature. Acceptors, rather than donors, are usually used for impurity diffusion into III–V compounds. This is because: (i) the solubility of acceptors, notably Zn, in most materials is high (typically $>10^{19}\,\mathrm{cm}^{-3}$ at diffusion temperatures of 800–900 °C). This readily provides type conversion in moderately doped n-type layers to form the desired p–n junction. (ii) Donors often interact with the host atoms during diffusion (e.g., Se diffused into GaAs can form a second phase of Ga_2Se_3). This interaction attacks the surface of the crystal, destroying the mirror-smooth finish necessary for subsequent processing; moreover, such inclusions may induce stresses and nonradiative states.

One important feature of impurity incorporation by *diffusion* rests in the fact that it can be localized in selected portions of the crystal to provide specific geometries (e.g., rectangular bars for a seven-segment numeric display). This is usually accomplished by gaseous diffusion through a dielectric mask. SiO_2 films deposited from silane and oxygen are the simplest such mask; however, at moderate temperatures SiO_2 is relatively porous to the Zn atoms often used for p–n junction formation. The addition of a few percent of phosphorus to the SiO_2 dramatically increases the masking capability of the film. Deposited layers of Si_3N_4 are even more inpenetrable, and have gained recent attention for this reason.

The dissociation of the III–Vs at elevated temperature requires maintaining a background pressure of the most volatile component (As for GaAsP, P for GaP, N for GaN, etc.). The crystal and the dopant are usually sealed in an ampoule, although open-tube flow systems also have been used. The background pressure can be derived from the impurity source itself (e.g., $ZnAs_2$ [2.117] or ZnP_2 [2.118]), however sometimes the group V element (As or P) is added individually to small amounts of elemental Zn.

The diffusion of Zn is dominantly an interstitial process [2.119]. At normal diffusion temperatures (800–900 °C), surface concentrations between 10^{19} and $10^{20}\,cm^{-3}$ usually occur. Many years ago, *Herzog* [2.120] found that the recombination in GaAs is predominantly nonradiative for acceptor concentrations exceeding about $3\times10^{19}\,cm^{-3}$. As a result, diffusion techniques have been developed that provide surface concentrations less than this value. One common technique involves the use of a Ga melt containing only a few percent of Zn. The partial pressure of Zn over the Ga–Zn melt is much less than the vapor pressure of a pure Zn source, thereby reducing the Zn concentration in the semiconductor during diffusion. Other techniques involve the use of Zn-containing glasses, and multiple-step diffusions in which the Zn is incorporated into the surface region of the crystal at a low temperature, and driven in during a subsequent diffusion at a higher temperature.

2.5.4 Glow-Discharge Decomposition

In the glow-discharge decomposition method, a volatile compound is ionized and disassembled by collisions in a high electric field; the resulting ions form an amorphous solid which can be crystallized or used in the amorphous state. It is conceivable that this technique may become practical for the manufacture of LEDs. Although amorphous materials are usually full of defects, it is found that hydrogen, by tying the dangling bonds, greatly reduces the number of gap-states and thus allows doping so that p–n junctions can be made. Injection luminescence has been demonstrated in amorphous Si p–i–n and Schottky barrier diodes emitting (efficiently at low temperature) 1.3 eV photons when forward biased [2.121]. The emission of visible light by photoluminescence in amorphous SiC has also been demonstrated [2.122]. Crystalline SiC has also been obtained by the glow discharge decomposition of methyltrichlorosilane [2.123]. The glow-discharge decomposition is a very new technology that currently concentrates on photovoltaic applications [2.124] where low cost is believed to more than compensate for low efficiency. As this technique is further developed, its application to low-cost LEDs may become possible.

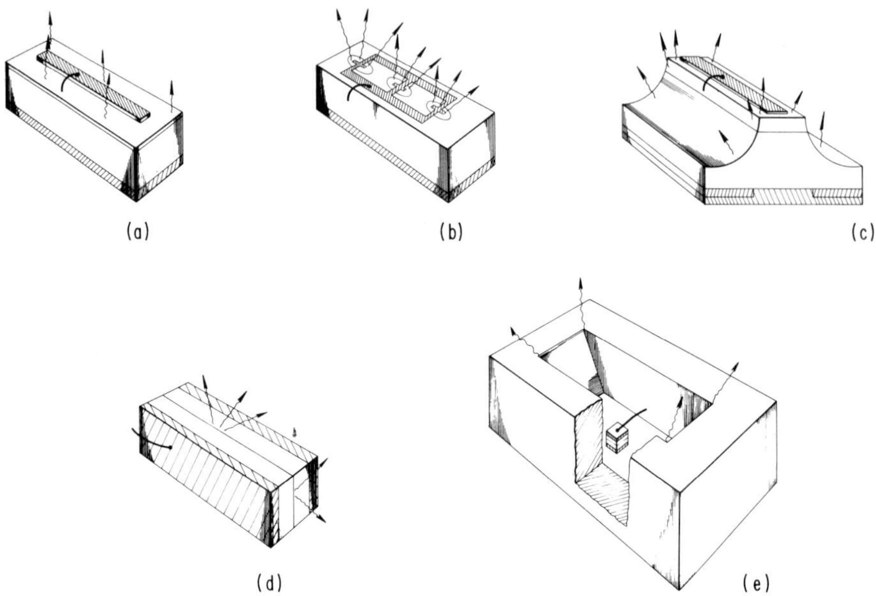

Fig. 2.36a–e. LED geometries used for the fabrication of seven-segment displays. (**a**)–(**c**) Surface emitters. (**d**) Edge emitter. (**e**) Cavity emitter

2.5.5 Device Structures

Light-emitting-diode materials are most widely used in the form of numeric displays consisting of seven segments. A variety of methods has been used to generate these segments. Some fabrication methods are more suitable for monolithic displays where all seven segments are produced at the same time on a common substrate, and others are used to generate individual diodes for panel lights, "on-off" indicators, etc. The choice of a particular technological approach to display fabrication is dictated primarily by the desired display size and the material used. In Fig. 2.36 we illustrate a number of designs for a LED segment, and will now describe how they are fabricated and under what conditions they are most useful.

Surface Emitters. The dominant emission here is directed out of one of the two large surfaces parallel to the p–n junction plane. In such structures, the uppermost layer is kept thin ($<5\,\mu m$) to reduce absorption losses, particularly in direct-band-gap materials. Current spreading from the Ohmic contacts to a thin layer can be a problem, and a compromise must be attained between small contact area (for optical transmission) and total junction coverage (for current spreading). Figure 2.36a illustrates the simplest type of surface emitter. Here, the contact bar can be simply evaporated through a metal mask, eliminating the need for photolithography.

To obtain higher brightness values or lower current requirements, a segment can be simulated by a line of small dots, as shown in Fig. 2.36b. This type of display is fabricated by diffusing Zn through openings in an SiO_2 diffusion mask into n-type material, and is obviously limited to those types of materials (e.g., $GaAs_{1-x}P_x$) where efficient electroluminescence can be obtained from diffused p–n junctions. The contacts in Fig. 2.36b are evaporated, the unwanted parts being subsequently removed by photolithographic techniques.

Figure 2.36c shows a surface-emitting segment designed for a monolithic GaP display [2.125], in which the high transparency of the material requires special techniques to avoid optical cross-coupling between segments. This is accomplished by etching a mesa structure, as shown, and applying the proper combination of absorbing and reflecting films on the bottom surface.

Edge emitters (Fig. 2.36d) are made by sawing the material into bars, and so positioning them that the exposed p–n junction faces the viewer. Light emitted from the sides and bottom is largely wasted, and the useful emission is concentrated into a very narrow region approximating a line. Since the brightness varies inversely with the emitting area (2.10), very bright, large-size segmented displays can be made in this way. The assembly process is not simple because of the mechanical problems in containing the sides of very small bars; nevertheless, red and green GaP displays using this approach have been reported [2.126].

A *cavity structure* [2.127] shown in Fig. 2.36e, uses little material and has low power consumption. The small diode chip in the center of the cavity radiates in all directions and thus operates at the peak light-utilization efficiency. Used with both GaP and $GaAs_{1-x}P_x$, it provides very pleasing segmented displays, but with somewhat lower brightness than the previous designs. Not shown in Fig. 2.36e is a plastic diffusing surface that is mounted on top of the cavity to trap the emitted radiation and form a rectangular image.

A newer structure in GaP LEDs was devised by *Inoue* et al. [2.128] utilizing the high refractive index of GaP and its crystalline properties to enhance the brightness profile for normal viewing of the display. Selective etching with phosphoric acid works fastest along (111) planes at 180 °C, producing very smooth mirror-like facets, as shown in Fig. 2.37. The gold electrode forms an etching mask so that an array of truncated pyramids can be generated. The GaP wafer is then bonded to a printed circuit so that the viewer sees only the n-type surface free of electrodes. The ohmic connection is obtained by applying a reverse bias across the junction, causing it to break down and to form a permanent short. The light emitted at the forward-biased p–n junction is reflected by the etched facets and transmitted by the viewing surface, with the brightness profile shown in Fig. 2.37.

A "double-base diode" can be used to control the area of the diode where injection luminescence will occur (Fig. 2.38). A control current along one layer adjusts the potential distribution in that layer, so that part of the junction is reverse biased and nonemitting. The series resistance of the second layer tends

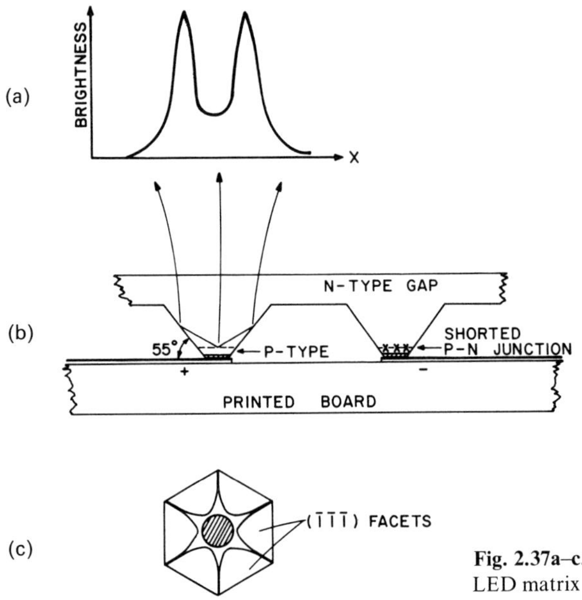

Fig. 2.37a–c. Cross-sectional view of GaP LED matrix

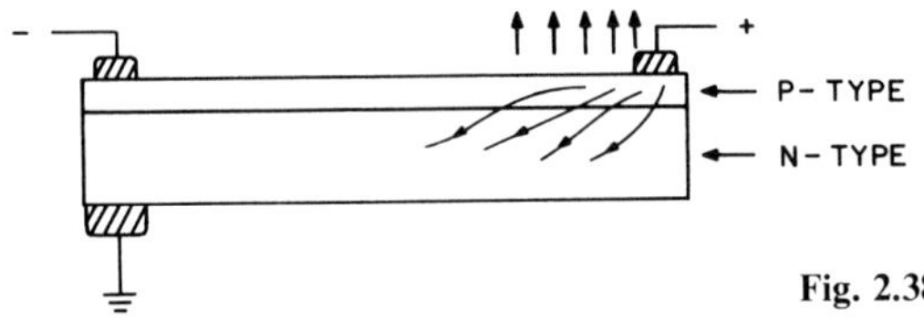

Fig. 2.38. Double-base bar-graph display device

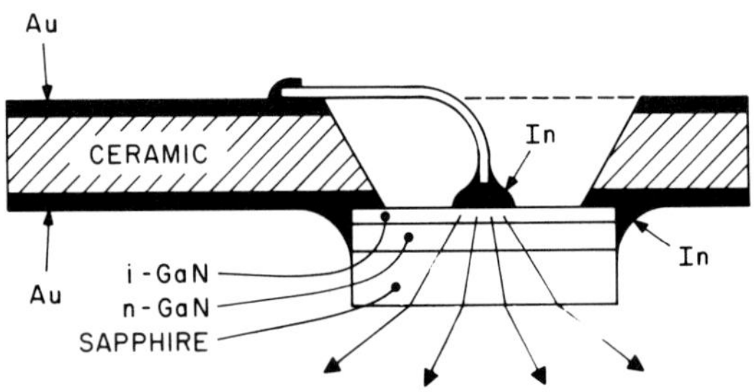

Fig. 2.39. Structure of the GaN LED

to limit the current density away from the bottom contact. Such a structure was fabricated using a SiC bar ($8 \times 0.2\,\mathrm{mm}^2$) [2.129]. The position of the light-to-dark boundary was resolvable with a 10% accuracy.

GaN. GaN LEDs have been made with the structure shown in Fig. 2.39. An undoped n-type conduction layer is grown on sapphire. Then a thin Zn-doped insulating layer is grown on the n-type layer. The wafer is cut into 1 mm diameter discs, which are then soldered with indium to a ceramic support. This

Fig. 2.40. Photograph of seven-bar numeric patterns of a GaN wafer viewed through the sapphire substrate. The wafer is about 2 cm in the long dimension

peripheral solder makes an ohmic connection to the n-type layer. A small indium dot soldered at the center of the insulating layer forms a $10^{-3}\,cm^2$ area of nonohmic connection.

To make an array of 7-bar numeric displays, metallic electrodes are evaporated through a mask on the Zn-doped layer to form the desired pattern (Fig. 2.40). The crystal is cut into rectangles, leads are soldered to the evaporated electrodes, and an ohmic connection is made to the n-type layer. The device is ready to operate. However, to secure the leads beyond the strength of the soldered contact, a layer of epoxy is poured on the GaN crystal, thus embedding the ends of the leads. Black epoxy is used to minimize the reflection of ambient light. The light is emitted through the sapphire substrate, which results in a high contrast ratio against the black background of the epoxy.

Compared to GaP green LEDs, an important advantage of GaN LEDs is that although the light is emitted under the contact, it is viewed through a very transparent substrate so that there is no shadowing by the contact. With the black background, the contrast is optimized without the need for color filters. A GaN display can be viewed in full sunlight, where the commercial red displays are barely visible.

Upconversion Phosphors. As an alternative to the use of high-energy-gap semiconducting compounds to obtain visible emission, one can employ a suitably doped "upconverting" phosphor, in which two or three infrared photons are converted into a single higher-energy photon in the visible portion of the spectrum [2.130]. For this technique, a highly efficient diode which emits in the proper infrared wavelength range (e.g., GaAs:Si with $\eta_{ext} \approx 10\%$ at 930–970 nm) is coated with a layer of phosphor doped with rare-earth impurities [2.131] (Fig. 2.41a). The combination of phosphor host and impurity is chosen for the particular emission wavelength desired.

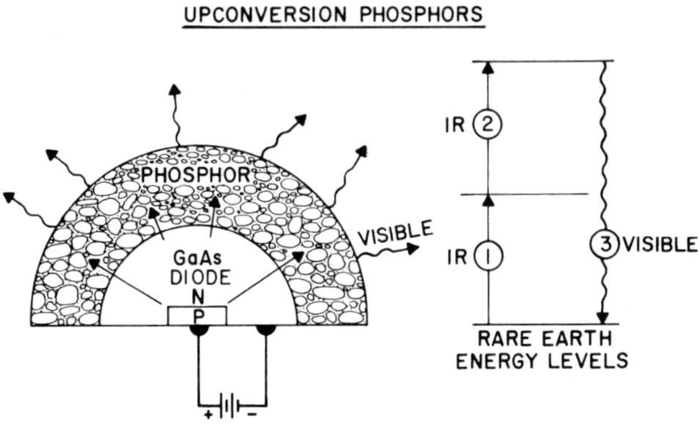

Fig. 2.41. Simplified representation of upconversion process for rare-earth-doped phosphors

The basic upconversion process is illustrated schematically in Fig. 2.41b. Here, the absorption of one infrared photon excites a rare-earth atom to level A, after which a second absorbed infrared photon can excite it to level B. The subsequent radiative decay of the excited state to its original ground state emits a visible photon. In this way, upconversion to the red, green, and blue has been achieved. Because the upconversion is a two-step process (or even a three-step process for blue emission), the intensity of the emitted visible radiation is approximately proportional to the second (or third) power of the infrared excitation intensity. Hence, for high-efficiency operation of the diode–phosphor system, the GaAs diode must be driven to very high current densities that are best achieved under pulsed conditions.

A fundamental limitation of the upconversion scheme lies in the difficulty of adequately coupling the infrared diode emission to the phosphor. Although the conversion efficiency of the phosphor itself is quite reasonable (30–55%) [2.132], only a small fraction of the diode emission is absorbed by the phosphor, which must be kept quite thin in order to reduce self-absorption of the visible radiation. This means that the diode–phosphor system produces a large amount of undesired GaAs infrared radiation as well as the visible emission.

2.6 Applications

The importance of LEDs is demonstrated by their widespread acceptance in commercial products. Large numbers of discrete LEDs are being used for pilot lights, often as a substitute for neon lamps. Most new high-fidelity audio equipment and industrial instruments have LED indicators. For such applications, the LED encapsulation can be shaped in several different ways to control the radiation pattern: a hemispherical lens (or the inclusion of light-scattering particles such as TiO_2) provides a wide angle of viewing; a directional lens is more suitable to send a bright signal to a pilot or to a driver, who is fixed in position relative to the LED.

The major application of LEDs, however, is in monolithic displays with seven bar segments (for numerals) or alphanumeric displays with a 35 dot matrix arranged in a 5×7 array. The small size and high contrast ratio of such displays fostered the growth of the digital wrist watch and portable calculator business. However, as these products have grown increasingly smaller, reduced battery size has made current consumption a critical parameter, and liquid-crystal displays are rapidly replacing LEDs in this application.

More versatile LED arrays have a larger number of elements, for example: the Hewlett Packard monolithic 30×36 LED array [2.133]. Four such 1.25 cm square chips can be mounted side by side on a printed circuit board to form a display 1.25 cm high and 5 cm wide, consisting of 144 columns of 30 diodes each. This panel can display 36 alphanumeric characters in up to three rows, or fewer rows of characters in any font or format. It also can display mathematical equations. Bar graphs can be generated by linear arrays of LEDs.

A more futuristic application of LEDs that has already been partly demonstrated by Sanyo Electric Co. is in the flat-panel TV [2.134]. In an early version, an automated tester selected individual GaP LED chips and bonded them in place in a 160×112 matrix on a ceramic support. The diodes are x–y addressed to present a broad-spectrum greenish picture with a 40 fL maximum total brightness. A gray scale of 16 levels was achieved by pulse-width modulation. In a more recent version, squared wafers of GaP comprising an x–y addressed matrix of diodes were assembled in a mosaic to form a larger area. The uniformity of the diodes over a wafer and from wafer to wafer was remarkable in the demonstrated panel. The concept of automated assembly was also employed by *Harris* [2.135] to fabricate two-color modules comprising x–y addressed red- and green-emitting LEDs in a 32×64 matrix. Each LED could be addressed individually to produce either red or green light or they could be turned on simultaneously to emit amber light. Peripheral interconnect electrodes allowed several modules (measuring $3.75 \times 7.5\,\text{cm}^2$) to be assembled into a larger area display.

References

2.1 O.V.Losev: Philos. Mag. **6**, 1024 (1928)
2.2 R.J.Haynes: Phys. Rev. **98**, 1866 (1955)
2.3 J.I.Pankove, M.Massoulie: Bull. Am. Phys. Soc. **7**, 88 (1962); J. Electrochem. Soc. **11**, 71 (1962)
2.4 N.Holonyak, Jr., S.F.Bevaqua: Appl. Phys. Lett. **1**, 82 (1962)
2.5 H.G.Grimmeiss, H.Scholz: Phys. Lett. **8**, 233 (1964)
2.6 P.J.Dean, M.Gershenzon, G.Kaminsky: J. Appl. Phys. **38**, 5332 (1967)
2.7 J.I.Pankove, E.A.Miller, J.E.Berkeyheiser: J. Lumin. **5**, 84 (1972)
J.I.Pankove: J. Lumin. **7**, 114 (1973)
2.8 S.Oda, H.Kukimoto: IEEE Trans. ED-**24**, 956 (1977)
2.9 R.J.Robinson, Z.K.Kun: Appl. Phys. Lett. **27**, 74 (1975)

2.10 O.V.Bogdankevich, S.A.Darznek, P.G.Eliseev: *Semiconductor Lasers* (Nauka, Moscow 1976) p. 333
2.11 W.Shockley: Bell Syst. Tech. J. **28**, 435 (1949)
2.12 C.T.Sah, R.N.Noyce, W.Shockley: Proc. IRE **45**, 1228 (1957)
2.13 C.H.Henry, R.A.Logan, F.R.Merritt: Appl. Phys. Lett. **31**, 454 (1977)
2.14 J.I.Pankove: In *Progress in Semiconductors*, Vol. 9, ed. by A.F.Gibson, R.E.Burgess (Heywood, London 1965) p. 48
2.15 H.Kroemer: Proc. IRE **45**, 1535 (1957)
2.16 A.G.Fisher: In *Luminescence of Inorganic Solids*, ed. by P.Goldberg (Academic Press, New York 1966) p. 541
2.17 H.Kressel: J. Electron. Mater. **4**, 1081 (1975)
2.18 A.S.Grove: *Physics and Technology of Semiconductor Devices* (Wiley, New York 1967) Part III
S.M.Sze: *Physics of Semiconductor Devices* (Wiley, New York 1969) pp. 425–504
2.19 C.N.Berglund: Appl. Phys. Lett. **9**, 441 (1966)
2.20 J.I.Pankove, P.E.Norris: RCA Rev. **33**, 377 (1972)
2.21 I.Giaever: Phys. Rev. Lett. **5**, 147 (1960)
2.22 M.D.Clark, S.Baidyaroy, F.Ryan, J.M.Ballantyne: Appl. Phys. Lett. **28**, 36 (1976)
2.23a Y.P.Varshni: Phys. Status Solidi **19**, 459 (1967)
2.23b D.A.Cusano: Solid State Commun. **2**, 353 (1964)
2.23c B.Tuck: J. Phys. Chem. Sol. **28**, 2161 (1967)
2.23d H.J.Queisser, M.B.Panish: J. Phys. Chem. Soc. **28**, 1177 (1967)
2.24 J.Starkiewicz, J.W.Allen: J. Phys. Chem. Sol. **23**, 881 (1961)
2.25 M.R.Lorenz, M.Pilkuhn: J. Appl. Phys. **37**, 4094 (1966)
2.26 A.Many, Y.Goldstein, N.B.Grover: *Semiconductor Surfaces* (North-Holland, Amsterdam 1965) p. 434
2.27 N.G.Davison, J.D.Levine: Solid State Phys. – Adv. Res. Appl. **25**, 1 (1970)
2.28a C.T.Sah, R.N.Noyce, W.Shockley: Proc. IRE **45**, 1228 (1957)
2.28b C.J.Nuese, J.J.Gannon, H.F.Gossenberger, C.R.Wronski: J. Electron. Mater. **2**, 571 (1973)
2.29 W.Heinke, H.J.Queisser: Phys. Rev. Lett. **33**, 1082 (1974)
2.30 A.Y.Cho, I.Hayashi: Solid-State Electron. **14**, 125 (1971)
2.31 J.S.Blakemore: *Semiconductor Statistics* (Pergamon Press, Oxford 1962) p. 214
2.32 M.I.Nathan, T.N.Morgan, G.Burns, A.E.Michels: Phys. Rev. **146**, 570 (1966)
2.33 J.I.Pankove, L.Tomasetta, B.F.Williams: Phys. Rev. Lett. **27**, 29 (1971)
2.34 J.P.Dismukes, W.M.Yim, J.J.Tietjen, R.E.Novak: RCA Rev. **31**, 680 (1970)
2.35 W.O.Groves, A.H.Herzog, M.G.Craford: Appl. Phys. Lett. **19**, 184 (1971)
2.36 R.H.Saul, J.Armstrong, W.H.Hackett, Jr.: Appl. Phys. Lett. **15**, 229 (1969)
2.37 R.Solomon, D.DeFevere: Appl. Phys. Lett. **21**, 257 (1972)
2.38 I.Ladany, H.Kressel: RCA Rev. **33**, 517 (1972)
2.39 W.Rosenzweig, R.A.Logan, W.Wiegmann: Solid-State Electron. **14**, 655 (1971)
2.40 R.Nicklin, C.D.Mobsby, G.Lidgard, P.B.Hart: J. Phys. C**4**, L3444 (1971)
2.41 R.N.Bhargava, C.Michel, W.L.Lupatkin, R.L.Bronnes, S.K.Kurtz: Appl. Phys. Lett. **20**, 227 (1972)
2.42 K.Lawley: Monsanto Co., Cupertino, California (private communication)
2.43 R.M.Potter, J.M.Blank, A.Addamiano: J. Appl. Phys. **40**, 2253 (1969)
2.44 H.P.Maruska, J.J.Tietjen: Appl. Phys. Lett. **15**, 327 (1969)
2.45 R.Madar, G.Jacob, J.Hallais, R.Fruchart: J. Cryst. Growth **31**, 197 (1975)
2.46 R.Juza, H.Hahn: Z. Anorg. Allgem. Chem. **239**, 282 (1938)
2.47 R.Paff: Private communication
2.48 J.I.Pankove, J.E.Berkeyheiser, H.P.Maruska, J.P.Wittke: Solid State Commun. **8**, 1051 (1971)
2.49 R.Dingle, K.L.Shaklee, R.F.Leheny, R.B.Zetterstrom: Appl. Phys. Lett. **19**, 5 (1971)
2.50 C.D.Wang, M.Gershenzon: Bull. Am. Phys. Soc. **24**, 343 (1979)
2.51 J.I.Pankove, S.Bloom, G.Harbeke: RCA Rev. **36**, 163 (1975)
2.52 M.Ilegems, H.C.Montgomery: J. Phys. Chem. Sol. **34**, 885 (1973)

2.53 D.L.Rode: Private communication
2.54 J.I.Pankove, E.A.Miller, J.E.Berkeyheiser: RCA Rev. **32**, 383 (1971)
2.55 J.I.Pankove, E.A.Miller, J.E.Berkeyheiser: J. Lumin. **6**, 54 (1973)
2.56 J.I.Pankove, M.A.Lampert: Phys. Rev. Lett. **33**, 361 (1974)
2.57 H.P.Maruska, D.A.Stevenson, J.I.Pankove: Appl. Phys. Lett. **22**, 303 (1973)
2.58 G.Jacob, D.Bois: Appl. Phys. Lett. **30**, 412 (1977); and personal communication
2.59 J.I.Pankove: RCA Rev. **34**, 336 (1973)
2.60 F.Porret: Personal communication
2.61 J.I.Pankove, P.E.Norris: RCA Rev. **33**, 377 (1972)
2.62 Y.M.Tairov, Y.A.Vodakov: In *Electroluminescence* ed. by J.I.Pankove, Topics in Applied Physics, Vol. 17 (Springer Berlin, Heidelberg, New York 1977) p. 31
2.63 G.F.Kholyano, Y.A.Vodakov: *Silicon Carbide*–1973, ed. by R.C.Marshall, J.W.Faust, C.E.Ryan (University of South Carolina Press, Columbia, South Carolina 1974) p. 574
2.64 G.B.Dubrovski: Private communication
2.65 A.G.Fischer: *Luminescence in Inorganic Solids*, ed. by P.Goldberg (Academic Press, New York 1966) Chap. 10
2.66 M.Aven: *II–VI Semiconducting Compounds*, ed. by D.G.Thomas (Benjamin, New York 1967) p. 1232
2.67 Y.S.Park, B.K.Shin: In *Electroluminescence*, ed. by J.I.Pankove, Topics in Applied Physics, Vol. 17 (Springer Berlin, Heidelberg, New York 1977) p. 133
2.68 Y.S.Park, B.K.Shin: J. Appl. Phys. **45**, 1444 (1974)
2.69 S.Wagner: In *Electroluminescence*, ed. by J.I.Pankove, Topics in Applied Physics, Vol. 17 (Springer Berlin, Heidelberg, New York 1977) p. 171
2.70 S.Wagner: J. Appl. Phys. **45**, 246 (1974)
2.71 A.G.Thompson, M.Cardona, K.L.Shaklee, J.C.Wooley: Phys. Rev. **146**, 601 (1966)
2.72 A.H.Herzog, W.O.Groves, M.G.Craford: J. Appl. Phys. **40**, 1830 (1969)
2.73 R.J.Archer: J. Electron. Mater. **1**, 128 (1972)
2.74 H.P.Maruska, J.I.Pankove: Solid-State Electron. **10**, 917 (1967)
2.75 H.Kressel, F.Z.Hawrylo, N.Almeleh: J. Appl. Phys. **40**, 2248 (1969)
2.76 A.Onton, M.R.Lorenz, W.Reuter: J. Appl. Phys. **42**, 3420 (1971)
2.77 R.J.Archer: Electrochem. Soc. Meeting, Los Angeles, Calif. (Spring 1970) paper 66
2.78 H.Kressel, H.F.Lockwood, H.Nelson: IEEE J. QE-**6**, 278 (1970)
2.79a M.G.Craford, W.O.Groves: Proc. IEEE **61**, 862 (1973)
2.79b C.J.Nuese: J. Electron. Mater. **6**, 253 (1977)
2.80a G.A.Antypas, R.L.Moon, L.W.James, J.Edgecomb, R.L.Bell: *Proc. 4th Intern. Symp. on Gallium Arsenide and Related Compounds* (Inst. of Physics, London 1973) p. 48
2.80b J.J.Coleman, N.Holonyak, Jr., M.J.Ludowise, P.D.Wright, W.O.Groves, D.L.Keune: IEEE J. QE-**11**, 471 (1975)
2.80c W.R.Hitchins, N.Holonyak, Jr., P.D.Wright, J.J.Coleman: Appl. Phys. Lett. **27**, 245 (1975)
2.80d R.Chin, H.Shichijo, N.Holonyak, Jr., J.A.Rossi, D.L.Keune, D.Finn: IEEE J. QE-**14**, 711 (1978)
2.81 G.B.Stringfellow, P.E.Greene: J. Appl. Phys. **40**, 502 (1969)
2.82 W.A.Brantley, O.G.Lorimor, P.D.Dapkus, S.E.Haszko, R.H.Saul: J. Appl. Phys. **46**, 2629 (1975)
2.83a M.Ettenberg: J. Appl. Phys. **45**, 901 (1974)
2.83b G.B.Stringfellow, H.T.Hall, Jr.: J. Electrochem. Soc. **123**, 916 (1976)
2.84 M.Ettenberg, G.H.Olsen, C.J.Nuese: Appl. Phys. Lett. **29**, 141 (1976)
2.85 C.E.E.Stewart: J. Cryst. Growth **8**, 259 (1971)
2.86 M.S.Abrahams, L.R.Weisberg, C.J.Buiocchi, J.Blanc: J. Mater. Sci. **4**, 223 (1969)
2.87a G.H.Olsen, M.S.Abrahams, C.J.Buiocchi, T.J.Zamerowski: J. Appl. Phys. **46**, 1643 (1975)
2.87b R.N.Bhargava: IEEE Trans. ED-**22**, 691 (1976)
2.88 F.Stern: Appl. Opt. **3**, 111 (1964)
2.89 W.N.Carr, G.E.Pittman: Appl. Phys. Lett. **3**, 173 (1963)
2.90 I.Ladany: J. Electrochem. Soc. **116**, 993 (1969)
2.91 J.M.Dishman, M.DiDomenico, Jr., R.Caruso: Phys. Rev. **2**, 1988 (1970)

2.92 C.Lanza, K.L.Konnerth, C.E.Kelly: Solid-State Electron. **10**, 21 (1967)
2.93 A.A.Bergh: IEEE Trans. ED-**18**, 166 (1971)
2.94 H.Kressel, N.E.Byer, H.F.Lockwood, F.Z.Hawrylo, H.Nelson, M.S.Abrahams, S.H.McFarlane: Metall. Trans. **1**, 635 (1970)
2.95 H.Schade, C.J.Nuese, J.J.Gannon: J. Appl. Phys. **42**, 5102 (1971)
2.96 S.Yamakoshi, T.Sugahar, O.Hasegawa, Y.Toyama, H.Takanashi: *Tech. Dig. Intern. Electron Devices Meeting* (IEEE, New York 1978) p. 642
2.97 M.Iwamoto, A.Kasami: Appl. Phys. Lett. **28**, 591 (1976)
K.Hirahara, Y.Iizuka, T.Beppu, A.Kasami: Presented at 1978 IEEE LED Specialist Conf., San Francisco (1978)
2.98 R.L.Hartman, B.Schwartz, M.Kuhn: Appl. Phys. Lett. **18**, 304 (1971)
2.99 G.H.Olsen, M.S.Abrahams, T.J.Zamerowski: J. Electrochem. Soc. **121**, 1650 (1974)
2.100 G.H.Olsen, M.Ettenberg, R.V.D'Aiello: Appl. Phys. Lett. **33**, 606 (1978)
2.101 J.J.Tietjen, V.S.Ban, R.E.Enstrom, D.Richman: J. Vac. Sci. Technol. **8**, 55 (1971)
2.102 W.C.Benzing: Applied Materials Technology, Inc., Santa Clara, Calif.
2.103 R.Dupuis, D.Dapkus: Appl. Phys. Lett. **31**, 201 (1977)
2.104 G.B.Stringfellow, H.T.Hall,Jr.: J. Cryst. Growth **43**, 47 (1978)
2.105 M.Illegems: J. Cryst. Growth **13/14**, 360 (1972)
2.106 H.M.Manasevit, F.M.Erdmann, W.I.Simpson: J. Electrochem. Soc. **118**, 1864 (1971)
2.107 T.L.Chu: J. Electrochem. Soc. **118**, 1200 (1971)
2.108 J.I.Pankove, D.Richman, E.A.Miller, J.E.Berkeyheiser: J. Lumin. **4**, 63 (1971)
2.109 H.P.Maruska, W.C.Rhines, D.A.Stevenson: Mater. Res. Bull. **7**, 777 (1972)
2.110 J.T.Kendall: *Silicon Carbide*, ed. by J.R.O'Connor, J.Smiltens (Pergamon Press, Oxford 1960) p. 67
M.L.Belle, N.K.Prokof'eva, M.B.Reifman: Sov. Phys. Semicond. **1**, 315 (1967)
2.111 A.Lely: Ber. Deutsch. Keram. Ges. **32**, 229 (1955)
2.112 H.Nelson: RCA Rev. **24**, 603 (1963)
2.113 H.Rupprecht, J.M.Woodall, G.D.Petit: Appl. Phys. Lett. **11**, 81 (1967)
2.114 H.Nelson: U.S. Patent 3,565,702 (1971), filed 1969
2.115 H.F.Lockwood, M.Ettenberg: J. Cryst. Growth **15**, 81 (1972)
2.116 J.M.Blum, K.K.Shis: J. Appl. Phys. **43**, 1394 (1972)
2.117 A.H.Herzog, W.O.Groves, M.G.Craford: J. Appl. Phys. **40**, 1830 (1969)
2.118 L.C.Luther, H.C.Casey,Jr., S.E.Haszko, A.S.Jordan, O.G.Lorimor, G.A.Rozgonyi: J. Electron. Mater. **1**, 54 (1972)
2.119 K.Weiser: J. Appl. Phys. **34**, 3387 (1963)
2.120 A.Herzog: Solid-State Electron. **9**, 721 (1966)
2.121 J.I.Pankove, D.E.Carlson: Appl. Phys. Lett. **29**, 620 (1976)
2.122 D.Engemann, R.Fischer, J.Knecht: Appl. Phys. Lett. **32**, 567 (1978)
2.123 J.I.Pankove, J.E.Berkeyheiser: RCA Tech. Note No. 523 (1962)
2.124 D.E.Carlson, C.R.Wronski: Appl. Phys. Lett. **28**, 671 (1976)
2.125 A.Kinnobu Kasami, M.Naito, M.Toyama: IEEE Trans. ED-**19**, 1093 (1972)
2.126 S.Usui, N.Watanabe: Intern. Electron. Devices Meeting, Washington, D.C. (1972) paper 22.6
2.127 A.A.Bergh, B.H.Johnson: Bell Lab. Rec. **47**, 320 (1969)
2.128 M.Inoue, H.Yamanaka, T.Uragaki, I.Teramoto: IEEE Trans. ED-**24**, 979 (1977)
2.129 A.A.Kal'nin, Yu M.Tairov: Prib. Tekh. Eksp. **I**, 143 (1972)
2.130 F.Auzel: CR Acad. Sci. (Paris) **262**, 1016 (1966); Proc. IEEE **61**, 758 (1973)
2.131 S.V.Galginaitis, G.E.Fennel: *Proc. 2nd Intern. Conf. Gallium Arsenide* (Inst. Phys. Phys. Soc., London 1968) Conf. Ser. 7, p. 131
2.132 J.P.Wittke, I.Ladany, P.N.Yocom: J. Appl. Phys. **43**, 597 (1972)
2.133 Electronics **49**, 22 (1976)
2.134 T.Niina, S.Kuroda, H.Yonei, H.Takesada: IEEE-SID-AGED Conf. Record **78 CH 1323-5ED**, 18 (1978)
2.135 R.L.Harris: IEEE-SID-AGED Conf. Record **78 CH 1323-5ED**, 20 (1978)
2.136 C.J.Nuese, J.J.Gannon, H.F.Gossenberger, C.R.Wronski: J. Electron. Mater. **2**, 571 (1973)

3. AC Plasma Display

T. N. Criscimagna and P. Pleshko

With 41 Figures

Abstract

This chapter discusses matrix-addressed ac plasma displays operated with memory. The discussion begins with an overview of the different possible modes of operation, followed by a review of the materials and processing options available in fabricating these devices. The voltage transfer characteristic is derived and then used to discuss the operation of these devices. The effect on operating margin of different waveforms, patterns and sequences is also discussed. Finally, panel limitations and extendibility of the technology is addressed.

3.1 Background Information

AC plasma displays currently manufactured by several companies, are finding wide acceptance where their slim profiles (see Fig. 3.1) are required as in banking applications. For applications where a high information content is required, the flickerless, distortionless image is ideal for those using a display in a dedicated mode for many consecutive hours without interruption.

Figure 3.2 shows a message presented on a 240-character ac plasma display. The illustrated message gives particulars regarding the display's characteristics.

3.1.1 Historical Evolution

The basic two-terminal ac plasma display device is shown schematically in Fig. 3.3. The driving point terminals are connected via capacitors to a gas filled cavity, which emits light in the visible range, when a discharge is formed. Because of the capacitive coupling to the gas an ac waveform is required for continuous emission of light pulses.

Driving each cell individually in a large array of cells would result in a prohibitively high cost for drivers and connections between the drivers and the cells. One way of reducing driver and interconnection costs is to interconnect the electrodes of each cell in a matrix array. For individual plasma cells to be operable independently when interconnected as a matrix array, an impedance in series with the gas discharge is required at each intersection of x–y lines.

Fig. 3.1. A physical comparison of a plasma panel display and a CRT display

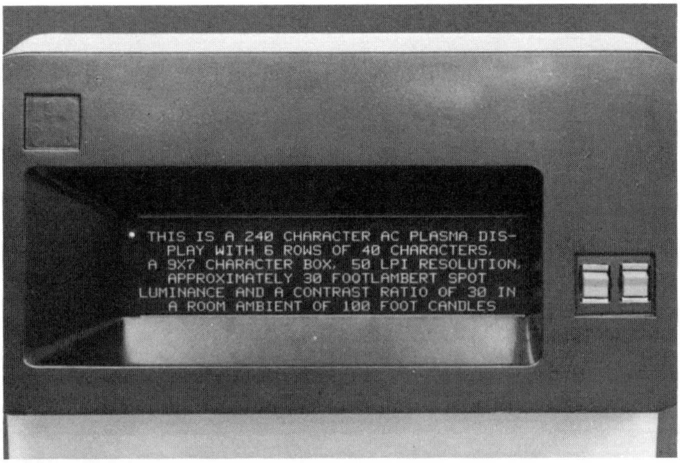

Fig. 3.2. A 240-character plasma display panel

Using resistors as the cell series impedance, it was not possible to fabricate the proper values of resistance with a continuous film deposition. The breakthrough, with respect to electrical isolation of each cell occurred at the University of Illinois in 1964 when it was realized that capacitors could provide the necessary electrical isolation [3.1]. The use of capacitors made integrated array fabrication feasible because it is more practical to fabricate capacitors with a continuous film of dielectric material, than it is to fabricate large value resistors. Resistor values are inversely proportional to a cell size – the smaller the cell, the larger the resistor, thus limiting resolution if one uses discretely formed resistors. Early experimental ac plasma panels had a honeycomb-like sheet between the two glass plates with the conductors and dielectric layer on them,

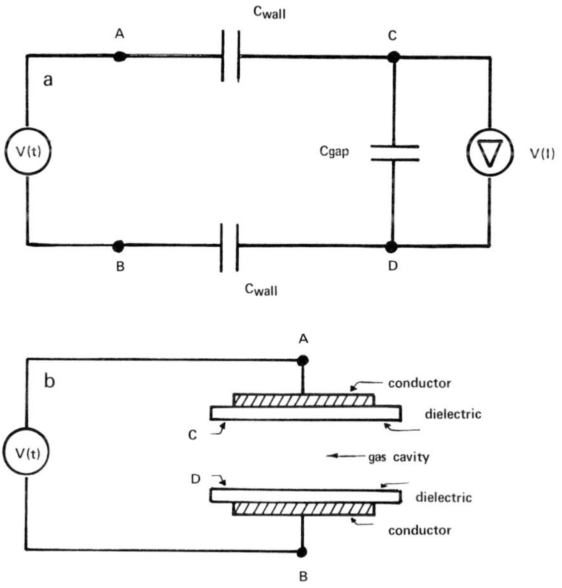

Fig. 3.3a, b. Simplified cell. (**a**) Electrical equivalent. (**b**) Physical representation

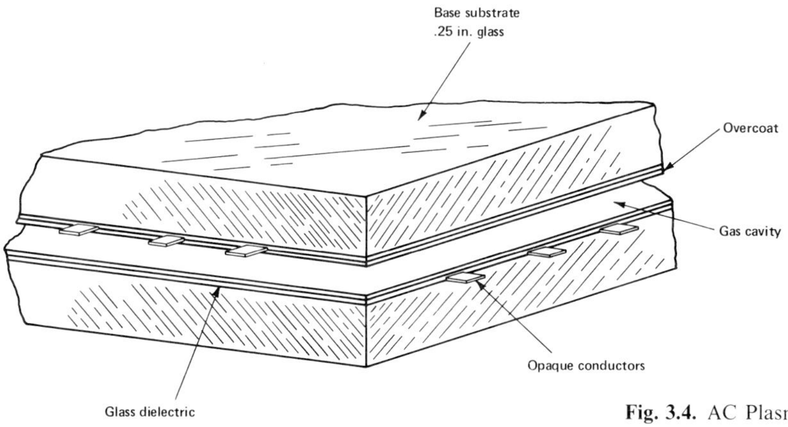

Fig. 3.4. AC Plasma panel

which would have resulted in high fabrication costs. Thus, another breakthrough occurred in panel fabrication at Owens-Illinois when it was realized [3.2] that one did not need to physically isolate each cell from its neighbors by constructing "walls" to contain each individual cell's gas discharge. The use of an open chamber greatly simplified panel construction eliminating the perforated dielectric sheet and the need for precise alignment between this sheet and the two plates. Figure 3.4 illustrates current panel fabrication used for any number of display cells (pels) from one element (used as indicators) to arrays incorporating a million pels (1024 × 1024).

3.1.2 Modes of Operation

The most common mode of operation for an ac plasma display is the bistable or "memory" mode of operation. In this mode the cell produces two levels of brightness, a bright ON level and an OFF level. The bistable operation yields cell brightness which is independent of the number of pels in the display panel. The brightness is determined by the intersection of the dynamic load line and plasma discharge characteristic. In the bistable mode of operation, a continuous ac square wave voltage called a "sustain" voltage is simultaneously applied to all cells (in parallel). This causes the ON cells to "fire" or discharge repetitively while OFF cells remain "unfired". Because all of the cells are sustained in parallel, using a single sustain driver for a large matrix array requires relatively large sustain currents.

The refresh mode of operation avoids the problem of high sustain currents because the maximum number of cells that can simultaneously be ON is the number of cells on a single scanned line of pels. Therefore, the bias generator used for the entire panel primarily supplies displacement current which is determined by the panel's chamber capacitance. The refresh mode of operation, with the same matrix addressed devices described above, is achieved by appropriately choosing a bias voltage below both of the bistable operating voltages [3.3, 4]. Selection voltages above the bias voltage produces cell operation without memory, requiring "refreshing" of the displayed pattern. Brightness depends on the drive voltages and the number of lines that are scanned. To maximize brightness, scanning is performed on the smaller matrix array dimension. Thus, high brightness can be achieved for small displays.

A shift panel operating in the bistable mode was reported by *Umeda* and *Hirose* [3.5]. In the shift configuration, use is made of the phenomenon that a cell in close proximity to another cell can act as input to that cell. In general, to shift information linearly in symmetric devices, directionality can only be achieved with 3 or more sequentially phased electrical drive voltages. This phased drive effects transfer of ON and OFF states unidirectionally along a chain of cells. Operation in the shift configuration results in some loss of operating margin as compared to the matrix configuration. For a three phase design, the required conductor pitch on one of the two plates comprising the panel is one-third that of a matrix display with the same pel resolution. Second-level conductors are required to interconnect all of the lines for a particular phase. To avoid this, a four phase drive is used in conjunction with a meander pattern of conductors [3.6] or a pattern which has two sets of parallel electrodes with each set on opposite substrates [3.7]. The need for second-level interconnection is thereby traded for the need to register or align the panel's two plates.

The shift line pattern is more complex to fabricate but requires less circuitry than a matrix-addressed panel. The shift panel configuration may or may not be more cost effective than the matrix alternatives depending on the relative cost of the shift panel and electronics as affected by the aspect ratio of the panel.

This chapter concentrates on the matrix-addressed ac plasma display operating in the bistable mode of operation.

3.1.3 Basic Cell Operation

As shown in Fig. 3.5, an ac plasma panel consists of an assembly of two substrates, spaced to form a chamber containing a neon gas mixture. Each substrate holds a set of parallel conductors, that are covered by a transparent dielectric, and the substrates are oriented such that the two sets of conductors are orthogonal. Selected intersections of these conductors become localized spots of neon-colored light when driven by suitable signals, thereby creating a display. A detailed description of the operation of a cell and a panel are included in this chapter, however, the following simplified description provides a foundation for the detailed description which follows.

In Fig. 3.5, the potential difference V_{ab}, that equals the firing (or ionizing) voltage, V_f, of the gas, is applied to the cell at time t_0. The gas ionizes, and electrons and ions move toward the anode and cathode respectively. This first discharge produces a short burst of neon-colored light, and a voltage (from the wall charge) that opposes the external potential difference, "quenching" the discharge.

At time t_1, the external voltage polarity is reversed and reduced in magnitude to a point below V_f. This reduced applied voltage is called the sustain voltage of the cell. Since the polarity of the applied voltage has been

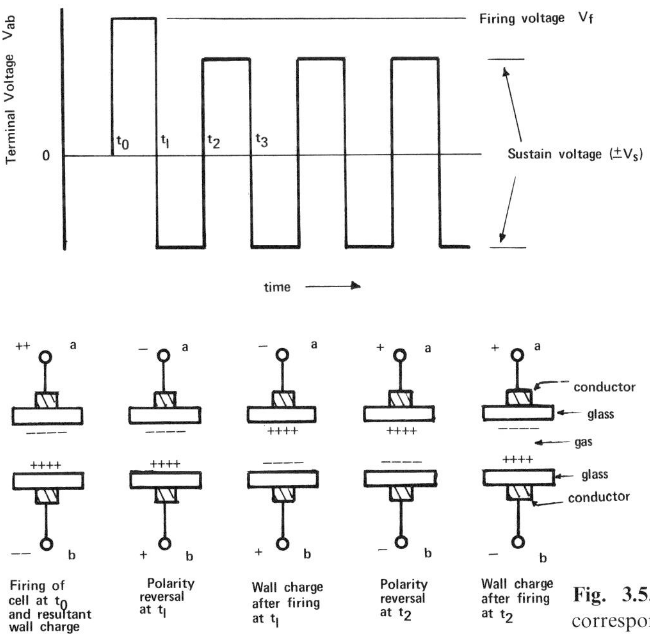

Fig. 3.5. Drive waveform and corresponding wall charges

reversed, it now aids the wall voltage produced during the first firing. The sum of these two voltages exceeds the firing voltage and a second firing occurs. This second discharge produces another pulse of light, and an opposing wall voltage. At time t_2 the polarity is again reversed initiating a third firing that results in a third burst of light and an opposing wall voltage.

In summary, a high potential difference, V_f, is used to initiate the first gas discharge. Subsequently a sustain drive of alternating polarity and lower amplitude is used to produce a burst of light and a sustaining wall voltage for every alternation. Thus the cell has been turned "ON" or "written" and is being sustained in this ON state. Typical values for V_f and for the sustain operation are 150 and 90 V respectively.

The sustain voltage cannot produce firings without the aid of the wall voltage, so the erase operation is simply one where a weak firing is forced that cannot produce enough wall voltage to sustain the subsequent firings. This reduction of the wall voltage to or close to zero, is generally accomplished by a low amplitude alternation of the applied ac waveform.

3.1.4 Cell Light Emission

The primary purpose of considering the characteristics and operation of an ac plasma cell is because the neon host gas used emits radiation at visible wavelengths. This phenomenon is what allows one to construct arrays of cells to form a visual display panel. The gas discharge that occurs in a cell, emits light from two regions: the negative glow and the positive column. For neon with 0.2% xenon dopant, the radiation from the negative glow region is at 5850 Å while the radiation from the positive column is predominantly at 6400 Å. Penning mixtures such as this, have reduced light output compared to the pure neon host gas.

Cell brightness is given by the time-averaged value of the light emissions from a cell. Thus the peak value of light emission, its variation with time and space and repetition rate are determinants of cell brightness. The instantaneous cell-light-output waveform follows the current waveform in time and varies directly with the cell current. Therefore, as cell current increases with wall capacitance and pressure, so does the brightness. In addition, increasing the sustain frequency, which increases the number of light pulses per unit time, increases the cell brightness.

Typical cell brightness of 30–75 fL are specified on ac-plasma-product data sheets with contrast ratios of 25 and cell luminance efficiencies of 0.1–0.5 lm/W.

3.2 Panel Materials and Processing

The ac plasma panel is fabricated by individually processing two substrates and then sealing a pair of substrates so that a chamber is formed which will contain the gas. The panel is then filled with the appropriate gas mixture through a fill

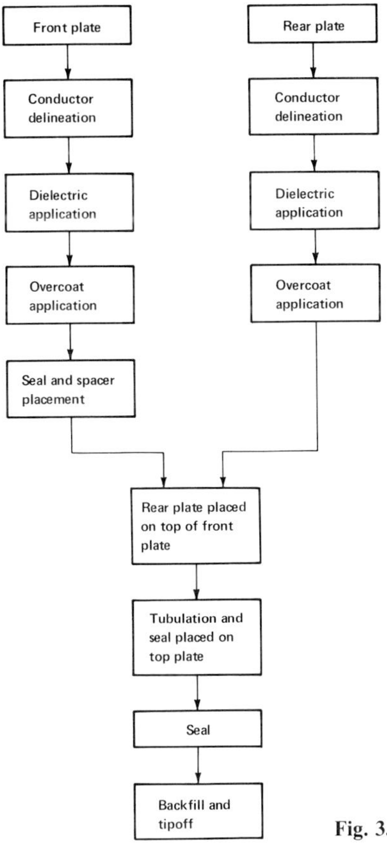

Fig. 3.6. Plate and panel processing flow

tube and sealed off. This processing sequence is illustrated in Fig. 3.6 and should be referred to as the detailed step by step construction is discussed in the following subsections. Panels that are fabricated using "hot" processes are restricted by the constraint that each subsequent step in the process be performed at a temperature such that it does not disturb the properties of the materials deposited in the preceding steps. With "cold" processing, this constraint on materials due to a requirement for a temperature hierarchy is alleviated.

In the following sections, the different materials and processes that are used to form the conductors, the dielectric layer, the overcoat material, and the gas mixtures, are discussed.

3.2.1 Substrate

The substrate is ordinary window glass (soda lime silicate) varying in thickness from 3 to 6 mm, depending on panel size.

3.2.2 Conductors

Conductors may be formed with either thick or thin films deposited through masks or as continuous layers. The latter are then covered with photoresist, which in turn is exposed and developed. The conductors are then etched in accordance with the resist pattern. Low-resolution panels allow the use of screened conductors, which have the lowest processing cost. Etched multilayer thin-film conductors, required for high resolution, result in the highest processing costs. Deposition and delineation of the conductors on the substrate require good adhesion of the conductors to the substrate, passivity with respect to subsequent processing steps, high conductivity, small tolerances of conductor dimensions, and a small number of opens, shorts, or other defects in the final conductor pattern.

a) Thick-Film Conductors

Screened-gold-paste conductors are used in panels [3.8], with relatively low resolution (less than or equal to 1.2 lines/mm) and are of the order of 150 μm wide and 12 μm high. Because of the high cost of gold paste, other metallic pastes have also been used for conductors, such as nickel [3.9]. Two requirements on the thick film conductors are 1) to have an acceptable value of resistivity (Ω/square), and 2) be nonreactive with the dielectric layer that is to be deposited over it.

The conductive pastes are formed with a glass frit which supplies the adhesion between the substrate and metallic particles. Metal conductors such as gold and nickel are usually chosen for their passivity. The resolution of thick-film conductors is better when a photosensitive binder material is used. This material is applied as a uniform layer, exposed and developed. Line resolutions better than 150 μm can be produced with uniform edges.

After the deposition and formation of the patterned conductors, the plates with the paste are put through a thermal cycle to remove the binder in the paste to achieve adhesion to the substrate, and to sinter the metallic particles into conductive lines. The pastes must have processing-temperature requirements suitably below that of the substrate's softening point. Even when thin-film conductors are used in a panel's active area, the conductor pattern at the edge of the substrate may be a thick-film metallurgy [3.8] which requires registering the screen with the thin-film pattern.

As a consequence of the electrical requirements of high conductivity, the conductors are required to have thicknesses which make them opaque. Fortunately, the glow discharge formed in the gas chamber extends beyond the width of the conductor (of the same order as the chamber gap). However, if the conductor width is an appreciable percentage of the total cell area, the percentage of available light visible to the viewer may be very low.

This condition can be alleviated to an appreciable degree by using slotted conductor patterns as shown in Fig. 3.7. This, design, however, may require

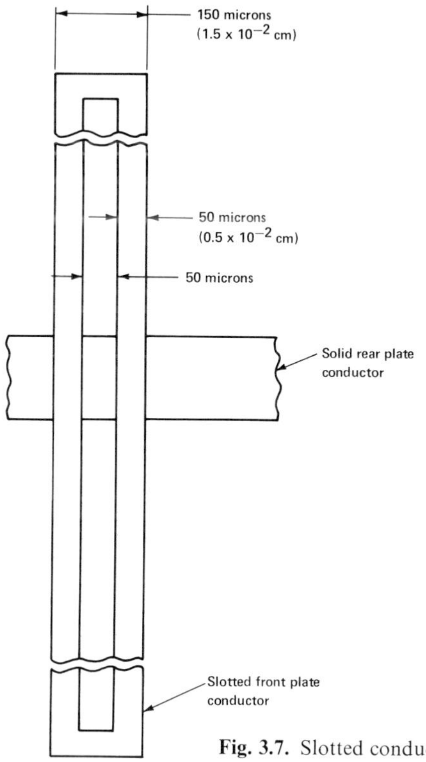

Fig. 3.7. Slotted conductor geometry

that a thin-film metallurgy be used even for low-resolution panels because the electrically parallel multiple conductors used for transmitting the light are quite narrow. A typical pattern for a 150 μm wide line would be a slotted pattern with two 50-μm-wide conductors separated by a 50-μm gap between the conductors. Thus a low-resolution solid-conductor pattern requires thin-film metallurgy, if a slotted-conductor pattern is used.

b) Thin-Film Metallurgy

Thin-film-metallurgy conductors are usually applied as a combination of different metals. The first layer, usually chrome, aluminium, or tantalum is applied for adhesion purposes and can be several thousand angstroms thick. The next layer, usually up to a few tens of thousands of angstroms thick, is for conductivity, with metals such as copper and gold used. The third layer is for passivity and adhesion and is similar to the first layer. Metals such as chrome and nickel can be used.

Deposition is done either by E-beam evaporation or by sputtering, the former being predominantly used because of its relatively high rate of material deposition. Delineation of conductor patterns via photolithographic techniques, using contact masking because of the large area of the substrate, is limited to a resolution of about 25 μm [3.10]. Aluminum sputtered through carbon rib supported masks has also been demonstrated with line resolution of 125 μm [3.11]. As resolutions become greater, and narrower lines are required, material conductivity requirements will become even more demanding. Therefore, new deposition methods applicable to relatively high-aspect-ratio (height-to-width) conductors will have to be developed. Again the melting (softening) point of the metals must be such as to withstand subsequent temperature cycles in the processing.

3.2.3 Dielectric Films

The dielectric film provides the capacitive coupling between the conductors and the gas in the chamber. The value of capacitance influences the peak current drawn by the cell (and hence brightness) and the amplitude of the voltages required to operate the cell. These electrical characteristics also control or affect "field spreading", i.e., coupling to adjacent cells. Thus, the range of thickness and the range of dielectric constants for this layer is important in optimizing a panel's performance.

a) Paste Dielectric Deposition

The dielectric can be applied to the metallized substrate by several methods. One way is as a paste, which is then dried and heated to obtain a uniform, glassy dielectric film. The reflow temperature of this glass has to be lower than the softening points of the substrate and the conductor films. The glass composition must be such that it does not excessively react with the conductor metallurgy during its temperature cycle.

Dielectric constants for typical glasses are in the range of 6–15 and film thicknesses vary from 12 to 50 μm, yielding a wide range of available cell capacitances. The lower bound on dielectric thickness seems to be set by achieving uniform metal coverage, and the upper bound by bubble formation during reflow. These glasses are primarily high lead oxide compositions used for low-temperature processing.

b) Evaporated Dielectric Deposition

Evaporation of thin film dielectric layers has also been reported [3.12]. Substrate heating may be required during deposition to minimize the stress in the deposited film. The lower bound on the dielectric thickness is fixed by conductor-edge-coverage requirements and by dielectric-breakdown considerations. Because of the method of deposition and relative thickness of the dielectric layer, the conductor thicknesses are required to be less than that possible with the paste dielectric processing. However, there is no severe constraint on the processing temperature of the dielectric relative to the substrate as there is with the paste dielectric processing method. Thicknesses of 3–15 μm of borosilicate glass have been deposited with a typical relative dielectric constant of approximately 4.5 [3.12].

3.2.4 Overcoat Materials and Gas Mixtures

The overcoat material interfaces with the gas mixture. As such, the operating voltage and stability of the panel are greatly influenced by the interaction of the overcoat and gas. Overcoat materials are generally oxides which are evaporated over the dielectric layer as a thin (~2000–3000 Å) layer.

To achieve a low operating voltage, a Penning gas mixture is generally used. In those panels in which the light output is generated by the gas discharge, neon is the host gas with the addition of an argon or xenon dopant of typically 0.01–0.3 %, respectively. Orange is the color of the light emission from a device when neon is used as the primary gas.

a) Operating Voltage

The firing voltage dependence on gas and overcoat parameters is given by *Murase* et al. [3.13]

$$V_f = \frac{A(pd)}{\log\left[\dfrac{B(pd)}{\log(1+1/\gamma)}\right]} \tag{3.1}$$

where A and B are constants determined by the gas mixture, p is the gas pressure, d is the panel chamber gap, and γ is the secondary emission coefficient of the overcoat material.

For a given gas mixture, a low operating voltage requires that the value of γ be large. Materials with large secondary-electron emissivities, γ, are considered to have small work functions or low electron affinities. For reliable "writing" (cell discharge) with single pulses, a fast discharge response time is also required. In addition, the overcoat must have good insulation characteristics and surface-discharge characteristics to minimize adjacent-cell interference.

The value of γ is dependent on the value of the electric field that is established between the cell's electrodes, divided by the pressure, i.e., E/p which is called the reduced field. Typical plots are shown in Fig. 3.8. A sharp drop of γ at low values of E/p, represented by curve 2, results in a more confined cell discharge. There is evidence that lead oxide and ytterbium oxide overcoats do confine the discharge more than magnesium oxide. Thus it might be inferred that they have a γ characteristic which falls off at low values of E/p. The 1024×1024, 3.3 line/mm, Owens-Illinois panel used an ytterbium oxide overcoat allegedly to reduce cell-to-cell interaction at the higher resolution.

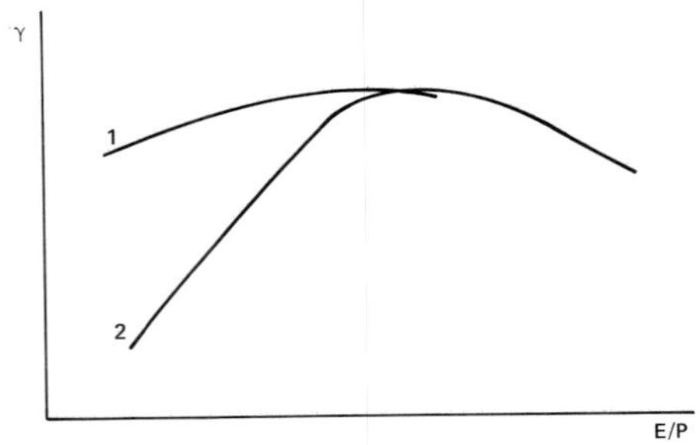

Fig. 3.8. Secondary emission characteristics

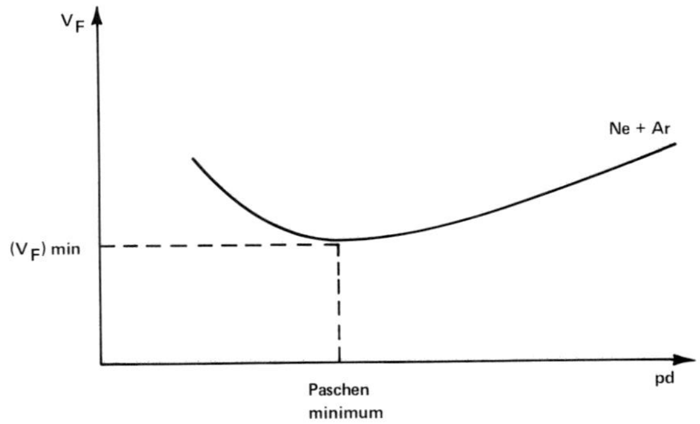

Fig. 3.9. Firing voltage vs pressure × distance

Table 3.1. Overcoat Performance Parameters

Material	Secondary emission coefficient γ	Firing voltage	Operating voltage	Gas	Discharge response time	Sputtering threshold [eV]	Ref.
MgO	0.57	100	90	Ne + Ar(?)	short	60	[3.14, 3.15]
Yb_2O_3			120				[3.15]
La_2O_3	0.55	130	110	Ne + Ar(?)	short	130	[3.14]
CeO_2		170		Ne + Ar(?)	long	125	[3.14]
MgO		100	90	Ne + 0.3Xe/400 Torr	short		[3.16]
CeO_2 La_2O_3 (90/10)	0.41	167	140	Ne + 0.3Xe/400 Torr	long		[3.16]

The dependence of the firing voltage on values of *pd* with a fixed overcoat and gas composition is shown in Fig. 3.9. Note that there is a value of *pd* which minimizes the firing voltage. This value of *pd* is different for different gas mixtures. In contrast to this effect, the memory margin of a single cell gets larger with larger values of *pd* [3.13]. Thus the value of the chosen *pd* is a design compromise.

The comparative performance of different relatively stable overcoats, published by *Urade* et al. [3.14], *Byrum* [3.15], and *Andoh* et al. [3.16], is summarized in Table 3.1.

b) Stability

Stability of operating voltages with time is affected by two types of phenomena. The first is the change in the cell's operating voltage dependent on whether or not the cell has been operated (ON) or nonoperated (OFF). Differences between operated and unoperated cells seem to be related to a "natural" contaminant for the refractory oxides due to the tendency of these materials to readily form hydroxides or hydroxyl groups at surfaces. Observed voltage changes, which are not always in the same direction for different overcoat materials are correlated to the transfer of these groups or simply the hydrogen from operated to nonoperated cells. This effect is much less pronounced in properly produced MgO devices than in rare-earth oxide devices. Vacuum heated MgO samples show a strong reduction of this effect. SEM photographs of surfaces show that the ac gas discharge has a "polishing" effect on the surface that, it is felt, essentially removes and decomposes the hydroxide-related outer layers which have a much lower sputtering threshold than the bulk film.

The other phenomenon has to do with sputtering of the overcoat material. The use of thin-film overcoats requires a knowledge of conditions under which sputter removal of the surface layer can occur, resulting in a voltage shift if the underlying layer has an appreciably different value of γ. First, the gas itself must be considered. The typical gas is predominantly neon with a very small percentage of either argon or xenon. When the gas is excited, similar quantities of Ne^+ and the minority ion are produced with significant amounts of Ne_2^+ also believed to be present. The kinetic energy of the impinging ions can be calculated from *Byrum* [3.15]

$$\text{Kinetic Energy} = 1/2m(K_0E)^2 \tag{3.2}$$

where m is the ion mass, K_0 is the reduced mobility at 760 Torr pressure and 273 K, and E is the field. The values of these parameters at low-field conditions are given in Table 3.2. These values are not readily extrapolated to high-field conditions (high-field to pressure ratios) but serve as an aid to gain insight into the effects of gas composition. From these values, it seems likely that the neon molecular ions and minority ions contribute most to surface damage due to kinetic energy. The values of mK_0^2 shown in Table 3.2 indicate that the higher-

Table 3.2. Low Field Ion Parameters

Host gas	Ion	Ion mass (m)	Reduced mobility (K_0)	$m\,K_0^2$
Neon	Ne^+	20	4.2	352.8
	A^+, Ne_2^+	40	7.8	2433.6
	Xe^+	131	6.5	5534.75

mobility minority ions can be expected to be the major contributors to sputtering.

The equation for calculating the mobility at different pressures p and temperatures T [3.15] is

$$K = K_0 \left(\frac{760}{p}\right)\left(\frac{T}{273}\right) \tag{3.3}$$

from which stronger sputtering action can be expected at lower pressures. Estimated and measured average values of the kinetic energy of the Ne^+ ions, including the effects of space-charge field distortion, range from 1 to 2 eV. This implies the presence of high-average-energy (>10 eV) minority ions in normal devices with Xe^+ ion energies greater than that of A^+ ions, primarily due to a larger mass.

For the operating voltage to remain constant, the overcoat-material surface must remain structurally stable under the conditions of gas excitation and, hence, should be resistant to sputtering. The overcoat-material parameter which is directly useful in calculating sputtering thresholds is the heat of sublimation, H. The energy threshold value for sputtering has been approximated by *Urade* et al. [3.14], as

$$E_s \approx \frac{10H}{\eta} \tag{3.4}$$

where

$$\eta = \frac{4mM}{(m+M)^2} \tag{3.5}$$

with H being the sublimation energy, m the mass of the surface atoms, and M the mass of the incident ions. Approximate values of sputtering thresholds for some of the overcoat-materials are shown in Table 3.1. These considerations suggest that the larger the heat of sublimation for these overcoat-materials and hence the strength of the chemical bonds within the material, the lower the extent of sputtering will be.

3.2.5 Seals, Spacers, and Backfill

When the plates are completely processed, as shown in the flow diagram of Fig. 3.6, two completed plates – a front plate and a back plate – are then assembled to form a panel. The front and back plate always differ, even if it is only because of the hole drilled in the back plate so that a gas fill tube can be affixed for filling this panel with the gas mixture.

The spacers are generally affixed to the front plate since the front plate must be on the bottom so that the back plate can have the fill tube put on it. Spacers which determine the panel's chamber gap are either put in place at the same time that the rim (peripheral) seal material is applied or, if the spacer is within the active panel area, prior to the application of the overcoat-material. The spacer may be of dielectric or metallic material shaped as a rod, bead, or some other shape. The spacer technology is very important because the operating voltage and margin of a cell depends on the chamber gap (at a given pressure) and on the distance from the spacer.

The rim seal may be either put in place as a cane or be screened on as a paste. The fill tube is also sealed to the plate with a seal material similar or identical to that used for the rim. The glass seals are cured by processing them through a furnace temperature cycle. The panel is then "baked-out" to remove contaminants, filled with the proper gas mixture to the desired pressure and, lastly, the fill tube is tipped off. The rim seal has to have a very low leak rate to sustain a long gas life because of the relatively large seal area to gas volume ratio (as compared to a cathode-ray tube).

3.3 Voltage-Transfer Characteristics of a Cell

The operational characteristic of an ac plasma cell is described by the voltage-transfer characteristic. This curve describes both the transient and steady-state behavior of a cell that is being driven by an alternating square-wave voltage. In this section, the derivation of this voltage transfer characteristic and its salient features will be described as a basis for understanding panel operation descriptions is subsequent sections.

3.3.1 Wall Charge and Wall Voltage

If it is assumed that the ionization level in the chamber gap prior to breakdown is negligible, the single ac plasma cell can be represented by an equivalent circuit consisting of three capacitors in series connection, as shown in Fig. 3.10. The chamber gap capacitance C_g has a value at this time close to that of an air capacitor of equal dimensions. The two wall capacitances C'_w can be represented by an equivalent capacitance C_w, as shown in Fig. 3.10.

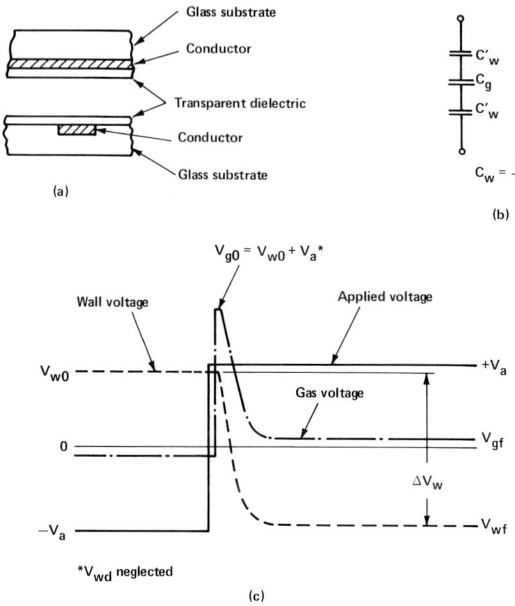

Fig. 3.10a–c. Simplified cross section of a single cell and its equivalent circuit

When the positive alternation of a simple rectangular voltage waveform is applied to a cell as in Fig. 3.10, the gas voltage V_{g0}, prior to the ionization of the gas can be expressed as

$$V_{g0} = V_{w0} + (V_a - V_{wd}) \tag{3.6}$$

where V_{wd} is the voltage developed across C_w by the displacement current produced by the voltage step V_a, and where V_{w0} is the initial voltage present across C_w as a result of a previous discharge current. Since V_a is typically 100 V (peak) and $C_w \gg C_g$, V_{wd} is negligible (typically 2 V). Consequently it will be neglected in the remainder of this analysis. If $V_{g0} \geqq V_f$, where V_f is the breakdown voltage of the gas, a discharge current i_g will flow through the gas developing a voltage across C_w, that opposes the initial gas voltage V_{g0}, such that the gas voltage V_g, as a function of time can be expressed as

$$V_g(t) = V_{w0} + V_a - \Delta V_w \tag{3.7}$$

where

$$\Delta V_w = \int_0^{t_1} i_g(t)dt/C_w(t) \tag{3.8}$$

and t_1 must be large enough to allow V_g to reach its final value V_{gf}. This opposing change in wall voltage quenches the discharge which decays slowly (typically 2–6 μs). After the cell current i_g ceases, V_{gf} can be expressed from (3.7) as

$$V_{gf} = V_{w0} + V_a - \Delta V_w \tag{3.9}$$

and from Fig. 3.10 the change in wall voltage ΔV_w can be written as

$$\Delta V_w = V_{w0} - V_{wf}\,. \tag{3.10}$$

Substituting (3.10) into (3.9) and simplifying yields

$$V_{wf} = -V_a + V_{gf}\,. \tag{3.11}$$

Schlig and *Stillwell* [3.17] have reported values for V_{gf} as low as a few volts, with V_a approximately 100 V. Therefore, to a reasonable approximation

$$V_{wf} = -V_a\,. \tag{3.12}$$

Generalizing (3.6) for a sequence of discharges (and neglecting V_{wd})

$$V_{g0i+1} = V_{ai+1} + V_{wi} \tag{3.13}$$

where i and $i+1$ refer to the ith and the ith $+1$ discharges, and V_{wi} is the final wall voltage developed on the ith discharge.

3.3.2 Stable Discharge Sequences

To simplify the material that follows, the convention adopted by *Slottow* and *Petty* [3.18], will be used. This convention suppresses sign changes of wall voltages, wall charge, and applied voltage, and considers magnitudes measured from a zero-volt reference.

With a rectangular ac voltage applied to a cell, a sequence of cell firings will be generated which will result in a steady-state operation at one of two stable operating points. Reference is made to Fig. 3.11 for generalized voltages that are applied across the cell and developed across the walls and the gas, as a function of time. The steady-state condition is one in which the final wall voltage is equal in magnitude from discharge to discharge, i.e.

$$V_{wi-1} = V_{wi} = V_{wi+1}\,. \tag{3.14}$$

Thus we can pick any discharge in a stable sequence of discharges and relate what happens from the end of one cycle to the beginning of the next cycle. From (3.14), (3.13) can be rewritten as follows

$$V_{g0i} = V_{ai} + V_{wi}\,. \tag{3.15}$$

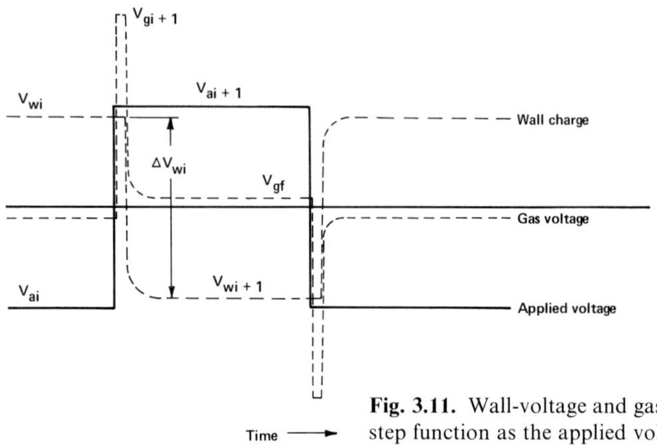

Fig. 3.11. Wall-voltage and gas-voltage waveforms with a step function as the applied voltage

Differentiating yields

$$\frac{dV_{g0i}}{dV_{wi}}=1. \tag{3.16}$$

In the general case the change in wall voltage on any given cell discharge is then given by

$$\Delta V_{wi}=V_{wi}+V_{wi+1} \tag{3.17}$$

which for the steady-state condition (3.14) becomes

$$\Delta V_{wi}=2V_{wi}. \tag{3.18}$$

Equation (3.17) can be expressed more generally as

$$\Delta V_{wi}=f(V_{g0i}). \tag{3.19}$$

This function, an internal-cell-voltage change due to an externally applied voltage, is called the voltage-transfer characteristic of the cell (Fig. 3.12) and will be discussed in the following section. However, to understand fully the transfer function, the conditions required to produce a stable sequence of discharges will be analyzed. Rearranging (3.17)

$$V_{wi+1}=-V_{wi}+\Delta V_{wi} \tag{3.20}$$

and differentiating yields

$$\frac{dV_{wi+1}}{dV_{wi}}=\frac{-dV_{wi}}{dV_{wi}}+\frac{d(\Delta V_{wi})}{dV_{g0i}}\cdot\frac{dV_{g0i}}{dV_{wi}} \tag{3.21}$$

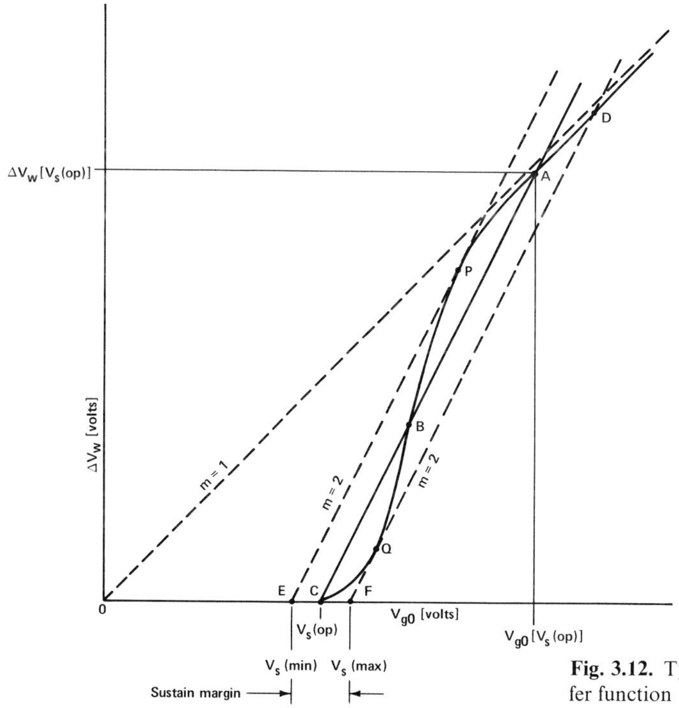

Fig. 3.12. Typical voltage transfer function

since $\Delta V_{wi} = f(V_{g0i})$. Therefore, using (3.16) and simplifying

$$dV_{wi+1} = (-1+m)dV_{wi} \tag{3.22}$$

where $m = d(\Delta V_{wi})/dV_{g0i}$, the slope of the voltage-transfer characteristic curve at equilibrium, and dV_{wi} and dV_{wi+1} are the deviations on the ith and ith$+1$ discharge from the equilibrium wall voltage. Convergence towards equilibrium requires that

$$\left|\frac{dV_{wi+1}}{dV_{wi}}\right| < 1 . \tag{3.23}$$

Substituting (3.23) into (3.22) and simplifying yields

$$|-1+m| < 1 \tag{3.24}$$

or

$$0 < m < 2 . \tag{3.25}$$

From (3.22, 25) equilibrium is achieved on the first discharge following the deviation from equilibrium if $m=1$ [3.18]. For values of m less than 1, equilibrium is achieved more slowly. When $m=2$, the deviation from equilibrium persists indefinitely. When $m=0$, the deviation alternates in sign and the operating point wanders, implying that $V_w \neq f(V_g)$, which is not physically possible.

3.3.3 Voltage-Transfer-Characteristic Curve

Techniques to experimentally obtain this functional curve have been reported by *Schlig* and *Stilwell* [3.17], *Johnson* [3.19], and *Petty* [3.20], and therefore will not be described here.

In accordance with (3.25) all points on the transfer-characteristic curve (Fig. 3.12), above P are capable of supporting a convergent sequence of discharges, for a cell in its ON state. Similarly, all points below point Q also constitute stably convergent operating points for a sequence of discharges of a cell in its OFF state.

Since $\Delta V_w = 2V_w$, any line with a slope $m=2$ between and including lines EP and FD by its intersection with the characteristic curve defines two stable operating points for a cell. For example, line ABC is one line (in the family of lines) that defines, by its intersection with the abscissa the applied operating voltage $V_s(\text{op})$. The line's intersections with the curve define a stable ON condition at point A, and the stable OFF condition at point C. When the cell is ON, the gas voltage at each alternation will be $V_{g0}[V_s(\text{op})]$ and the voltage transfer at each discharge will be $\Delta V_w[V_s(\text{op})]$ since they are the coordinates of point A. When the cell is OFF, the gas voltage at each alternation will be slightly larger than $V_s(\text{op})$ and the voltage transfer will be almost zero, as indicated by the coordinates of point C.

A bistable condition exists because the curve exhibits two regions that satisfy (3.25). These regions are connected by a line where the slope is greater than two. A stable ON condition exists above point D, but the cell is not bistable since it cannot be switched to its OFF state. Likewise, a stable condition exists below point E, but the cell cannot be switched to its ON state. Accordingly then, the bistable condition is bounded by points E and F – point E defines the minimum sustain voltage $V_s(\text{min})$, and point F the maximum sustain voltage $V_s(\text{max})$. Thus, the voltage range or margin, $V_s(\text{mar})$, for bistable operation is defined as

$$V_s(\text{mar}) = V_s(\text{max}) - V_s(\text{min})\,. \tag{3.26}$$

If a single cell is in its ON state at point A and a momentary perturbation of its wall voltage occurs, causing the operation of the cell to shift but still remain above point B, the discharge sequence will gradually resume its equilibrium state at point A. If the momentary perturbation results in an operation below point B, the discharge sequence will decay over several cycles of the sustain voltage to point C, the OFF stable operating point. Likewise, if a cell is in its OFF state at point C and a discharge brings the cell's operation up to a point below point B, the initiated discharge sequence will decay, and the cell will resume its equilibrium OFF state at point C. However, if the disturbance results in operating above point B on the curve, the discharge sequence will gradually intensify until point A is reached.

Point B is a point of equilibrium since it lies on the intersection of the voltage-transfer curve and a line whose slope m is equal to two. However, the slope of the voltage transfer curve at B is greater than two. Therefore point B is an unstable equilibrium point. If a cell operates at point B, *any* voltage disturbance, no matter how small, either causes the cell's operation to move up to point A or down to point C on the curve. It should be noted that the voltage-transfer curve does not quite reach the slope $m=1$ line drawn through the origin. This was accounted for in (3.11) by the gas voltage V_{gf}.

3.3.4 Writing Operation

A typical sustain wave shape with a write pulse and an erase pulse is shown in Fig. 3.13. The alternation rising above V_s by a voltage increment V_{wr}, is the write voltage; the alternation that only rises to V_e (below peak sustain) is the erase voltage. During the first two sustain cycles the wall voltage (dotted line) is about zero. Each sustain alternation produces a small current flow and, subsequently a small wall voltage. However, it is not enough to exceed a threshold and switch the cell to its stable ON state [V_s is assumed to be below V_s(max)].

As shown in Fig. 3.14 when (V_s+V_{wr}) is applied to the cell, a gas discharge will occur, producing a voltage transfer of $\Delta V_w(V_{wr})$. Since V_{wO} was zero, the wall voltage produced by the write-pulse discharge was $V_w(V_{wr})$.

a) Minimum Write Voltage

If the next discharge (after the write pulse) produces a change in wall voltage greater than $\Delta V_w(V_B)$, eventually a stable sequence of discharges around point A will be reached. Minimum write voltage V_{wr}(min) can then be defined as

$$V_s+V_w[V_{wr}(\text{min})]>V_B \tag{3.27}$$

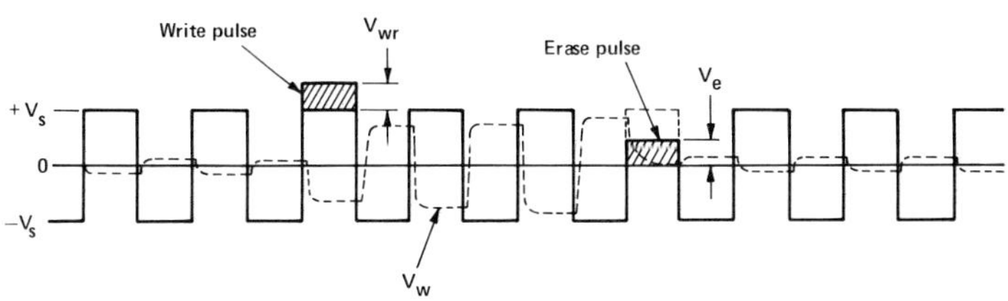

Fig. 3.13. Simple write, sustain, and erase waveforms to illustrate wall-voltage control

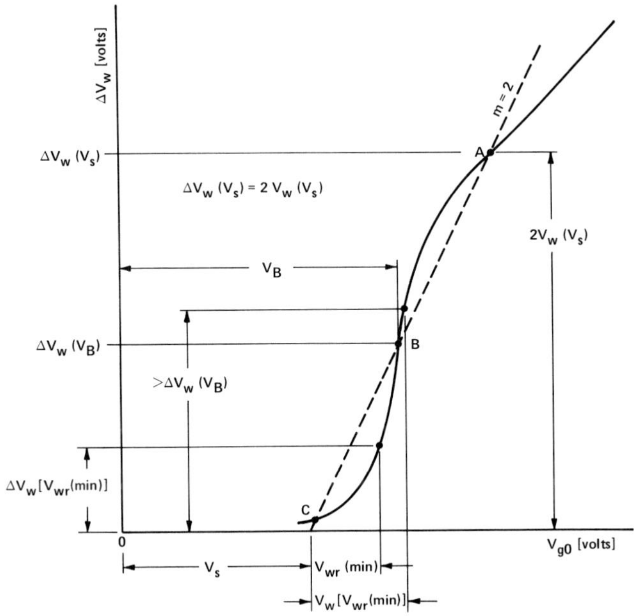

Fig. 3.14. Minimum write-pulse amplitude

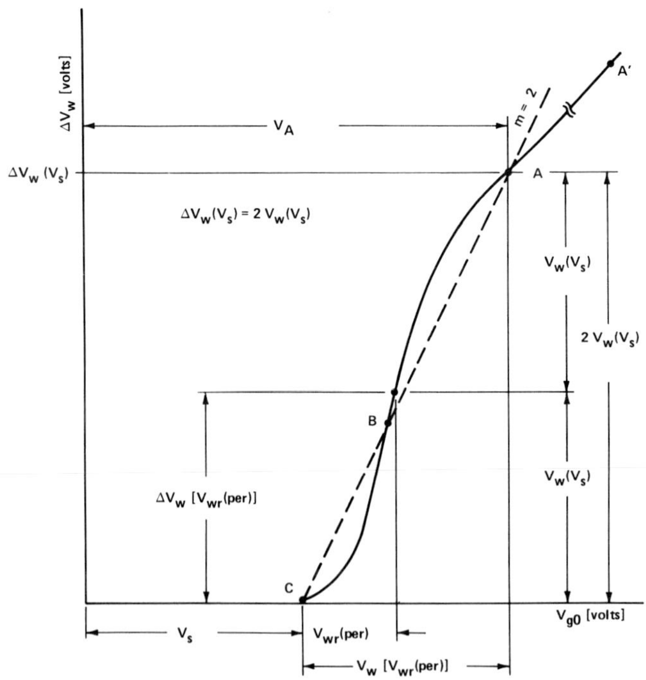

Fig. 3.15. Perfect write-pulse amplitude

where V_B is defined by the unstable equilibrium point B, on the voltage-transfer-characteristic curve. Since the write voltage was the minimum voltage required to initiate a sequence of discharge that lead to operation at point A, the discharges start off quite low in intensity and gradually (over a few cycles) grow to the equilibrium intensity.

b) Perfect Write Voltage

If the write voltage was increased such that

$$V_s + V_w[V_{wr}(\text{per})] = V_A , \tag{3.28}$$

then the first sustain discharge after the write pulse is of equilibrium intensity, and $V_{wr}(\text{per})$ is called the perfect write voltage. Figure 3.15 illustrates $V_w[V_{wr}(\text{per})]$.

c) Maximum Write Voltage

A write voltage pulse greater than $V_{wr}(\text{per})$ also results in a stable sequence of discharges since the first sustain discharge is above point A, but the succeeding discharges decay to equilibrium at point A.

If the voltage transfer curve, as shown in Fig. 3.15, continues to increase monotonically beyond point A to A′ and if only single-cell effects are considered, there would be no maximum write-pulse amplitude for the write-pulse wave shape illustrated in Fig. 3.13. However, as discribed in Sect. 3.6.2b), a transition step (return-to-zero) on the trailing edge of the write-pulse wave shape can limit the maximum write-pulse amplitude. This phenomenon, is called "self erase" and occurs if a transition step on the trailing edge of the write pulse produces a discharge that reduces to less than $V_w[V_{wr}(\text{min})]$, the wall charge just created by the write pulse. The end result is that the high-amplitude write pulse (equal to about V_s) is effectively lower in amplitude than $V_{wr}(\text{min})$, and the cell would not be written. Methods used to correct this problem are described in Sect. 3.6.2b).

3.3.5 Erase Operation

At equilibrium point A, Fig. 3.16, each discharge results in a wall voltage of $V_w(V_s)$. If one of these discharges is altered by a voltage disturbance in the sustain drive, such that $\Delta V_w = V_w(V_s)$, then the discharge would result in a wall voltage of zero, and the cell could not maintain its ON state. The sustain waveform is changed in some suitable manner to reduce the wall voltage to or near to zero, causing "erasure" of the cell. Waveforms commonly used (Sect. 3.6.3) are either a lower-amplitude pulse as shown in Fig. 3.13, or a narrow pulse.

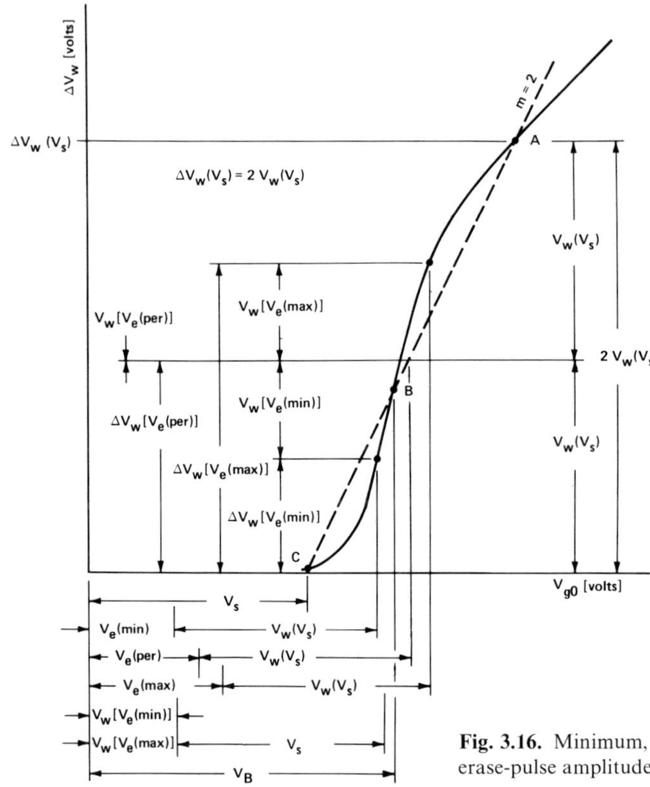

Fig. 3.16. Minimum, perfect, and maximum erase-pulse amplitudes

The low-amplitude erase pulse is used for explanation here as its operation is largely dependent on the value of voltage only and can thus be described using the voltage-transfer-characteristic curve.

a) Minimum Erase Voltage

As illustrated in Fig. 3.16, the gas voltage V_{g0} prior to the erase discharge is $V_e + V_w(V_s)$ where V_e is the erase-pulse amplitude. The change in wall voltage during the erase discharge is $\Delta V_w(V_e)$. Therefore, the wall voltage remaining after the erase discharge $V_w(V_e)$, can be expressed as

$$V_w(V_e) = V_w(V_s) - \Delta V_w(V_e)\,. \tag{3.29}$$

Since $\Delta V_w(V_e)$ varies directly with V_e, as V_e increases, $V_w(V_e)$ decreases. When

$$V_w(V_e) + V_s < V_B \tag{3.30}$$

the sequence of discharges in the cell will gradually decay to a new stable state at C, the OFF state. The minimum amplitude of V_e that satisfies (3.30) is called the minimum erase voltage, $V_e(\text{min})$.

b) Perfect Erase Voltage

As V_e is increased above $V_e(\text{min})$, a point will be reached where $\Delta V_w(V_e) = V_w(V_s)$ and then from (3.29), $V_w(V_e) = 0$. With $V_w(V_e) = 0$, (3.30) will be satisfied as shown in Fig. 3.16 and the cell will be erased. The sequence of discharges in this case does not decay gradually following the erase pulse, but instead ceases abruptly, leaving the cell in its stable OFF state at C. The perfect erase voltage $V_e(\text{per})$ is attained when $V_w(V_e) = 0$.

c) Maximum Erase Voltage

As V_e is further increased above $V_e(\text{per})$ a point is reached where

$$V_w(V_e) + V_s > V_B, \tag{3.31}$$

and the sequence of discharges, although momentarily reduced in intensity, will gradually grow and regain equilibrium at point A. The erase amplitude is then too high to erase. The maximum amplitude of V_e that satisfies (3.30) is called the maximum erase voltage, or $V_e(\text{max})$.

d) General Aspects of the Erase Operation

It should be noted that although the erase operation is often thought of as being complementary to the write operation, there is one essential difference. With ideal conditions in mind, the write voltage has no maximum voltage, or at least, the ideal voltage-transfer-characteristic curve shows none. As the write voltage increases, the discharges at write time, and immediately after write time, become more and more violent but gradually decay to equilibrium point A. The erase operation however is bounded, as indicated by the voltage-transfer-characteristic curve. If the erase-pulse discharge is either too weak, or too strong, the remaining wall voltage (either polarity) is sufficient enough to initiate a sequence of discharges that will reach the equilibrium ON state. Depending on the polarity of $V_w(V_e)$, either in the first positive or first negative alternation of the sustainer will reinitiate the sequence.

In practical applications, both the write and the erase-pulse amplitudes are bounded. Careful wave-shape design is necessary to keep these bounds far enough apart to insure practical operation.

3.4 Nonhomogeneous Cell Characteristics in a Panel

If all the cells of a display panel had identical operating characteristics and if cell-to-cell interactions were nonexistent, panel operation could be described by a single cell's operation. However, panel cells do not have identical characteristics and, depending on resolution, cell-to-cell interaction can be appreciable.

Manufacturing tolerances in panel conductor widths, chamber gap, dielectric thickness, dielectric constant, and the secondary-electron-emission coefficient of the dielectric overcoat result in nonuniform cell characteristics. In addition, cell-to-cell interactions also result in nonuniform cell behavior.

Since a display panel is an x–y matrix of nonhomogeneous cells, the description of a panel's operation can be complex as compared to that of a single cell. However the operation of a single cell can still serve as a reference point for describing the effects of variations in cell parameters.

3.4.1 Cell-to-Cell Interaction

There are three basic forms of cell-to-cell interaction or coupling. Two of these manifest themselves through the gas as the coupling medium, while the third interaction manifests itself through the dielectric layer.

One of the gas couplings takes the form of interactive priming, while the other takes the form of wall voltage resulting from the spread of discharges.

a) Priming

According to the definition adopted by *Holtz* [3.21], priming refers to the action of metastable neon atoms, free electrons, and charged particles which assist the initiation of a gas discharge. When a suitable concentration of these starting particles is present, the probability of initiating a discharge with a single pulse of a given amplitude and duration is equal to 1. As the level of priming decreases, the amplitude of the write pulse has to increase to maintain the statistical probability of a successful write operation at 1.

In an ac plasma display panel, a minimum priming level is supplied to the gas chamber by the use of pilot cells. These pilot cells are border cells that are ON and continuously sustained, usually by a driving waveform synchronized with sustain. The center cells of a panel are supplied with the minimum level of priming because of their distance from the pilot cells, while the cells near the panel border are supplied with the maximum priming level. Therefore, the V_{wr}(min) of the panel's center cells are higher than the V_{wr}(min) of the border cells, unless an equalizing source of priming is provided for the center cells.

This dependency of write voltage on pilot-cell distance was reported by *Ngo* [3.22] along with a proposed solution, to control the timing of pilot-cell firings, as a function of distance from the selected cell. This reduces the variable priming action and the spread in V_{wr}(min) and V_{wr}(max) over the entire panel.

Neighboring cells interact by priming each other. This priming by neighboring cells has more effect where pilot priming is lowest, i.e., at or near the center of the panel. The edge cells, however are very strongly primed by the adjacent pilot cells, and therefore are not affected as much by the priming action of their neighboring cells.

In addition to priming effects varying with distance, priming also varies as a function of the elapsed time after the neighboring cells are turned OFF.

Consider a group of cells that are ON and are being sustained. During sustain, they generate very high priming levels in their immediate vicinity. When the cells are erased or turned OFF, this priming level decays. However, if a single cell in this group of cells or even a nearby cell, is written before the decay process reduces the priming level to the minimal pilot level, the V_{wr}(min) of that cell will be a function of 1) the time that has elapsed since the group of cells was erased and 2) the number and proximity of cells in the group. The combined effects of these two will subsequently be referred to as self-priming. The self-priming levels attained in this manner are equal to those levels found near the border pilots.

Thus, there are two types of priming interactions, spatial and temporal. Spatial priming interaction is a function of a cell's distance from the pilots and the display image content. Temporal priming interaction is a function of the erase/write frame rate and the density of the image just erased.

b) Effects of Priming Interaction

All of the characteristics of a cell as described in Sects. 3.3.4, 5 are affected by variable priming levels. In some cases, the effect is almost negligible. However, all affected characteristics vary inversely with the priming level and are described as follows in order to present a complete picture of interactive priming.

V_{wr}(min) – The wall voltage transferred by the write pulse varies with the priming level, because of the difference created in the discharge formative delay [3.17]. Therefore, for a given change in the priming level, the change in $V_w[V_{wr}(\text{min})]$ for a narrow write pulse is larger than the change in $V_w[V_{wr}(\text{min})]$ for a wider pulse.

V_{wr}(max) – As mentioned earlier, the voltage-transfer-characteristic curve does not indicate the existence of a maximum write voltage. However, there are practical reasons why it is a bounded function. One reason is that all selection systems are not perfect and, therefore, cells that are not selected for writing can experience small voltage disturbances called partial-select pulses. A cell's sensitivity to these disturbances is directly proportional to the priming level.

Another reason for the write operation being bounded is that high amplitude write pulses can significantly add to the priming level of the sustain discharge to create a temporarily high priming level. This high priming level tends to limit V_{wr}(max) by causing adjacent nonselected cells to be turned ON, a condition not reflected by the voltage-transfer-characteristic curve which is for a single cell. This effect is most pronounced near V_s(max).

V_e(min)/(max) – As described in Sect. 3.3.5d), the erase amplitude is bounded, and an erase-amplitude window exists at any sustain level. The magnitude of the window $[V_e(\text{max}) - V_e(\text{min})]$ is inversely proportional to the priming level. In other words, if the priming level is high, $V_w(V_e)$ is more critical. With an increase in the priming level, $V_w(V_e)$ that satisfied (3.30), will, instead, allow a cell to continue to discharge and gradually resume its normal ON state equilibrium – if the ambient priming level is higher than normal. Therefore, erasing a cell in a field of ON cells is a more critical operation than erasing a single isolated cell, especially at or near V_s(max).

V_s(max) – The sustain amplitude at which a cell switches to its ON state, without a write pulse or any other sustain waveform disturbance, is reduced by an increase in the priming level.

V_s(min) – When determining V_s(min), the cell is ON, and is therefore self priming. Consequently, V_s(min) is affected only slightly by priming interaction.

It should be noted that if the discharge time lag or discharge formative delay is a strong function of the priming level [3.16], when an operation such as the erase operation is implemented by a narrow-width erase pulse (that relies on prematurely stopping the discharge to erase a cell), the operation of the erase can be strongly affected by changes in the ambient priming level.

c) Wall Voltage Interaction

Gaseous interaction in the form of priming has been described. A second form of gaseous interaction is a wall-voltage interaction. The mechanism for the interaction is a wall charge, distributed or spread, as shown in Fig. 3.17. A cross section of a small portion of an ac plasma display panel is illustrated. Panel cells are defined by the intersections of two orthogonal sets of parallel conductors.

Consider the center cell M, and the wall charges that have been collected on the dielectric surface areas associated with cells L and R during the last sustain discharge. The top conductors A and C (the anodes) have attracted electrons, creating a negative charge, while the common bottom electrode D (the cathode) is charged positive by neon ions. The positive charges are laterally spread out along the conductor length, and it is this lateral spreading of wall charges along each conductor that causes wall-voltage interaction.

The positive charges spread beyond the midpoints between conductors A and B, and B and C. As such, cell M has a small wall charge within its cell area,

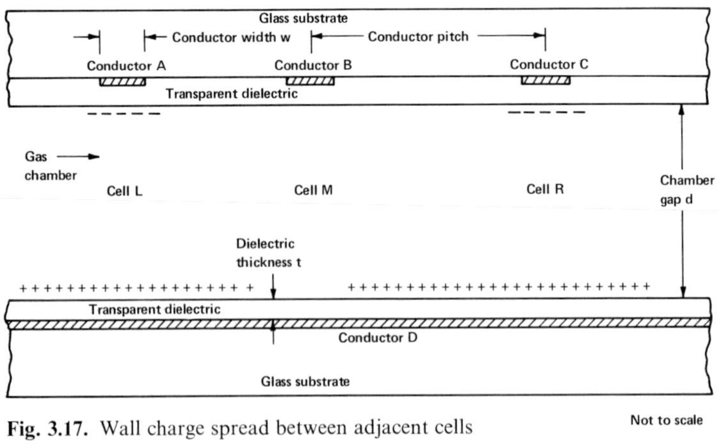

Fig. 3.17. Wall charge spread between adjacent cells

even though it is in its OFF state. If the wall voltage produced by this wall charge is large enough, cell M will turn ON just as if it had been written by a write pulse. Thus the V_s(max) of cell M is a function of the state and proximity of adjacent cells, as well as the intensity and glow-spread of the adjacent cells' discharges.

Where cells are physically isolated by glass walls or barriers, wall charge spread can be more readily controlled. However one serious drawback is that pilot priming can be substantially reduced, making the write operation more difficult and less reliable. In addition, as mentioned in Sect. 3.1.1 these barriers complicate panel construction, and are therefore only used when crosstalk cannot be controlled by other means.

Cell-to-cell wall voltage interaction can be minimized by using geometric constraints that affect the electric field distribution around the cell. For example reducing pel resolution or increasing cell pitch is very effective. However, pel resolution is usually dictated by the display-panel application and usually cannot be reduced. Fortunately, the line width w, the chamber gap distance d, and the dielectric thickness t can be reduced without the above restriction. The lower limits of w and d are primarily determined by manufacturing tolerances.

Reducing nominal dimensions to reduce crosstalk can also increase the percentage deviations from the nominal dimension, if manufacturing tolerances remain unchanged. At some point the improvement achieved by using smaller dimensions is offset by cell-to-cell nonuniformity produced by these tolerances. Dielectric thickness is limited by the process used to form the layer and the voltage breakdown of the material. Reducing w increases sustain and write voltages while decreasing d and t tends to increase the cell discharge current. Both of these trends complicate the design and increase the cost of panel drive circuitry, especially for very large panels.

Optimizing the physical parameters of a panel to reduce wall-voltage crosstalk is a compromise that, at best, requires careful design and coordination between panel and circuit designers.

Wall-voltage crosstalk tends to lower the amplitude of V_s(max), but increases the amplitude of V_{wr}(min) as reported by *Ngo* [3.23]. To understand this, consider a positive alternation for the write operation of a selected cell as shown in Fig. 3.13. The adjacent ON cells are also being driven at write time with a positive alternation of the sustainer, and are therefore developing a negative wall voltage. The selected cell will then have a negative wall voltage when a positive pulse is being applied to write. The effect is obvious, a higher-amplitude pulse is required to make up for the bucking wall voltage. This effect is maximized if the write pulse's leading edge is delayed so as to come after the leading edge of the sustain alternation – a common practice in waveform design which is discussed in Sect. 3.6.2d).

The combined effect of gas-priming interaction and wall-voltage interaction is qualitatively illustrated in Fig. 3.18 as reported by *Ngo* [3.23]. The V_{wr}(min) of a cell gradually decreases as its distance from ON cells decreases (spatial

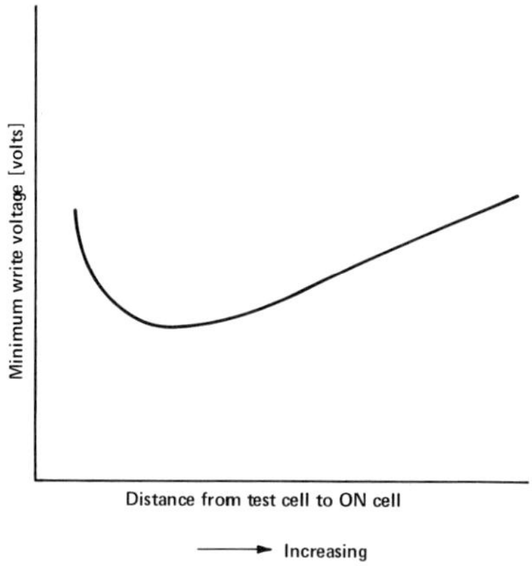

Fig. 3.18. Minimum write-pulse amplitude vs conductor position with varying chamber gap and dielectric constant

priming effect), until this distance becomes so small (a few cells) that wall-voltage interaction starts to dominate. Then V_{wr}(min) actually increases as explained above. *Ngo* [3.23] reports that this effect can increase the write voltage by approximately 15 V.

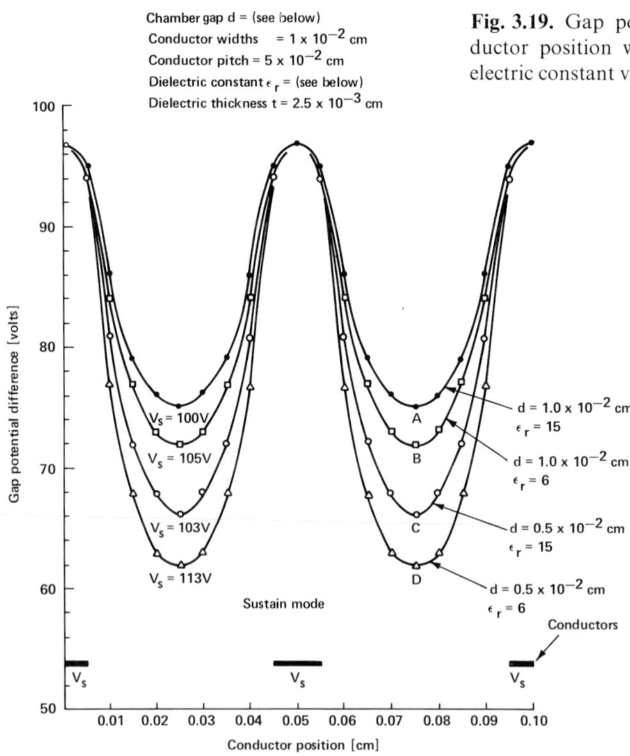

Fig. 3.19. Gap potential difference vs conductor position with chamber gap and dielectric constant varying

d) Dielectric Interaction

Computer programs have been written to simulate the dynamic operation of a cell. However, to simulate the dynamic operation of a panel, and to include all the known cell-to-cell interactions, and the write/erase operation with worst case image patterns, is a very complex task. To date such a program has not been reported. However, two-dimensional electric-field-plot programs have been written that account for voltage interaction between cells prior to their discharge. These programs calculate the difference in potential across the two dielectric surfaces, neglecting the nonlinearity of the voltage gradient across the chamber gap. To simplify these programs the gas is assumed to be a simple dielectric medium and, therefore, wall voltage and charge are not accounted for. Although these simplifications appear to be quite limiting, experience has verified the utility of such a program as reported by *Criscimagna* et al. [3.10], where dielectric interaction during the write and erase operations was maximized to provide a panel internal-addressing function.

The physical parameters that most affect dielectric interaction are the dielectric thickness, the dielectric constant, conductor widths, conductor pitch, and the chamber gap. The substrate thickness and dielectric constant also affect

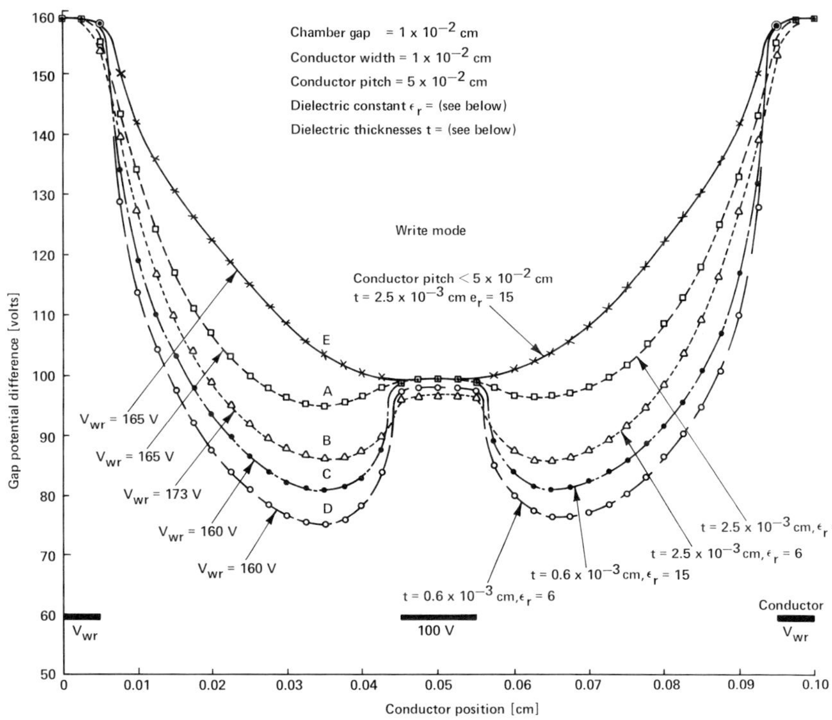

Fig. 3.20. Gap potential difference vs conductor position with varying dielectric thickness and dielectric constant

dielectric interaction, but to a much lesser degree, and will therefore not be considered.

Dielectric interaction manifests itself during two different modes of panel operation – the sustain operation and the write/erase operation. A graph for the sustain operation is shown in Fig. 3.19, where the gap potential (component normal to the dielectric surface) vs conductor positions is plotted. The physical parameters of the panel are included in the drawing, the bottom conductors serving as positional references.

The gap potential in between adjacent panel lines prior to the gas discharge is proportional to the dielectric constant ε_r and the chamber gap d. Experience has shown that the physical parameters resulting in curve A will produce more glow-spread and, consequently, more wall-voltage interaction than the parameters that result in curves B, C, and D.

In designing an ac plasma panel, especially one with high resolution, dielectric materials with the lowest ε_r should be used whenever possible, and the chamber gap made small. The extent to which one can lower ε_r and reduce d, however, is a compromise, since in each case the voltage requirement for V_s increases, as illustrated by curves B, C, and D.

The write/erase operation is also affected by dielectric interaction as shown in Fig. 3.20, where the same basic geometry is used. Only the write operation is illustrated, but similar relationships exist for the erase operation. The case shown is one where a middle or nested cell is driven by the sustain potential (100 V), while adjacent cells are driven by the write voltage (160–173 V). The gap potential between cells is definitely affected by the adjacent selected cells. The goal of the panel designer is a voltage curve like that of curve D where the gap potential between the cells is actually lower than the gap potential at the middle cell. Desirable cell-to-cell isolation can be achieved by reducing the dielectric thickness t, and the dielectric constant ε_r. Isolation can also be achieved by reducing the chamber gap d as implied by Fig. 3.19. It should be noted that a lower ε_r requires a higher drive voltage, as illustrated by curve B. Curve D, however, shows that this drawback can be reduced by using a thinner dielectric with a lower value of ε_r. This combination not only maximizes isolation, it also reduces the current required by the thinner dielectric.

Curve E illustrates the effect of using all of the same parameters of curve A, but decreasing the conductor pitch and/or increasing the conductor width. Under higher resolution conditions the gap potential between conductors can reach amplitudes high enough to help turn the center or half-selected cells ON. The more frequently the write pulse and the resulting high voltage between conductors occurs in the train of sustain cycles, the more likely it is that the center will be turned ON. As an example with a write pulse for every sustain cycle (synchronous writing), the cell turns ON just as if it were being driven by a sustain amplitude greater than V_s(max).

In summary, three types of cell-to-cell interactions have been described. Separately, or in combination, they can affect the operating characteristics of the cells of a panel, causing the cells to appear nonuniform. The next section deals with the effects of tolerances in physical parameters of a panel and how these tolerances actually change the cell characteristics so that they are in fact nonuniform.

3.4.2 Physical Parameters and Their Effects

Some of the physical parameters of interest are conductor or line width, chamber gap, and gas pressure. Their variations produce changes in cell characteristics.

The effect of conductor width on cell characteristics is shown by *O'Hanlon* [3.24], where V_s(max) and V_s(min) are plotted vs line width, at the Paschen minimum (Fig. 3.21). The operating voltages of a cell are shown to vary inversely with the line width of the conductors making up that cell. Note that the operating voltages increase quite rapidly for line widths of less than 0.01 cm. This creates a problem in high-resolution panel design because higher voltage circuit components and tighter manufacturing tolerances on line width are required.

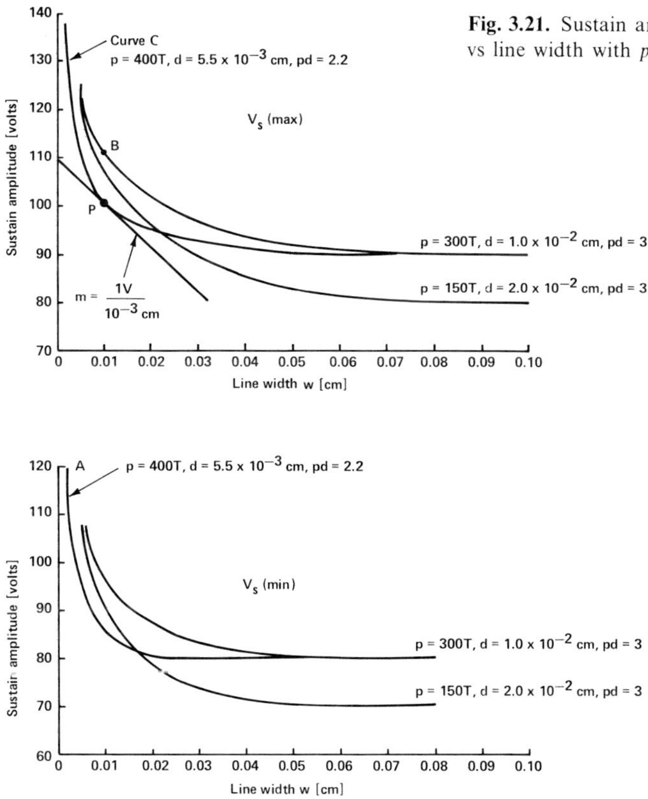

Fig. 3.21. Sustain amplitude (min and max) vs line width with $pd = 2.2$ and 3.0 Torr. cm

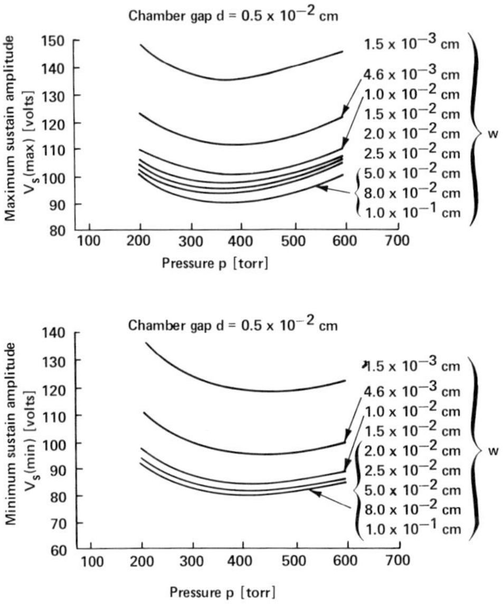

Fig. 3.22. Sustain amplitude (min and max) vs pressure and line width for a chamber gap of 0.5×10^{-2} cm

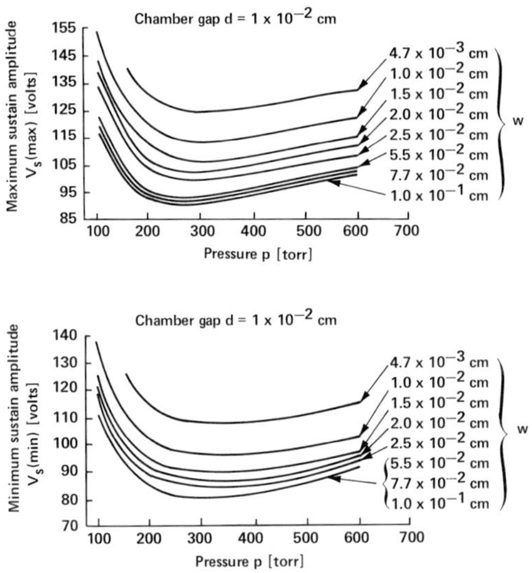

Fig. 3.23. Sustain amplitude (min and max) vs pressure and line width for a chamber gap of 1.0×10^{-2} cm

The smaller chamber gap (Fig. 3.21) is the most suitable one for a high resolution display since the sharp increase in sustain voltages start at a lower point (curve C) than for the larger chamber gaps. This is consistent with Fig. 3.19 which indicates a more confined field (and glow) for the smaller-chamber-gap panels. From the tangent to curve C at point P, it can be seen that a change in line width of 0.001 cm will change the sustain voltages of a cell by approximately 1 V.

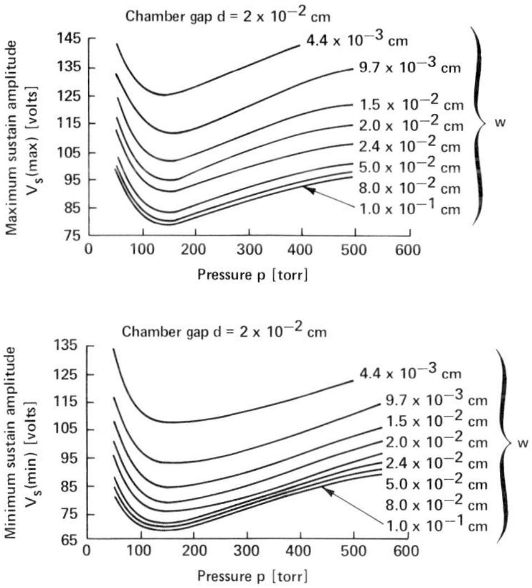

Fig. 3.24. Sustain amplitude (min and max) vs pressure and line width for a chamber gap of 2.0×10^{-2} cm

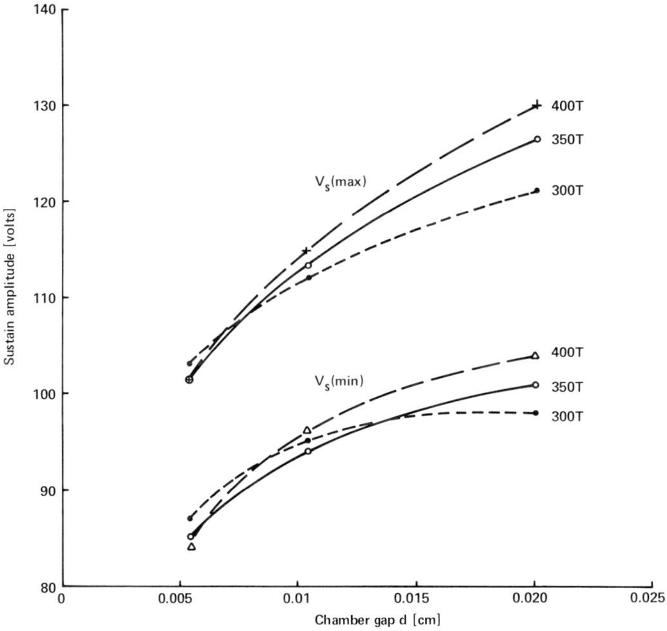

Fig. 3.25. Sustain amplitude (min and max) vs chamber gap and pressure for a line width of 1.0×10^{-2} cm

It was mentioned in Sect. 3.4.1d) that the curve E of Fig. 3.20 indicated the strong possibility of a nested half select cell being spuriously written because of dielectric interaction. The curves in Fig. 3.21 confirm this fact as the V_s(max) for a line width of 0.01 cm is in the 100–110 V range, and the gap potential between cells from Fig. 3.20 is also in that range.

In Figs. 3.22–24 *O'Hanlon* shows the effect on the sustain voltages of changing the chamber pressure, for different line widths. For a given chamber gap there is a pressure, that gives the lowest operating sustain voltages, and is referred to as the Paschen minimum. *O'Hanlon* observed that the Paschen minimum varies slightly with the line width and is not constant; also the Paschen minimum does not occur at the same *pd* product for different chamber gaps.

Of primary interest in panel design, is the sensitivity of cell characteristics to variations in chamber gap. *O'Hanlon* shows this in Fig. 3.25, where V_s(min) and V_s(max) are plotted against the chamber spacing, with the gas pressure as a parameter. The operating voltages of a cell increase with an increase in chamber gap.

It is interesting to note that while Fig. 3.25 indicates that a single cell margin $[V_s(\max) - V_s(\min)]$ increases with a larger chamber gap spacing, Fig. 3.19 indicates that crosstalk will increase between cells. Since an increase in crosstalk will decrease the overall panel margin, increasing the chamber gap will not necessarily increase the overall panel margin.

In panel design, there are many interacting factors that must be simultaneously considered, i.e., the effects of changing one parameter of a design may completely negate the gain achieved by optimizing a different parameter.

Section 3.4 has dealt with the causes of cells having nonuniform operating characteristics. Cell-to-cell interaction and intrinsic cell nonuniformity have been described. Both of these factors contribute in decreasing the overall panel operating margins. The next section will deal with graphing of interrelated panel parameters to determine the operating margins of a panel.

3.5 Write, Erase, and Sustain Margins

The following is a description of commonly used voltage plots that graphically reflect the combined effects of all the physical parameters of a panel as well as the waveforms used to drive it. If worst-case test patterns are used, a realistic measure of the panel's operating-voltage margins will be realized.

3.5.1 Panel Operating-Voltage Margins

Figure 3.26 is a representative plot of the write, erase, and sustain voltage margins of an ac plasma panel. Since the write and erase pulses are used in conjunction with the sustain waveform to selectively manipulate test patterns, the margins are referred to as dynamic margins. This mode of testing, similar to actual use of the panel, is different from the mode referred to as static testing where only the sustain waveform is used to manipulate simple and less rigorous test patterns.

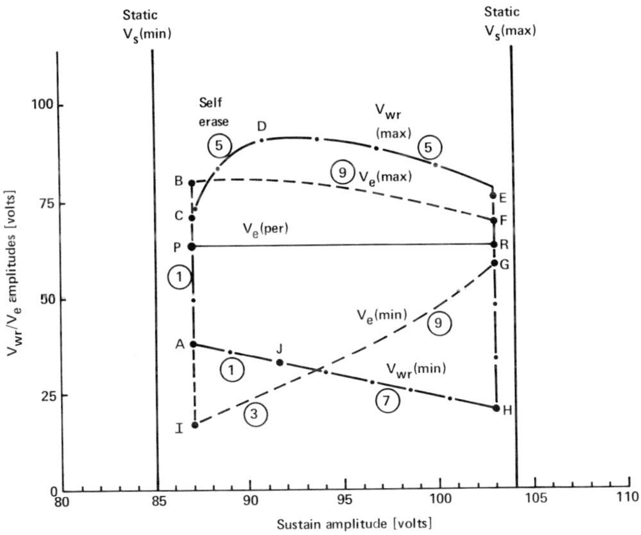

Fig. 3.26.
Panel operating range

Simple test patterns (e.g., checkerboard) can be implemented by "group switching" of the sustain waveform. The two sustain voltage bounds V_s(min) and V_s(max) (static test) in Fig. 3.26 represent the sustain voltage range where all of the cells of the panel exhibit bistability without the disturbing effects that can accompany selective writing and erasing. Therefore, the static sustain margin represents the limit of the dynamic margin which is seldom, if ever, attained.

For the static test, a simple pattern of ON cells is used to create a maximum priming level, while the sustain voltage is gradually increased until the first cell (not in the priming pattern) turns ON. This is the static V_s(max) for the test cells. Next, with a different pattern of ON cells used to provide a minimum local priming level, the sustain voltage is gradually increased until the first cell in that priming pattern turns OFF. This is the static V_s(min) for the cells in that pattern. The test can be repeated, rotating the roles of the cells, until all cells have been tested. The more extensive the coverage of the panel, the more accurate is the measurement. Generally, static testing is used for a quick test and maximum accuracy is not needed.

There are two operating areas plotted in Fig. 3.26 – erase and write. The erase operating area shown is typical for an erase pulse of 4–6 μs in width. The decrease in erase margin at the higher sustain voltages is expected (Sect. 3.4.1b), since a smaller value of $V_w(V_e)$ is required at the higher sustain amplitudes. The write operating area for an improperly designed write-pulse wave shape exhibits severe self-erase characteristics at the lower sustain amplitudes. Generally, the write-pulse amplitude required to write successfully decreases as the sustain amplitude increases.

3.5.2 Image Test Patterns

Worst-case test patterns are difficult to define since they vary, if only slightly, for different panel geometries. However, worst-case patterns do have common characteristics that can be described.

Worst-case test patterns are display image patterns that produce extremes of spatial priming, wall-voltage coupling, and dielectric interaction. The sequencing rate of these patterns are also varied to produce extremes of temporal priming.

a) Test Patterns and Test Panels

A pattern of 64 × 64 pels is a satisfactory test area, and therefore can be used on small test panels as well as large or full-size panels. Because large extremes in pilot priming cannot be satisfactorily simulated on small test panels, their use is usually limited to "rough cut" testing. In addition, for small test panels a few lines on the borders of the test pattern must be used for guard cells as in a big panel. These (OFF) guard cells are driven by sustain waveforms only (no write or erase).

On full-size panels small areas can be tested in selected positions and then a composite operating-area graph can be plotted for the entire panel. For full-load circuit tests, the small test patterns can be repeated until the entire panel is ON.

b) Worst-Case Test Patterns

The checkerboard, or even the alternate-checkerboard pattern (alternating the checkerboard with its inverse pattern), is not a worst-case test pattern because it is devoid of image discontinuities and image-content extremes. Therefore, the pattern regularity that makes it so easy to implement static tests detracts from its effectiveness as a worst-case test pattern.

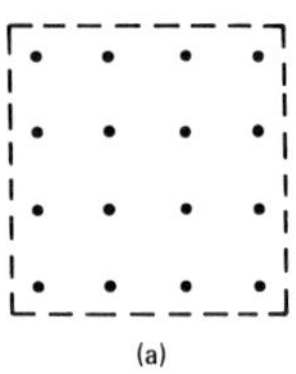

Fig. 3.27a–c. (**a**) Worst-case test pattern. (**b**) Abnormally shaped gap potential profile for a selected line (sel) caused by dielectric interaction of unselected ($\overline{\text{sel}}$) adjacent lines on both sides. (**c**) Abnormally high gap potential between an unselected ($\overline{\text{sel}}$) line and the adjacent selected (sel) lines, due to dielectric interaction

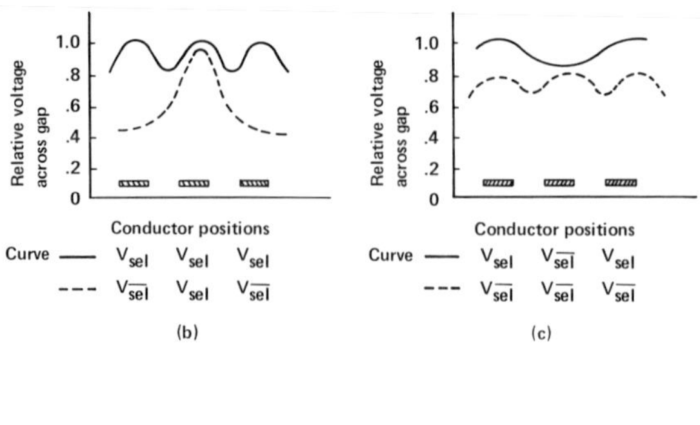

One rather simple worst-case test pattern is illustrated in Fig. 3.27a. A test pattern referred to as A, consists of a group of isolated ON cells in a field of OFF cells, thereby qualifying it as a pattern that produces an extremely low image content and, therefore, a low priming level. It also creates a high level of dielectric interaction as shown in Fig. 3.27b where the gap potential for the selected line (sel) is effectively reduced. Pattern B is the complement of pattern A and, therefore, produces a high level of priming as well as a high level of dielectric interaction, as shown in Fig. 3.27c. As a result of these levels of priming and interaction, the write and erase operations take place under worst-case conditions.

c) Worst-Case Test-Pattern Sequences

A typical test procedure can be created as follows:

1) Write pattern A (isolated ones)
2) Sustain cycles
3) Erase pattern A
4) Sustain cycles
5) Write pattern B (isolated zeros)
6) Sustain cycles
7) Write pattern A (nested ones)
8) Sustain cycles
9) Erase pattern A
10) Sustain cycles
11) Erase entire test area
12) Return to Step 1.

By extreme changes in the number of sustain cycles in Sequence Steps 2, 4, 6, 8, and 10 and in between repetitive write pulses in Sequence Steps 1, 5, and 7 repetitive erase pulses in Sequence Steps 3, 9, and 11, large changes in temporal self-priming and write/erase waveform disturbances can be achieved.

The circled numbers shown in Fig. 3.26 refer to the Sequence Steps in which the image failure occurs. The lower limit of the write-pulse amplitude is noted during Sequence Steps 1 and 7, where the cells fail to write because of insufficient amplitude (Fig. 3.27b), and priming, or because of the inhibiting effect of wall-voltage interaction *Ngo* [3.23]. Failures noted during Sequence Step 5 establish the upper bound of the write-pulse amplitude, when extra cells are written because of the effects of high priming levels and dielectric interaction illustrated in Fig. 3.27c.

The lower limit of the erase-pulse amplitude is noted during Sequence Step 3, where cells fail to erase because of insufficient amplitude (Fig. 3.27b), and during Sequence Step 9, where nested cells turn ON again after the erase because of the effects of high priming levels and dielectric interaction (Fig. 3.27c). The upper limits for the erase amplitude are primarily established by the dense pattern of Sequence Step 9.

Worst-case test patterns and sequences should always be used to evaluate changes in the physical parameters of panel, and changes in drive waveforms.

3.6 Drive Waveforms

This section deals with the composite waveforms that appear across every cell of a panel during the sustain operation and across the selected cells during the write and erase operations.

3.6.1 Sustain Waveforms

The ac plasma panel is best sustained by square waves that have rise and fall transitions of less than 1.0 µs. The voltage reference (external to the panel) of these waveforms changes with different circuit design philosophies, but since the panel is an ac device, the sustain operation does not. Therefore only the ac component of the sustain waveforms is described.

a) Non-Return-To-Zero Waveforms

A non-return-to-zero (NRZ) sustain waveform is illustrated in Fig. 3.28a. Its sustain amplitude is referred to as V_s (peak from midpoint). This amplitude convention will be used throughout this and later sections.

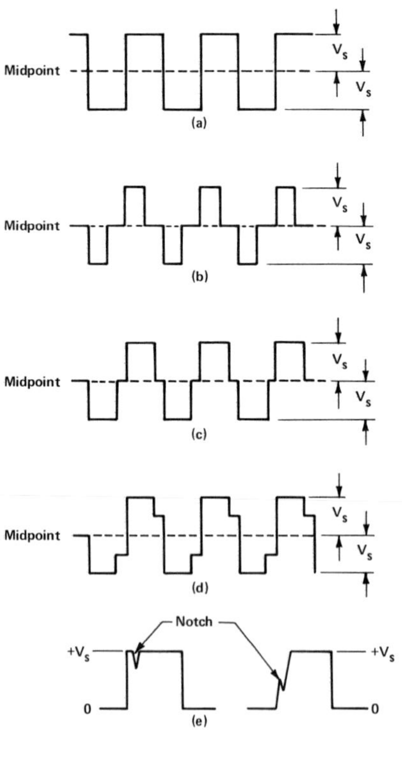

Fig. 3.28a–e. Sustain wave shapes and leading-edge distortion due to circuit loading

The characteristic of a panel driven by an NRZ waveform is affected by the absence of any transition delays (steps). One such characteristic of a panel is its ability to maintain a relatively high sustain margin at high sustain frequencies (above 50 kHz), thus producing a brighter display image. The reason for this is that with an NRZ waveform one has maximum pulse width at a given frequency. Thus there is sufficient time in each alternation for an adequate wall-charge collection and wall-voltage buildup at higher frequencies.

Panels driven by the NRZ sustain waveform exhibit slightly higher peak discharge currents and significantly higher displacement currents because of the larger transitions. These two factors can become serious obstacles for large high-resolution panels.

b) Symmetrical Return-To-Zero Waveforms

The symmetrical return-to-zero (RTZ) sustain waveform, illustrated in Fig. 3.28b produces an extreme in transition delays. Each alternation is reduced in width by a transition delay equal to 25% of the total period.

The characteristics of a panel driven by this symmetrical RTZ waveform are affected by the transition delays and the resultant narrow alternation (one-fourth of the period). Its frequency cutoff is lower than that of the NRZ waveform and its sustain voltages are slightly higher. More significantly, panels driven by an RTZ waveform require less peak discharge current as well as less displacement current. Therefore, the RTZ waveform tends to be more suitable for large or high-resolution panels.

c) Modified Return-To-Zero Waveforms

There are many variations between the symmetrical RTZ and the NRZ sustain waveforms that are designed to minimize the disadvantages and retain the advantages of the two generic forms. These commonly used variations, shown in Fig. 3.28c, d, are derived by minimizing the transition delays either in amplitude or in time duration. The end results improve the high-frequency sustain margins of the RTZ waveform and decrease the peak sustain and displacement currents of the NRZ waveform. These variant waveforms are commonly used in ac plasma display terminals. Other variations that can be made fall into the categories already described and will not be further discussed.

d) Distortion in Sustain Waveforms

The leading-edge transition and the first few microseconds of the square-wave peak level are the most critical parts of any sustain waveform. Changes in these areas produce first-order effects on panel operating characteristics, while changes to the trailing-edge shapes have second-order effects. Therefore, care must be taken in circuit design to prevent distortions, Fig. 3.28e, in the sustain waveform due to high current loads. Such distortions can cause excessive losses in sustain margin if they exceed about 5 V.

3.6.2 Write Waveforms

Earlier in this chapter (Sect. 3.3.4) the write operation and a simplified write pulse were described. The write-pulse waveshape illustrated in Fig. 3.13 was simplified to facilitate the description of the write operation. In this section practical composite waveshapes for the write operation will be described and their characteristics noted. The write-pulse amplitude in a write operation is the result of a voltage addition of two components generated on two orthogonal panel lines. The partial-select components are discussed in a later section.

a) Selection States of Cells

A selected cell, either for a write or an erase operation, is defined as the intersection of one selected horizontal and one selected vertical line. All remaining cells on either selected line are partially selected cells, and all the cells on neither of the two selected lines are unselected cells.

b) Write Waveforms for Selected Cells

The write pulses shown in Figs. 3.29a–c all perform the same but are referenced to different voltage levels. In Fig. 3.29a the write pulse uses the sustain peak amplitude V_s as its reference. In Fig. 3.29b a pedestal between V_s and the sustain midpoint, Fig. 3.28d, is used as a reference, and in Fig. 3.29c the sustain midpoint is used as the reference for the write pulse. Aside from the circuit design implications, these three forms of write pulses are operationally equivalent. The write pulse in Fig. 3.29d is slightly different in that it no longer is contiguous to the sustain alternation. This change is usually a result of circuit considerations; however, the RTZ step shown is of a shorter duration than for Fig. 3.29a–c. This reduction of the time allowed for the relatively weak discharge that normally occurs during this transition-step time does affect the write operation. The wall charge transfer is reduced and the self-erase characteristics are modified.

Self-erase has a tendency to occur with write pulses that have transition delays. Consider a write pulse with a peak amplitude equal to approximately $2V_s$ that produces a $\Delta V_w \approx 2V_s$. The transition from the write-pulse peak amplitude to the sustain midpoint will produce a similar change in wall voltage. The end result is $V_w(V_{wr})=0$.

The write pulse shown in Fig. 3.29e is an attempt to eliminate this weak firing by eliminating the transition step, creating an NRZ transition. The self-erase tendency is eliminated by this change but there is a negative byproduct. The NRZ transition from $+V_{wr}$ to $-V_s$ produces a very intense discharge that loads the drive circuits, producing extremely high priming levels that tend to turn adjacent cells ON. (It is interesting to note that instead of eliminating the self-erase tendency, *Stillwell* and *Schlig* [3.25] put it to use by employing a high-amplitude write pulse ($2V_s$) as an erase pulse. This erase pulse is reported to have relatively high-amplitude margins.)

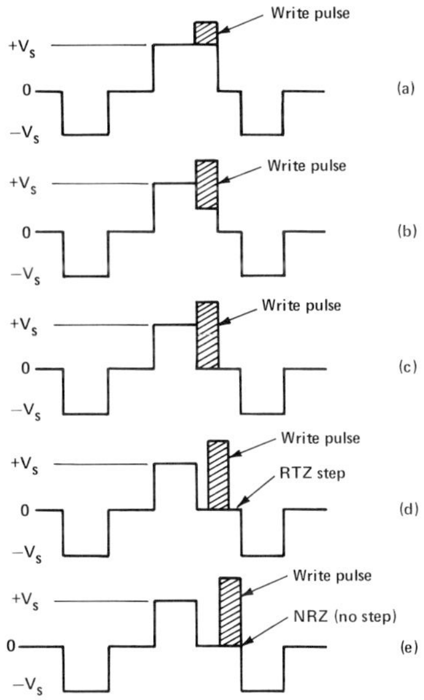

Fig. 3.29a–e. Different types of write waveforms

c) Write-Pulse Width and Rise Time

A narrow-width write pulse (generally less than 5 μs) requires a higher V_{wr}(min). Since this leads to more expensive circuitry as well as the self-erase tendency, wider write pulses are desirable. Sustain wave shapes must be designed to accommodate wide write pulses if possible, and where it is not possible the sustain wave shape has to be literally interrupted to make time for the wide write pulse. Either alternative can work, depending on the physical characteristics of the panels.

Wherever possible, slower-rise-time write pulses are used because they have been generally found to yield larger write-pulse-amplitude margins [3.26]. Since $\Delta V_w(V_{wr})$ is inversely proportional to the rise time of the write pulse, the longer-rise-time write pulses minimize the discharge intensities and crosstalk.

d) Delayed Position of the Write Pulse

In all the cases shown in Fig. 3.29, the write pulse follows a sustain alternation of like polarity. The importance of this delay is twofold [3.27]. First, writing between sustain discharges avoids creating an increase in self-priming due to the intense discharge that would be generated by a write pulse on the leading edge of the sustain, which would reduce V_s(max). Secondly, placing the write pulse towards the trailing edge of the sustain alternation results in removing the half-select cancellation perturbation from the leading edge of the sustain pulse, ensuring that all cells are regenerated by an undistorted sustain pulse.

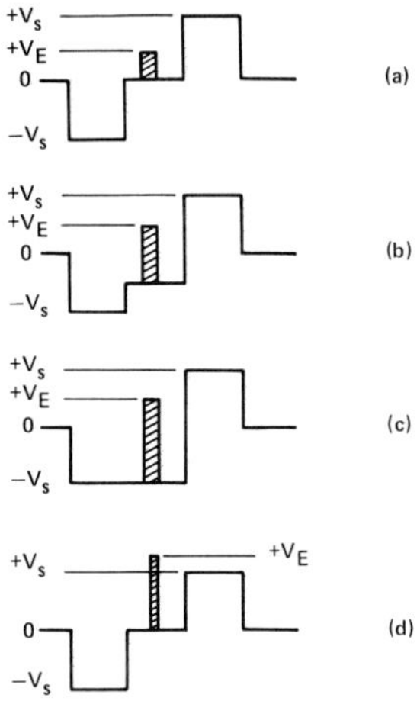

Fig. 3.30a–d. Different types of erase wave shapes

3.6.3 Erase Waveforms for Selected Cells

As for the write operation, only the composite erase-operation waveforms (for selected cells) will be described. The selection states for the erase operation are defined as they were for the write operation.

The erase pulse can be referenced to different voltage levels in the sustain waveform as shown in Figs. 3.30a–c. As before, the choice of reference levels, is based on circuit considerations since the voltage requirements for the pulses change with the different reference levels used. The relatively wide pulses shown are about 6 μs wide.

The narrow erase pulse shown in Fig. 3.30d, which is about 1 μs wide, can be used just as effectively as the wide erase pulse. The narrow pulse requires relatively high amplitudes that can even exceed the sustain amplitude – if they are narrow enough. This can become a circuit drawback that must be balanced against the relatively high erase-voltage margin that is attainable by using a narrow erase pulse.

In general, the wide erase pulse is more sensitive to amplitude variations than is the narrow erase pulse. Conversely, the narrow erase pulse is more sensitive to width variations and priming levels than is the wide erase pulse (Sect. 3.4.1b). The wide erase pulse, therefore, is not as sensitive to variations in rise time or leading-edge distortions as is the narrow erase pulse. This places more stringent circuit design constraints on the use of the narrow erase pulse.

The amplitude sensitivity of the wide erase pulse can be minimized by using slow-rise-time wave shapes [3.26]. The erase pulse comes in between two consecutive sustain alternations as does the write pulse. However, unlike the write pulse, the erase pulse follows a sustain alternation of the opposite polarity. As a result, waveform disturbances that inhibit the erasure of partially selected cells, will have little, if any, effect on other cells.

3.7 Panel Operating Drive Systems

This section describes methods of implementing the following sustain drive, selection pulses, cancellation pulses used to prevent erroneous writing and erasing of cells, and the electronics involved with two typical drive systems.

3.7.1 Sustain Drive

There are two basic methods used to sustain an ac plasma panel. One is to drive both axes (double-sided drive) using phased square waves having a peak amplitude of V_s. The other method is to drive one axis (single-sided drive) with the composite wave shape having a peak-to-peak amplitude of $2V_s$, while the other axis is maintained at a fixed voltage level, usually ground for circuit convenience. Both of these methods are illustrated in Fig. 3.31 for a single cell intersection. In both cases the cell is driven by the same ac voltage waveform, therefore the choice of the two methods depends on only circuit considerations. Any of the sustain waveform illustrated in Fig. 3.28 can be generated as shown in Fig. 3.31. The remainder of this section, therefore, will deal with double-sided drive systems only.

3.7.2 Half-Select Pulses

Half-amplitude write or erase pulses that are synchronous with the sustain drive are used to selectively turn cells ON or OFF. A simplified block diagram with all the essential elements required to operate a cell is shown in Fig. 3.32. Upon command, at some time t, each write/eraste (W/E) driver generates one of two half-select pulses, with different amplitudes and widths if necessary, that are added to the instantaneous sustain drive output voltage level. This results in a write or an erase pulse positioned as shown in Fig. 3.32. At all other times, the W/E drivers must bidirectionally pass the sustain drive to the panel line without distortion. Since the half-amplitude pulses must be additive across the cell, the W/E drivers for each axis must generate pulses of opposite polarities.

If the cell is only half-selected, the half-select pulse, being less than V_s in amplitude, will not cause a cell to be "written", Fig. 3.32. Neither will a half-select erase pulse cause a cell to erase, because its amplitude is insufficient to cause the polarity reversal required for an erase discharge.

The half-select write pulse follows the sustain operation and, therefore, has no effect; the half-select erase pulse also follows a sustain operation and, likewise, has no effect. The positions chosen for the write and erase pulses produce nondisturbing half-select-condition waveforms. This is absolutely necessary if the maximum margins are to be realized in operating the panel.

The system just described has one drawback, that is related to the voltage references used for the half-select pulses. The pulse amplitudes necessary for

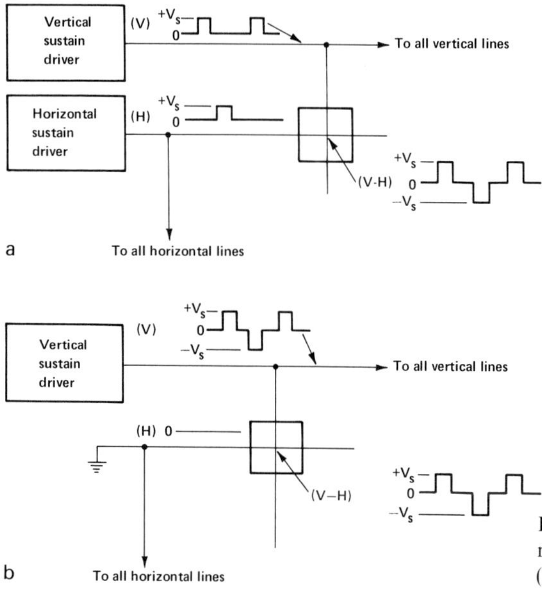

Fig. 3.31a, b. Sustain drive configuration. **(a)** Double-sided drive. **(b)** Single-sided drive

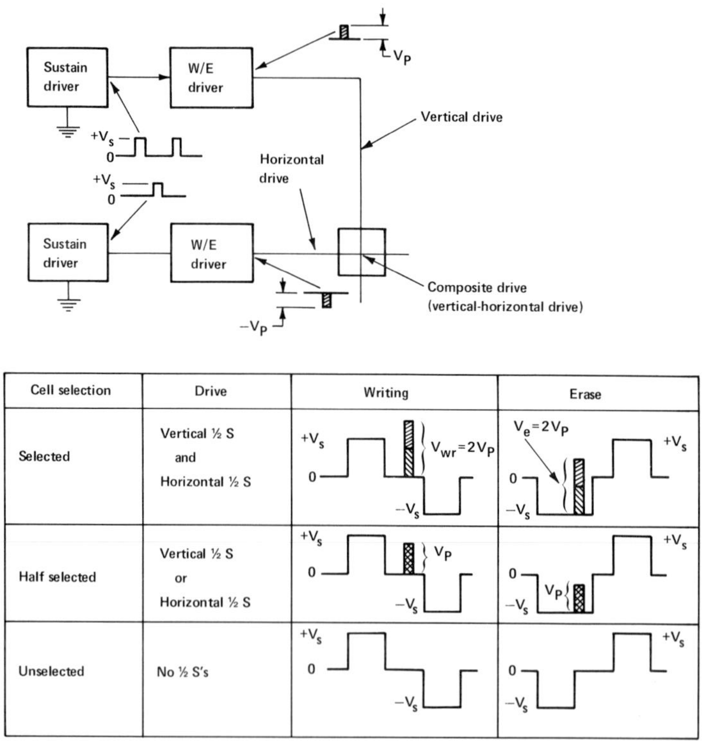

Fig. 3.32. Circuit elements essential to the operation of a cell as a display pel

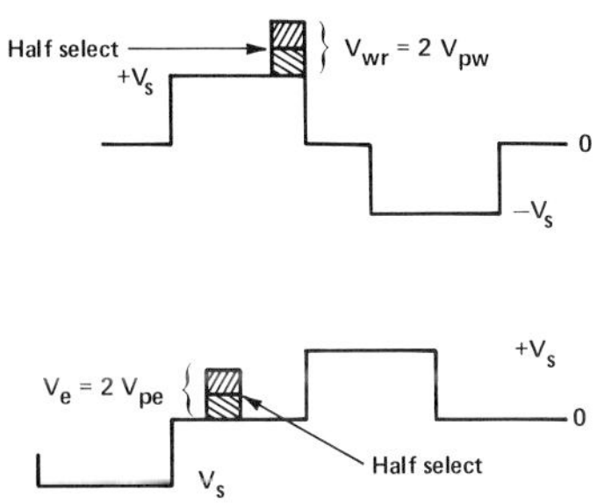

Fig. 3.33. Half-select pulses

writing and erasing are quite high and under certain conditions can approach, and even equal, V_s (typically 100 V). To use lower amplitude pulses and the associated less-expensive circuitry, a higher voltage-level reference or pedestal must be used. The highest readily available reference level is the sustain peak itself, V_s. When the peak and the midpoint sustain voltage levels are respectively used as pedestals for write and erase pulses, severe half-select problems occur (see Fig. 3.33). The pulse amplitudes V_p are clearly lower in amplitude, but the half-select write pulse exceeds V_s and will write cells, especially if it occurs every sustain cycle.

The half-select erase pulse will erase cells since it does produce a polarity reversal (above sustain midpoint). These uncancelled half-selects must be inhibited for reliable panel operation, but without resorting to the higher pulse amplitudes of Fig. 3.32.

3.7.3 Half-Select Cancellation Pulses

The waveforms illustrated in Fig. 3.34 include half-select cancellation pulses for the write and erase operations, as well as the half-select pulses. Waveforms (A) and (B) are the respective drive waveforms for the selected and unselected horizontal panel lines during a sustain cycle with a write or an erase operation. Waveforms (C) and (D) are the drive waveforms for the unselected and selected vertical panel lines respectively during a sustain cycle with a write or an erase operation. At write time, the selected horizontal panel lines are driven by a positive half-select pulse, $V_{wr}/2$, referenced to $+V_s$, while the unselected horizontal panel lines are driven by a level that is below $+V_s$ by $V_{wr}/2$ $+(V_s - V_{wr}/2)$. Similarly, the selected vertical lines are driven by a negative half-select pulse, $-V_{wr}/2$, referenced to ground potential while the unselected vertical lines are driven by a level that is above the ground potential by $+V_{wr}/2$. At a time appropriate for the erase operation similarly phased pulses modulate the sustain waveforms.

Waveform (A–D) is the composite drive waveform applied to selected cells, where a full V_{wr} or V_e pulse is applied to the cell. Waveform (A–C)/(B–D) is the composite drive waveform applied to all half-selected cells. Since the half-select pulses on waveforms (A) and (D) have been, respectively, cancelled by the

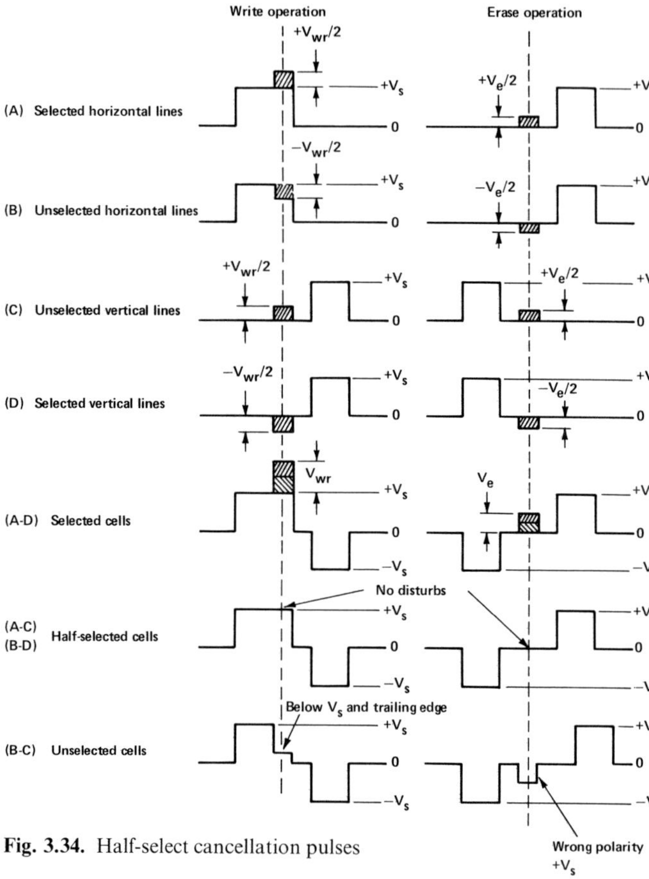

Fig. 3.34. Half-select cancellation pulses

pulses on waveforms (C) and (B), the half-selected cells remain totally undisturbed and are driven by a normal sustain waveform, thereby maintaining their prior ON/OFF state.

Waveform (B–D) is the drive waveform for all unselected cells. At write time the cancellation pulses cause a trailing-edge deviation in the sustain waveform. However, the state of a cell is not disturbed since the deviation's amplitude is below V_s and the remaining sustain alternation is full-width. In the case of the erase, the cancellation pulses result in a full-amplitude erase pulse that has the same polarity as the previous sustain, making an erase operation impossible. Therefore, all unselected cells will also maintain their prior ON/OFF state.

Higher-amplitude reference pedestals can be used with half-select cancellation pulses, reducing the required write and erase amplitudes without the adverse effects illustrated in Fig. 3.33. A similar approach to half-select cancellation is illustrated in Fig. 3.35 where the write and erase pedestals are built into the sustain waveshape (Fig. 3.28d). The write and erase pedestals are voltage levels somewhere between the sustain midpoint and the sustain peak.

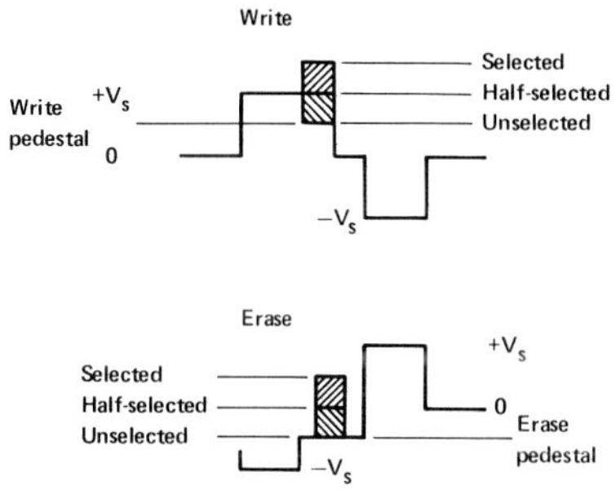

Fig. 3.35. Half-select cancellation pedestals

The end result is identical to that of using cancellation pulses – i.e., a pedestal is formed that effectively cancels half-select pulses.

3.7.4 Full-Select Pulses

If a pulse is added to the half-select and half-select cancellation pulses of the horizontal and the vertical drive waveforms (Fig. 3.34), a subtle difference in the selection process occurs. For the write operation $+V_{wr}/2$ is added to the select and cancellation pulses for both axis, and for the erase operation $+V_e/2$ is likewise added. The result is the waveforms shown in Fig. 3.36 where the last three (composite) waveforms are identical to those shown in Fig. 3.34. The subtle difference is that unipolar pulses instead of bipolar pulses can now be used as full-select pulses. This simplifies the design of the selection circuits, resulting in a circuit cost reduction.

To summarize this full-select system, the select condition (A–D) for both write and erase operations results from full-select pulses in waveform (A) and the absence of a pulse in waveform (D). The half-select state (A–C) (B–D) results from a full-select pulse in waveform (A) and a full-select cancellation pulse in waveform (C), or the absence of any pulse in both waveforms (B) and (D). The unselected state (B–C) results from the absence of a pulse in waveform (B) and a full-select-cancellation pulse in waveform (C). The selection pulses are of the same polarity and only two of the four waveforms need pulses, which is another cost reducing circuit simplification.

3.7.5 Panel Drive Systems

The purpose of the W/E drivers used in drive systems is two-fold. First, they serve as multiple distribution paths from the sustain driver to the panel lines. Second, they provide each panel line with write and erase pulses in accordance with their selection states. One of the systems described in this section employs W/E drivers in a matrixed configuration to reduce the circuit component count. The other system uses a nonmatrixed configuration with its higher component count and resorts to circuit integration to reduce cost.

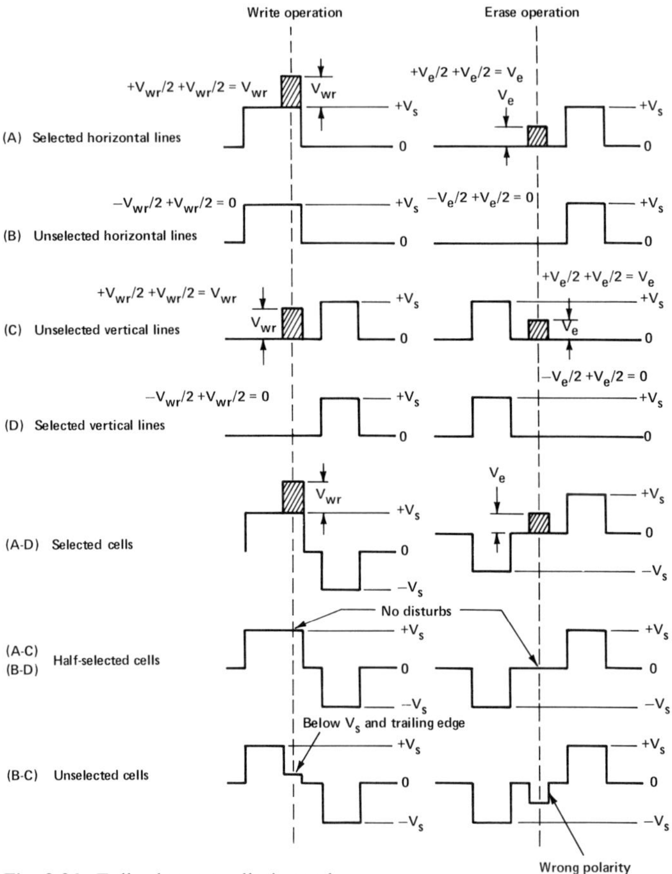

Fig. 3.36. Full-select cancellation pulses

a) Diode-Matrix Drive Array System

A diode-matrix drive system using a diode-resistor multiplexing circuit for a W/E driver was reported by *Johnson* and *Schmersal* [3.28]. Using active drivers with discrete components on every panel line can be bulky and costly. However the multiplexing circuit illustrated in Fig. 3.37 consists of only three discrete components (two diodes–one resistor) for each panel line, thus providing a practical and relatively inexpensive drive system. Low-impedance source and sink paths are respectively provided for the discharge and displacement currents by diode D 1, and by diode D 2 with SW 1 (closed during sustain).

For the write/erase operations, a selection pulse is applied to the upper end of resistor R. If R is pulsed with a selection pulse AND if SW 1 is open (selected condition), the panel line will receive a pulse that is higher than the sustain-driver level. Diode D 1 back biases under these conditions to disconnect the pulsed panel line from the sustain driver.

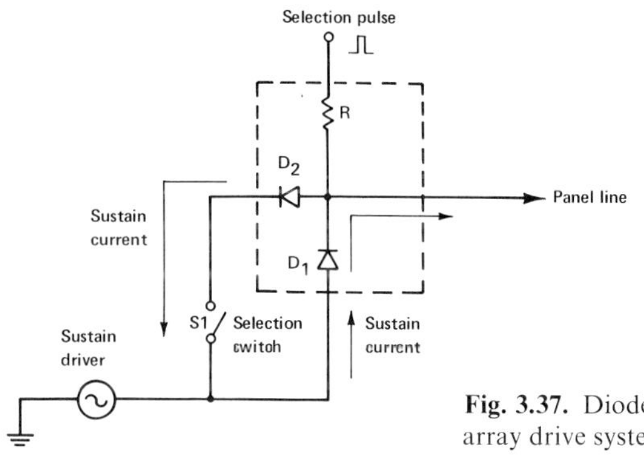

Fig. 3.37. Diode-resistor circuit used in diode-matrix array drive system

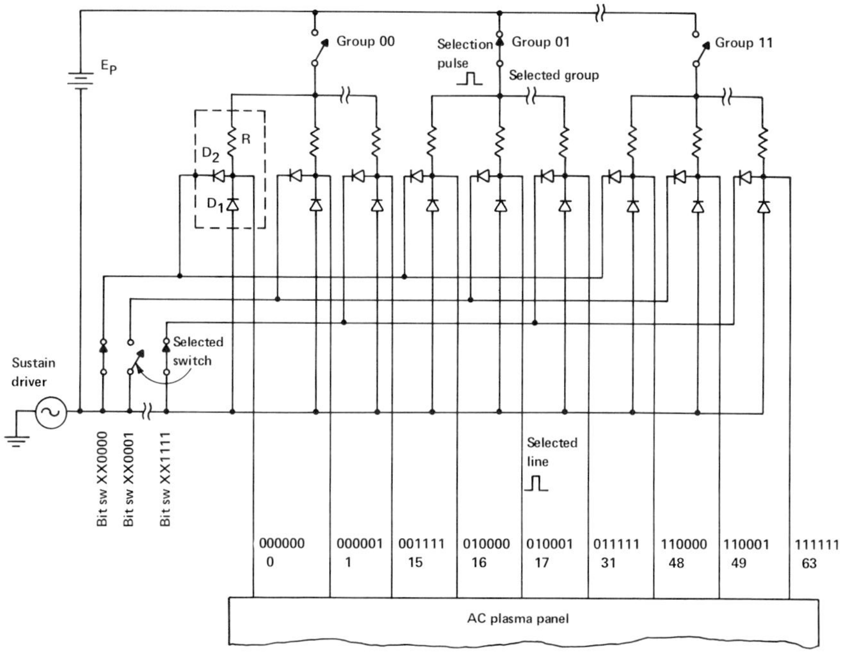

Fig. 3.38. Diode-matrix array drive system

The diode-resistor circuit provides the AND function required for use in a matrix configuration as illustrated in Fig. 3.38, the matrix is a 4×16 matrix of 64 W/E drivers. There are four group switches (actually transistor switches) 00 through 11 (binary). When momentarily closed, each group switch applies a selection pulse to the resistors of the selected group of diode-resistor circuits. As shown in Fig. 3.38 group switch 01 is momentarily closed while bit switch $\times \times 0001$ is open. Therefore, panel line 010001 (17) is the selected panel line and will, therefore, be pulsed. The remaining 15 bit switches are closed, holding the remaining 15 panel lines in the selected group down to the sustain level (no pulse). All the resistors in a selected group that are held down by the "not-selected" bit switches dissipate useless power during the selection pulse time. This can be a problem on very large panels where the resistors must be made small enough to drive high interelectrode capacitances.

Twenty switches compared to the otherwise required 64 is a significant reduction in circuitry. However, the 20 switches are high-current and high-voltage (75–100 V) circuits since they must handle the currents for multiple lines. The number of pels that can be written or erased with one selection pulse is determined by the number of panel lines associated with one group switch. For a 512×512 line panel, 16 groups of 32 lines per group is a reasonable partitioning of circuits. Therefore, 32 cells on a panel line could be selectively written or erased with one operation.

This drive system is a practical example of diode-matrix-drive array using discrete components. However it has limitations that should be noted. First, its power effeiciency is limited by the power dissipated in the resistors. Second, its high-current–high-voltage switches cannot be highly integrated because they dissipate more power than LSI modules can readily handle (without heat sinks). Third, the use of discrete components, even with the integrated diode modules, requires extensive board space and interconnection cabling. Last, and perhaps most serious from a users point of view, is the limited write/erase speed that results from the 32-bit-slice capability. With 32 cells addressed every 20 μs (50 kHz sustain frequency), the bit addressing rate would be 1.6 Mbit/s. This is a high bit rate, but not as high as for the next system to be described. Writing on every sustain cycle to attain this bit rate imposes limitations on the panel's dynamic margins because of the high priming levels and the high average voltage between cells (due to dielectric interaction) during-worst case patterns.

A similar but improved diode-matrix array system was reported by *Dick* [3.29] where charge-storage diodes (CSD) were used to replace the resistors. The total use of diodes in this system makes integration practical and effective. However, even this approach still has the problems of the high-current discrete-component switches and the limited write/erase bit rate peculiar to matrixed arrays.

b) Active Circuit Driver System

A driver system that uses half-select cancellation pulses, as shown in Fig. 3.34 and an active driver W/E circuit for each panel line was reported by *Criscimagna* [3.27]. The W/E drivers, called selection circuits, use active components as shown in the dotted lines in Fig. 3.39. Diodes D 1 and D 2 provide low-impedance sink and source paths for the discharge and displacement currents. Since switch 1 (S 1) is closed at all times (except write- or erase-pulse time) it provides D 1 with a path for the sink current to the sustain driver.

The selection circuit switches the panel line to the selection-phase source point A, or to the sustain drive point B, depending upon the state of the logic-latch input at point D. The state of the logic latch is dependent on the

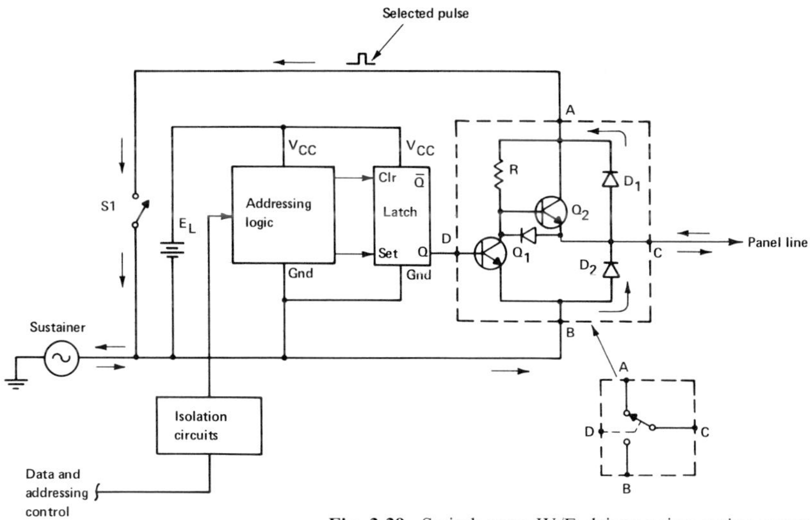

Fig. 3.39. Switch-type W/E driver using active components

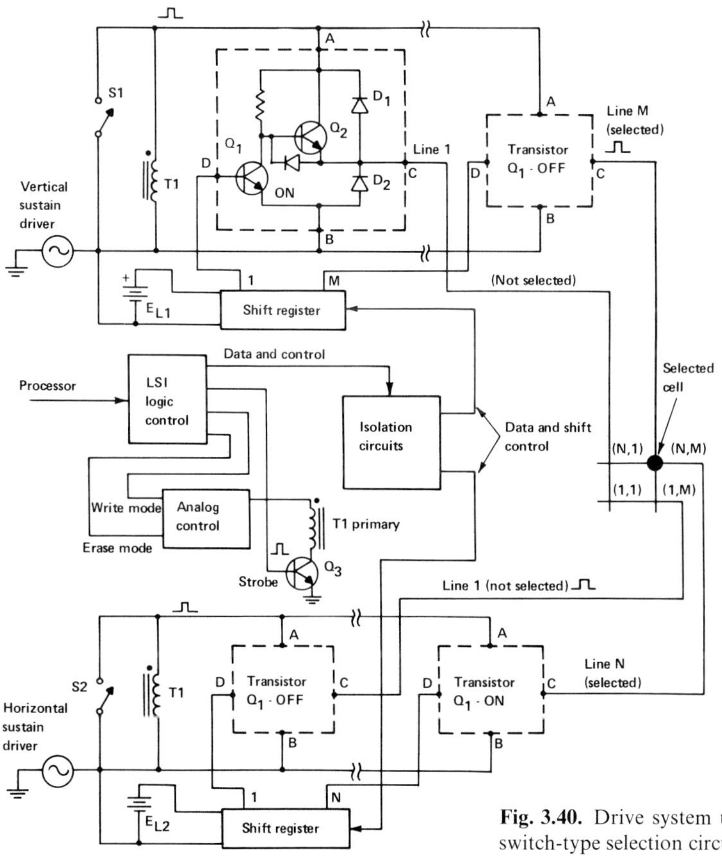

Fig. 3.40. Drive system using integrated switch-type selection circuits

addressing logic output that is controlled by the data transmitted via the ioslation circuits. The isolation circuits are required since the selection logic is floating on the sustain drive, and the control logic (LSI) is ground referenced in order to communicate with the controlling processor.

The configuration was not designed to reduce component counts as was the diode-matrix array configuration. Instead, it was designed to be implemented in low-power integrated circuits that could be supported by just a few discrete circuits.

Figure 3.40 is a complete, but simplified, system circuit diagram involving both axes of a panel that is designed to operate with the full-select pulse waveforms illustrated in Fig. 3.36. Since both axes operate in the same manner, only the vertical axis circuitry and the functions common to both axes will be described. Each axis has one switch to provide diodes D 1 with a path to the sustain driver, one pulse generator (illustrated as a transformer secondary). A shift register and a logic power supply referenced to the sustain bus (floating) are also provided. The shift register provides selection-state information for the selection switch at points D to control the switch transistor Q 1. When transistor Q 1 is ON, the panel line is connected to the sustain bus at point B, and when it is OFF the panel line is connected to the common pulser bus at point A. Th3 LSI control logic communicates to the floating shift register through isolation circuits that can be implemented by pulse transformers or diode switching circuits. The use of a shift register minimizes the number of isolation circuits required since the data is serial. However, where speed is essential, parallel data can be transferred to a different logic configuration by using more isolation circuits.

Although Fig. 3.40 does not illustrate the integrated selection:circuit partitioning as reported by *Criscimagna* [3.30], it should be understood that 32 selection circuits and a 32-bit shift register are easily integrated as one module. The shift register can be loaded as one long shift register (512 stages, for example) by providing shift carry lines, or, each register can be shifted independently. (The selection-logic architecture is almost totally dependent on specifications established by applications requirements.)

During the sustain operation, diodes D 2 and diodes D 1, along with S 1 provide sustain current for each panel line. This provision is totally independent of the state of the selection switch. Therefore, the shift register can be loaded during the sustain operation until the bit image of an entire panel line is loaded (only one line on the other axis is selected).

The circuits are now ready to write or erase an image for an entire panel line using only one selection pulse. S 1 is opened and transformer T 1 is synchronously triggered with a pulse that has the amplitude, width, and timing appropriate for the desired panel operation (Fig. 3.36). A full-select pulse is routed to panel line M, while panel line N is not pulsed. Cell (N, M) is, therefore, selected and either erased or written. All the other selection states are as shown in Fig. 3.36. After the selection pulse, S 1 is again closed and the sustain operation resumes.

During sustain operations the voltage is zero across the selection switches, as is the power dissipation. During the duration of the write or erase pulse, the switches perform as low impedance switching circuits. Thus, the power dissipation is again, very low. The current requirements for each circuit are also low since each circuit drives only one panel line. All of these factors make these circuits very suitable for medium scale integration. Thus, the entire system is virtually integrated, the circuit package is cool and compact, and the bit rate is high.

Considering a 512 × 512-line panel as before, and assuming a shift register speed of 5 MHz, the data for one panel line can be loaded in approximately 100 μs. Therefore a, write pulse or an erase pulse to control a full panel line can occur every five sustain cycles (50 kHz sustain frequency). This is then a write/erase bit rate of about 5 Mbit/s, or three times as fast as the diode-matrix array system. In addition, writing every fifth cycle improves the panel's dynamic margins under worst-case pattern writing and erasing by reducing the priming levels and the dielectric interaction effect on the voltage in between cells.

Speeds of 26 Mbit/s can be attained with this system (at the expense of margins) by writing every sustain cycle as with the diode-matrix array. Faster, parallel entry floating selection logic would be required.

This active circuit drive system outperforms the diode-matrix array system.

c) Integration of the Sustain and the Selection Functions

A promising approach to panel drive is the method of combining the sustain and selection function into one integrated circuit driver each panel line. This would eliminate the need for a bulk sustain driver and discrete pulser circuits; and a totally integrated panel drive could be realized. However, there is one problem. The voltage requirements of the circuits are quite high. V_s and V_{wr} in Fig. 3.41 are typically 100 V and 180 V, respectively. In Fig. 3.41a for a full-select pulse system, one axis would require 180-V circuits, while the other axis could manage with 100-V circuits.

The half-select pulse system illustrated in Fig. 3.42b would require 140-V circuits for both axes. Medium scale integrated circuits with rated voltage breakdowns exceeding 100 V are not economically available today.

Accordingly, *Miller* and *Schermerhorn* [3.31] reported a hybrid drive system where waveforms similar to those shown in Fig. 3.41a were used. The lower voltage axis was driven by integrated circuits (four per module) generating 60-V sustain and deselect square waves. The high-voltage axis used the same integrated circuits, but in a configuration as reported by *Criscimagna* [3.27]. The integrated circuits were used as floating selection circuits with floating logic and isolation circuits for communication to them. The bulk sustainer was operated at a peak greater than V_s to make up for the 60-V sustain drive on the other axis which was less than V_s. When 140–180-V medium-scale integrated circuits become economically available, this drive scheme will result in total circuit integration for panel drives.

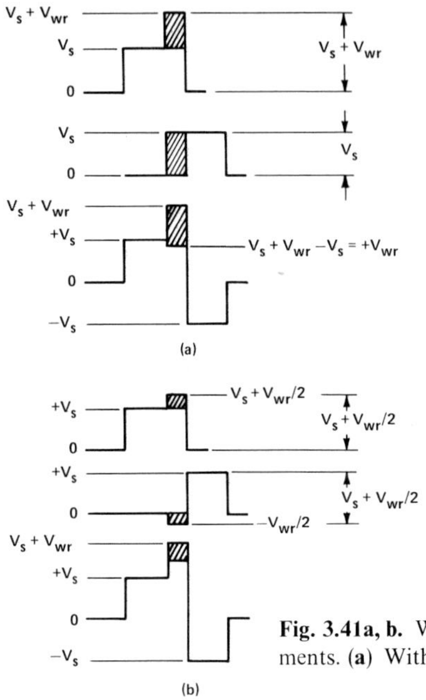

Fig. 3.41a, b. Write waveforms to illustrate circuit voltage requirements. (**a**) With full-select pulses. (**b**) With half-select pulses

3.8 Panel Limitations and Extendibility

In discussing limitations and extendibility of the ac plasma technology, considerations will be restricted to resolution, color, and image generation. In the following, only the highlights in each of these areas will be covered with a discussion of the type of limitation on extendibility indicated for each area considered. The imposed limits are set by manufacturability, overall cost, technology capability, or a combination of these.

3.8.1 Resolution

Today's ac plasma high-resolution panel in large-quantity manufacture, is a 512×512 panel with a resolution of 2.4 lines/mm (60 lines/in.). This panel is typically fabricated with a 40 μm dielectric thickness, 85 μm conductor width, a 100 μm chamber gap (d) and a pressure times chamber-gap (pd) value of approximately 6 Torr-cm with a neon–0.3% xenon gas mixture [3.13].

One approach to extending the ac plasma technology to higher resolution is to use the similarity principle, [3.32]. To double the current high-resolution technology, or design for 4.8 lines/mm, linear cell dimensions would have to be reduced by a factor of two and the gas pressure increased to 1200 Torr to keep the same value of pd as required by the similarity principle. This pressure is a problem since it is above atmospheric pressure, causing the panel plates to lift off the spacers, thus making the panel gap less uniform. Thus, because of atmospheric pressure values, the panel has an upper limit on its pressure at around 600 Torr. The penalty for using this lower value of *pd* is reduced margins, as shown in [Ref. 3.13, Fig. 5.0]. In addition, the tolerance on a 50 μm chamber gap (as required by the scaling) with the same panel process would be twice as high since the variation in gap does not scale with the scaled linear dimensions. This in turn affects voltage margin tolerance, current, spot size, and brightness. Thus going to a high resolution (4.8 lines/mm) may not be achievable by merely "scaling" according to the similarity principle, but will require changes in panel materials, processing and construction to achieve higher operating voltage margins and tighter tolerances.

The highest resolution demonstrated to date with a full panel implementation has been at 3.3 lines/mm (83 lines/in) [3.33], although with a fairly low margin of 3 V as compared to 10 V for a 512×512 panel [3.13]. Some margin improvement may be achievable on the high-resolution panels by optimization of drive waveforms, which was not attempted for the 1024×1024 panel described by *Czajkowski* et al. [3.33].

3.8.2 Color

Both monochromatic and multicolor ac plasma development has been reported in the literature. The development of a monochromatic panel with a color different from the neon-orange color generated by the primary gas, neon, was reported by *Yamashita* et al. [3.34]. A green-phosphor panel was developed in which uniformly distributed microencapsulated phosphor particles were sparsely deposited over the panel plates. A specially developed overcoat material requiring lower panel operating voltages than panels with an MgO overcoat material was also incorporated into the panel. Because of the uniform distribution of the phosphor particles, some were directly in the path of the discharge. However, in spite of the electronic-particle bombardment of the phosphor, panel life was quoted as being greater than 5000 h. This long life was attributed to the newly developed green phosphor and microencapsulation of the phosphor particles.

Multicolor ac plasma panels were investigated at Owens Illinois [3.35, 36]. A three-color, 128×128, 2.4 line/mm-resolution display panel was fabricated. Xenon was the primary gas and uv-excited tricolor phosphors were employed. The phosphors were deposited as bar strips outside the cell discharge area, i.e., parallel to the conductors. A 20 V static sustain margin and dynamic sustain margins greater than 10 V were reported.

In this panel, uv optical crosstalk was a problem due to the reflection of uv radiation from the front and rear plate dielectric surfaces. This crosstalk was minimized by depositing uv-absorbing black material on both substrates which reduced the OFF/ON ratio to less than 8.2% for any one of the three color phosphors.

The actual ON luminances of the phosphor bars were approximately 45, 18, and 6 mL for the green, red and blue phosphors, respectively. The reduced amount of uv crosstalk is still a problem, however, because the OFF luminance for the green phosphor calculates to be 3.7 mL, which is quite high, being more than half of the ON luminance of the blue phosphor.

3.8.3 Image

The ac plasma device, as discussed, yields two bistable levels of brightness for a single cell operated with a single frequency, single amplitude, continuous sustain waveform. Multiple levels of brightness with a single cell have been achieved with a sacrifice in operating-memory margin, by using a more-complex sustain waveform. A maximum of three levels of brightness was reported by *Petty* and *Slottow* [3.37].

Another approach to generating multiple levels of brightness at each cell is to erase and rewrite all the cells in the panel, effectively operating in a refresh mode, with multiple levels of brightness produced by modulating the length of time a cell is ON (within a given frames time). This variable-duty factor achieves a gray scale and was reported by *Anderson* and *Fowler* [3.38]. In their scheme, bursts of 8, 16, 32, 64, 128, and 256 pulses were generated, which, in combination yielded 64 levels of intensity. This was implemented on a standard 512×512-line panel. The panel image information was stored in a random-access memory (RAM) with one bit of RAM storage required for each picture element.

Techniques of generating half-tone images in which a number of bilevel cells are used to form one picture element are discussed by *Knowlton* and *Harmon* [3.39]. However this approach sacrifices display resolution in achieving gray scale. Therefore for matrix displays in which resolution cannot readily be sacrificed (i.e., without incurring appreciable additional cost), these schemes are not very applicable.

In their survey paper, *Jarvis* et al. [3.40] compare techniques in which the display spatial resolution is equal to the sampled resolution of the original image. In the representation of a continuous tone image by a bilevel display, the basic processing organization is as follows. An original-image picture element (pel) is represented by a multibit signal corresponding to the pels' intensity. This signal is then compared to a threshold. If the signal exceeds the threshold, the corresponding cell is set ON, otherwise it is set OFF.

The different schemes developed differ mainly in the processing algorithm used for producing the threshold values for comparison with the original-image pel intensities. Some methods use fixed thresholds, and others determine the threshold on the basis of neighboring-image pel intensities. The latter type of algorithm requires more auxiliary storage and more processing than the former. *Jarvis* et al. [3.40] compared three types of images generated by four different basic algorithms. They reported that the ordered dither technique, which is a fixed-threshold type of algorithm using a set of position-dependent thresholds determined only by the coordinates of the element being processed, yields excellent image renditions with a relatively simple implementation.

3.9 Summary

The ac plasma display is a flat-panel matrix display technology, manufacturable in sizes from a single pel to one million pels, with resolutions of up to 3.7 lines/mm (83 lines/in). The panel is compact and yields a display image that is bright, has high contrast, and is without distortion or jitter.

References

3.1 H.G.Slottow: IEEE Trans. ED-**23** (7), 762 (1976)
3.2 J.F.Nolan: "Gas Discharge Display Panel", Conf. Digest, IEEE Int. Elect. Dev. Meeting, Wash. D.C. (1969)
3.3 W.E.Coleman, J.P.Gaur, J.H.Hoskinson, J.L.Janning, R.C.Smith: Proc. SID, **13**, 34–36 (1972)
3.4 A.Yano, I.Inomata, T.Iwakawa: "Plasma Display", NEC (Nippon Electric Co.), Res. Devel. No. 30, 54–63 (1973)
3.5 S.Umeda, T.Hirose: "Self-Shift Plasma Display", SID Symp., San Francisco, Calif., June 1972, p. 38
3.6 S.Andoh, K.Oki, K.Yoshikawa: "Self-Shift PDP with Meander Electrodes", SID Symp., Boston. Mass. April 1977, p. 78
3.7 W.E.Coleman, D.G.Craycraft: "A Serial Input Plasma Charge Transfer Display Device", SID Symp., Washington, D.C., April 1975, pp. 114, 115
3.8 H.J.Hoehn: "Plasma/Display Memory Panel Fabrication", SID Symp., Boston, Mass., April 1977, pp. 18, 19
3.9 Circuits Manufacturing Magazine: *Thick Film Nickel Conductor for DC Gas Discharge Displays* (Benwill, Valley Stream, New York 1977)
3.10 T.N.Criscimagna, J.R.Beidl, M.Steinmetz, J.Hevesi: "Coupled-Matrix Threshold-Logic AC Plasma Panel", SID Symp., Beverly Hills, Calif., May 1976
3.11 R.Hammer: "Obtaining Gas Panel Metallization Patterns Simply and Directly by Vacuum Deposition Through Rib-Supported Mask Structures", IBM Res. Report RC-6631 (1977)
3.12 K.C.Park, E.J.Weitzman: J. Vac. Sci. Tech. **14**, 1318 (1977)
3.13 K.Murase, Y.Shirouchi, H.Yamashita: Fujitsu Sci. Tech. J. **12**, (1), 159–184 (1976)
3.14 T.Urade, T.Iemori, M.Osawa, N.Nakayama, I.Morito: IEEE Trans. ED-**23** (3), 313–318 (1976)
3.15 B.W.Byrum, Jr.: IEEE Trans. ED-**22** (9), 685–689 (1975)
3.16 S.Andoh, K.Murase, S.Umeda, N.Nakayama: IEEE Trans. ED-**23** (3), 319–324 (1976)
3.17 E.S.Schlig, G.R.Stilwell, Jr.: *Characterization of Voltage and Charge Transfer in AC Gas-Discharge Displays* (to be published)
3.18 H.G.Slottow, W.D.Petty: IEEE Trans. ED-**2**, 650 (1971)
3.19 R.R.Johnson: "The Application of Plasma Display Techniques to Computer Memory Systems", Univ. of Ill., CSL Report No. R-461, Appendix (1970)
3.20 W.D.Petty: "Multiple States and Variable Intensity in the Plasma Panel", Univ. of Ill., CSL Report No. R-497, Appendix A (1970)
3.21 G.E.Holtz: Proc. SID, **13**, 2–5 (1972)
3.22 P.D.T.Ngo: IEEE Trans. ED-**22**, 676–681 (1975)

3.23 P.D.T.Ngo: "Charge Spread and its Effect on AC Plasma Panel Operating Margins", IEEE Biennial Display Conf., New York, N.Y., Oct. 1976, p. 118
3.24 J.F.O'Hanlon: "A Phenomenological Study of AC Gas Panels Fabricated with Vacuum-Deposited Dielectric Layers", IBM Res. Report RC-6590 (1977)
3.25 G.R.Stillwell, Jr., E.S.Schlig: IEEE Trans. ED-**24**, 1125–1127 (1977)
3.26 K.Murase, S.Anhoh, S.Umeda, N.Nakayama: "A Driving Technique to Stabilize AC Plasma Display", SID Symp., Washington, D.C., April 1975
3.27 T.N.Criscimagna: Electro-Opt. Syst. Des. **3** (9), 34–36 (1971)
3.28 W.E.Johnson, L.J.Schmersal: Proc. SID **13** (1), 56–60 (1972)
3.29 G.W.Dick: Proc. SID **13** (1), 6–13 (1972)
3.30 T.N.Criscimagna: Proc. SID **17** (3), 124–129 (1976)
3.31 J.W.V.Miller, J.D.Schermerhorn: "Bipolar Integrated Circuit Plasma Panel Drive System", SID Symp., Beverly Hills, Calif., May 1976
3.32 H.J.Hoehn, R.A.Martel: IEEE Trans. ED-**18**, 659–663 (1971)
3.33 R.Czajkowski, W.Coates, P.Fitzhenry, R.L.Johnson, M.Stone: "A Microcomputer Driven Graphic Display System Using the 1024 × 1024 Line (83 lpi) AC Plasma Display Panel", SID Symp., Boston, Mass., 1977, pp. 26, 27
3.34 H.Yamashita, S.Andoh, T.Shinoda: "A Green AC Plasma Display", SID Symp., Beverly Hills, Calif., May 1976, pp. 80, 81
3.35 F.H.Brown, M.T.Zayac: Proc. SID **13** (1), 52–55 (1972)
3.36 H.J.Hoehn, R.A.Martel: IEEE Trans. ED-**20** (11), 1078–1081 (1973)
3.37 W.D.Petty, H.G.Slottow: IEEE Trans. ED-**18**, 654–658 (1971)
3.38 B.C.Anderson, V.J.Fowler: "AC Plasma Panel TV Display with 64 Discrete Intensity Levels", SID Symp., San Diego, Calif., May 1974, pp. 28, 29
3.39 K.Knowlton, L.Harmon: Comput. Graphics Image Process. **1**, 1–20 (1972)
3.40 J.F.Jarvis, C.N.Judice, W.H.Ninke: Comput. Graphics Image Process. **5**, 13–40 (1976)

4. Liquid-Crystal Displays, LCD

D. J. Channin and A. Sussman

With 10 Figures

Liquid-crystal displays (LCDs) are light-controlling devices, as opposed to light-emitting displays such as cathode-ray tubes, light-emitting diodes, and plasma displays. They rely on the optical properties of liquid-crystal materials to control the transmission of light from an external source, be it ambient illumination or lighting provided as a part of the display. There are several kinds of liquid-crystal materials that have been used in displays and a multitude of configurations for constructing addressing, and optically presenting information using these materials. This chapter will attempt to survey this diversity of technology, but will concentrate on the configurations most used at present or that have particular promise of future applications.

4.1 Background Information

All liquid-crystal displays consist of a thin layer of liquid-crystal material together with means for changing the optical properties of the layer in accordance with the information to be displayed. Liquid crystals are organic materials that possess an intermediate phase, or mesophase, between the solid and isotropic liquid phases. In the liquid-crystalline phase, they have the ability to flow and conform to their surroundings as do ordinary liquids, but the molecules of liquid crystals interact with each other to produce various kinds of ordered states. The simplest of these is the *nematic* phase, in which the molecules tend to have their major axes parallel to each other. The *cholesteric* phase has a helicical structure of molecular ordering, and the *smectic* phases show various kinds of layered structures with the long molecular axes parallel to each other within each layer. (The properties of these phases have been reviewed in several recent books and articles [4.1–4].)

With proper treatment of the surfaces bounding the liquid-crystal layer, the ordering can be made uniform throughout the entire layer. Because the materials are liquids, relatively little external energy need be applied to produce large changes in the molecular ordering. In addition, since the liquid crystals are anisotropic, these changes can be used to produce visible optical effects through polarization rotation, light scattering, or pleochroic light absorption.

Conveniently, the simplest type of liquid crystal has also turned out to be the most useful for display applications. All the devices currently marketed use a nematic layer sandwiched between glass plates on whose inside surfaces patterned conductive films are used to apply electric fields to the liquid crystal and induce changes in the direction of molecular orientation. The principal characteristics of this kind of device that have led to its rapidly growing commercial acceptance are its extremely low power drain at low voltage, visibility at high ambient-light levels, and low manufacturing cost.

4.2 Device Physics

4.2.1 Nematic LCD

Nematic liquid crystals consist of rod-like organic molecules. The mean axial direction of these molecules at any point is given by a unit vector $\boldsymbol{d}(x, y, z)$, called the director. The ordering of the liquid-crystal molecules gives rise to an elastic free-energy density W_e, which is expressed in terms of the director as

$$W_e = \tfrac{1}{2}K_{11}(\nabla \cdot \boldsymbol{d})^2 + \tfrac{1}{2}K_{22}(\boldsymbol{d} \cdot \nabla \times \boldsymbol{d})^2 + \tfrac{1}{2}K_{33}(\boldsymbol{d} \times \nabla \times \boldsymbol{d})^2 \,. \tag{4.1}$$

In this expression, K_{11}, K_{22}, and K_{33} are called the liquid-crystal elastic constants, and have typical values $\sim 10^{-11}$ J/m. W_e is proportional to the square of the gradients of $\boldsymbol{d}$, and is minimized when the molecular orientation is uniform. In the absence of external forces, this is the equilibrium configuration of the liquid crystal.

The anisotropic structure of the nematic molecules combined with the ordered nature of the mesophase is the source of a corresponding anisotropy in many of the macroscopic properties of a liquid crystal such as a dielectric constant, index of refraction, and the conductivity. The dielectric tensor has two components; $\varepsilon_{\|}$ for an electric-field component parallel to $\boldsymbol{d}$ and $\varepsilon_{\perp}$ for a far field perpendicular to $\boldsymbol{d}$. The difference between these two constants, $\varepsilon_a = \varepsilon_{\|} - \varepsilon_{\perp}$, is called the dielectric anisotropy and can be positive with values approaching 20.0 or negative with values of up to -5.0, depending on the structure of the organic compound. The electrostatic free-energy density due to the application of an electric field $\boldsymbol{E}$ is given by

$$W_d = -\tfrac{1}{2}\boldsymbol{D} \cdot \boldsymbol{E} = -\tfrac{1}{2}\varepsilon \boldsymbol{E} \cdot \boldsymbol{E} = -\tfrac{1}{2}\varepsilon_{\perp} E^2 - \tfrac{1}{2}(\varepsilon_{\|} - \varepsilon_{\perp})(\boldsymbol{d} \cdot \boldsymbol{E})^2 \,. \tag{4.2}$$

If $\varepsilon_a > 0$, the electrostatic free energy is minimized when $\boldsymbol{d}$ points parallel to $\boldsymbol{E}$, and when $\varepsilon_a < 0$, it is minimized when $\boldsymbol{d}$ points perpendicular to $\boldsymbol{E}$. When a field is applied to a sample of arbitrary orientation, a torque is created that acts to rotate $\boldsymbol{d}$ toward the orientation of minimum electrostatic free energy.

The total free-energy density of the liquid crystal is $W = W_e + W_d$, and the equilibrium configuration of the liquid crystal is that for which the free energy of the entire material will be a minimum, i.e.,

$$\delta\left[\int_v W(x, y, z)\,dx\,dy\,dz\right] = 0\,. \tag{4.3}$$

The liquid crystal is subject to additional constraints where the material is in contact with the surfaces of its container. Generally, the liquid-crystal molecules are given a particular orientation that depends on the liquid-crystal material and the nature of the surface. The physical basis of this phenomena is not well understood at present, but various techniques have been developed to give particular kinds of alignment. The term homeotropic alignment refers to the director being perpendicular to the surface, while homogeneous alignment describes the situation in which the director is parallel to the surfaces and points in a particular direction. The surface alignment provides boundary conditions that the director $\boldsymbol{d}$ must satisfy in addition to (4.3).

If at a given time the director configuration $\boldsymbol{d}(x, y, z)$ is not the equilibrium one dictated by the solution to (4.3) commensurate with the boundary conditions, internal torques are set up that cause rotation of $\boldsymbol{d}$ toward the equilibrium configuration. The viscous force within the liquid crystal opposes this rotation. In the absence of flow of the liquid-crystal material itself, this torque is given by

$$\boldsymbol{\Gamma}_{\text{vis}} = -\gamma_1 \boldsymbol{v} \times \frac{\partial \boldsymbol{d}}{\partial t} = -\gamma_1 \boldsymbol{d} \times (\boldsymbol{\Omega} \times \boldsymbol{d})\,. \tag{4.4}$$

Here, $\boldsymbol{\Omega}(x, y, z)$ is the angular velocity of the director, $\boldsymbol{v}$ the flow velocity, and γ_1 a viscosity coefficient. The viscosity is strongly temperature dependent, but a typical range of values at room temperature is 10^{-1}–10^{-2} kg/m $\cdot$ s^{-1}.

4.2.2 Field-Effect LCD

Liquid-crystal displays based on the properties described are called field-effect devices. A simple geometry that displays the phenomena resulting from these phasical properties is shown in Fig. 4.1. A liquid crystal of positive dielectric anisotropy is confined between two sources, which are treated to give homogeneous alignment in the x–y plane and parallel to the y axis. An electric field $\boldsymbol{E}$ parallel to the z axis is produced by a voltage applied to electrodes on the surfaces of the liquid crystal. This field acts to rotate the direction in the y–z plane by the angle $\phi(z)$ as shown. Using

$$\begin{aligned} d_x &= 0 \\ d_y &= (\cos\phi) \\ d_z &= (\sin\phi) \end{aligned} \tag{4.5}$$

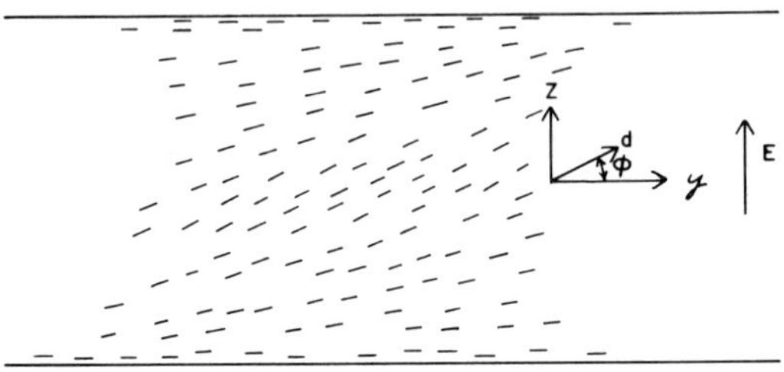

Fig. 4.1. Orientation of nematic molecules parallel to layer surface, with electric field perpendicular to surface

(4.1, 2) yield

$$\begin{aligned} W &= W_e + W_d \\ &= \tfrac{1}{2}(K_{11}\cos^2\phi + K_{33}\sin^2\phi)\frac{\partial^2\phi}{\partial z^2} \\ &\quad -\tfrac{1}{2}\varepsilon_\perp E^2 - \tfrac{1}{2}\varepsilon_a E^2 \sin^2\phi . \end{aligned} \tag{4.6}$$

The function $\phi(y)$ that satisfies (4.3) is that which satisfies the Euler-Lagrange equation

$$\frac{\partial W}{\partial \phi} - \frac{d}{dz}\frac{\partial W}{\partial \phi_z} = 0 \quad \left(\phi_z = \frac{\partial \phi}{\partial z}\right). \tag{4.7}$$

This leads us to the following equation and boundary conditions for the equilibrium directional configuration:

$$\begin{aligned} &(K_{11}\cos^2\phi + K_{33}\sin^2\phi)\frac{\partial^2\phi}{\partial z^2} \\ &\quad + \left[(K_{11}-K_{33})\left(\frac{\partial\phi}{\partial z}\right)^2 + \varepsilon_a E^2\right]\sin\phi\cos\phi = 0 \\ &\phi(z=-l/2) = \phi(z=+l/2) = 0 . \end{aligned} \tag{4.8}$$

Equation (4.8) represents the equilibrium between elastic and dielectric torques. The viscous torque may be added directly to (4.8) to give the dynamic behavior of the director orientation. It is convenient to make two approximations at this point. First, we observe that usually $K_{11} \approx K_{33}$, and therefore the term with coefficient $(K_{11} - K_{33})$ is dropped. Second, we consider for the moment only small values of ϕ. The linearized equation that results is

$$\begin{aligned} &K_{11}\frac{\partial^2\phi}{\partial z^2} + \varepsilon_a E^2 \phi - \gamma_1 \frac{\partial\phi}{\partial t} = 0 \\ &\phi(-l/2, t) = \phi(l/2, t) = 0 . \end{aligned} \tag{4.9}$$

Equation (4.9) has the following solution:

$$\begin{aligned} \phi(z,t) &= \phi_0 e^{-t/\tau}\cos(\pi z/l) \\ \phi_0 &= \phi(z=0,\ t=0) \\ \tau &= \gamma_1\left[K_{11}\left(\frac{\pi}{l}\right)^2 - \varepsilon_a E^2\right]. \end{aligned} \tag{4.10}$$

One sees that there is a critical field $E_c=(\pi/l)\sqrt{K_{11}/\varepsilon_a}$, below which any initial disturbance ϕ_0 relaxes back to $\phi(z)=0$, and above which the initial disturbance will grow. The device response has a threshold at $E=E_c$. Above the threshold, thermal fluctuations always provide an initial disturbance that grows under the action of the field. This is called a Freedericksz transition and occurs in other geometries as well. The corresponding critical voltage

$$V_c = lE_c = \pi\sqrt{K_{11}/\varepsilon_a} \tag{4.11}$$

is independent of the thickness l of the liquid crystal layer. In a typical case, $l=10\,\mu\text{m}=10^{-5}$ m, $K_{11}=10^{-11}$ J/m, $\varepsilon_a=20\,\varepsilon_0$, yielding $V_c=0.75$ V.

With zero field, the initial response ϕ_0 decays exponentially with time constant

$$\tau_{\text{relax}} = \gamma_1/K_{11}(\pi/l)^2 \tag{4.12}$$

Using $\gamma_1=5\times10^{-2}$ kg/m·s^{-1} and the previous values for the other parameters, we find that $\tau_{\text{relax}}=5.1\times10^{-2}$ s.

When a field much greater than the critical field is applied, the elastic torques are negligible compared with the dielectric and viscous torques throughout most of the liquid crystal, and the response rate becomes

$$\begin{aligned} \tau_{\text{response}}^{-1} &= 1/\phi\,\partial\phi/\partial t = \varepsilon_a E^2/\gamma_1 \\ &= \varepsilon_a V^2/\gamma_1 l^2 . \end{aligned} \tag{4.13}$$

With large fields applied, this rate can be much faster than the natural relaxation rate, τ_{relax}^{-1}. For example, the application of 50 V to the sample of the previous example causes a response time $\tau_{\text{response}}=1.1\times10^{-5}$ s. Operation at microsecond rates is possible in devices incorporating electric-field rotation [4.5–6].

Liquid-crystal devices are generally operated with large director rotations, $\phi\sim\pi/2$, so the linear-response theory is not sufficient to fully describe the device behavior. Nevertheless, the linear theory correctly describes the dependence of basic device behavior on device parameters such as cell thickness

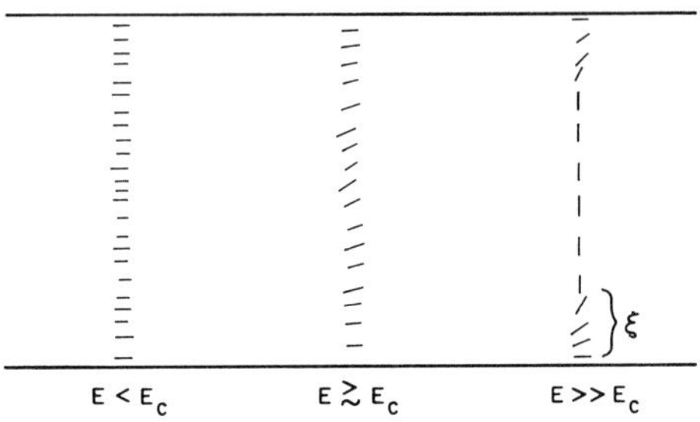

Fig. 4.2. Response of nematic molecules to electric field. E_c is the critical field magnitude for molecular reorientation (Freedericksz Transition)

and applied voltage. In particular, the critical voltage is a fundamental parameter of liquid-crystal devices and is given without approximation by the linear theory.

When fields greater than the critical field are applied, (4.9) fails to reach any equilibrium value for ϕ when $\partial\phi/\partial t=0$. The nonlinearity of (4.8) determines the equilibrium director configuration. Figure 4.2 shows the liquid-crystal orientation at various fields. For $E \gg E_c$ the director rotation is almost $\pi/2$ (***d*** parallel to ***E***) except for a boundary layer of width ξ adjacent to the surfaces. For large fields, the director configuration near the lower surfaces is determined from (4.8) to be given by

$$\begin{aligned}&\tan(\phi/2\pm\pi/4)=\exp[\pm 1/\xi(y\pm l/2)]\\&\qquad\xi=1/E\sqrt{K_{11}/\varepsilon_a}\,.\end{aligned} \tag{4.14}$$

It should be noted that solutions for positive or negative ϕ are equally admissible. This degeneracy is removed in an actual device by nonuniformities of the surface conditions, which give a preferred direction ϕ. However, this preferred direction may be different at different locations, giving domains of alternate sign of ϕ. In practical devices it is desirable to control the surface conditions to give a single domain throughout the device.

The assumption of a static fluid is in general not correct. In most geometries the director rotation Ω includes motion of the liquid crystal fluid. This motion will then affect the director rotation. This process is called "back flow", and can influence the detailed dynamics and optical properties of liquid-crystal devices. The hydrodynamical equations involved in this process are presented in the next section on electrohydrodynamic devices.

A variety of field-effect LCDs have been demonstrated, and the most important of these will be discussed in Sect. 4.5. Because these have different configurations for director orientation and electric-field geometry, the operational equations use different combinations of elastic and dielectric constants from the example discussed above. Nevertheless, the same principles and magnitudes of operational variables (such as response time and critical voltage) hold for all field-effect devices.

4.2.3 Electrohydrodynamic LCD

Liquid-crystal materials generally have high resistivities (10^8–10^{11} Ω-cm), but even low values of electrical conductivity can have major effects on the director orientation. These effects occur because current flow induces fluid flow, thus initiating various types of hydrodynamical structures. To describe the flow requires the specification of the fluid velocity throughout the liquid crystal. The equation for momentum transport represents the equilibrium of inertial, electrical, and viscous forces on every volume element of the material

$$\varrho \frac{\partial \boldsymbol{v}_i}{\partial t} + q\boldsymbol{E} + \sum_j \frac{\partial t_{ij}}{\partial x_j} = 0. \tag{4.15}$$

Here ϱ and q are the mass and charge densities. The viscous stress tensor t_{ij} is given by the equation

$$\begin{aligned} -t_{ij} = {} & \alpha_1 d_k d_n d_{kn} d_i d_j + \alpha_2 d_j m_i + \alpha_3 d_i m_j \\ & + \alpha_4 d_{ij} + \alpha_5 d_i d_k d_{kj} + \alpha_6 d_{ik} d_k d_j \end{aligned} \tag{4.16a}$$

$$d_{ij} = \tfrac{1}{2}\left(\frac{\partial v_i}{\partial x_j} + \frac{\partial v_j}{\partial x_i}\right) \tag{4.16b}$$

$$m_i = [(\boldsymbol{\Omega} - \boldsymbol{\omega}) \times \boldsymbol{d}]_i \tag{4.16c}$$

$$\boldsymbol{\omega} = \tfrac{1}{2} \operatorname{curl} \boldsymbol{v}. \tag{4.16d}$$

The coefficients α_1–α_6 are viscosity coefficients. $\boldsymbol{\omega}$ is the angular velocity of the liquid-crystal fluid. In ordinary circumstances the inertial term in (4.15) is negligible in comparison with the electrical and viscous terms.

The gradients in velocity couple to the director orientation by means of the complete viscous torque

$$\boldsymbol{\Gamma}_{\text{vis}} = \boldsymbol{d} \times [\gamma_1(\boldsymbol{\Omega} - \boldsymbol{\omega}) \times \boldsymbol{d} + \gamma_2 d_{ij} \boldsymbol{d}]. \tag{4.17}$$

Equation (4.17) reduces to (4.4) when the fluid velocity is zero. The coefficients γ_1 and γ_2 are related to the α_i's by $\gamma_1 = \alpha_3 - \alpha_2$ and $\gamma_2 = \alpha_2 - \alpha_5 = \alpha_2 + \alpha_3$.

To complete the picture, we have Maxwell's equations and a tensor relationship between the electric current density and the electric field

$$\boldsymbol{j} = \sigma \boldsymbol{E}. \tag{4.18}$$

The conductivity anisotropy incorporated in (4.18) can cause charge separation when current flows in liquid crystals. The tensor σ has the same form as ε, with one component $\sigma_{\|}$ for current flow parallel to the v and two equal components $\sigma_{\perp}$ for current flow perpendicular to $\boldsymbol{v}$. Generally, $\sigma_{\|} > \sigma_{\perp}$, with $\sigma_{\|}/\sigma_{\perp} \approx 1.5$.

Solutions of the complete Eqs. (4.15–19) have been discussed by various authors [4.7–11]. Here we will present only a physical description of the phenomena and those results especially pertinent to display devices.

We consider again the geometry of Fig. 4.1 with the surfaces of the liquid-crystal layer treated to give homogeneous alignment of ($\boldsymbol{d} \| y$) and perpendicular to the applied field $\boldsymbol{E}_0 \| z$. However we now take $E_a < 0$. In this case the torque on $\boldsymbol{d}$ due to $\boldsymbol{E}_0$ is such as to favor the existing alignment. A uniform current flow has no effect on this situation.

Now suppose the director orientation develops some spatial variation in the y–z plane. Equation (4.18) shows that there will then be components of current j_y flow parallel to y. Inhomogeneities in $j_y(y)$ will cause space charge $q(y)$ to develop, and this will in turn cause an electric-field component $E_y(y)$ to exist. This field component may then cause additional director rotation in the y–z plane, according to processes described below. Thus the uniform orientation $\boldsymbol{d} \| y$ can, under conditions outlined below, become unstable as a result of anisotropic conductivity and develop an inhomogeneous pattern of alignment in the y–z plane.

There are two means by which the space-charge field can cause director rotation away from $\boldsymbol{d} \| y$. One is through the electrostatic torque due to E_y (4.2), which acts to cause rotation towards the condition $\boldsymbol{d} \perp \boldsymbol{E}$. The other is through hydrodynamic processes. Equation (4.15) shows that the combination of space charge q and field component E_z causes fluid flow v_z. However, this flow is blocked by the surfaces of the liquid layer, and to maintain continuity of flow, circulation cells are set up in the y–z plane with velocity components v_y and v_z and gradients $\partial v_y/\partial z$ and $\partial v_z/\partial y$. Equation (4.17) in turn shows how these velocity gradients give a viscous torque Γ_X which causes director rotation in the y–z plane. The relative importance of the electrostatic and hydrodynamic processes depends on the frequency of the applied field $\boldsymbol{E}_0$.

Physically, the significant factor is the dielectric relaxation time $\tau_d = \varepsilon_{\|}/\sigma_{\|}$, the characteristic time for the space charge to come into equilibrium with the electric field. At frequencies low compared to $(2\pi\tau_d)^{-1}$ the space charge oscillates at the same frequency f as $\boldsymbol{E}_0$, and hydrodynamic processes dominate. This is usually called the "circulation regime". A threshold voltage V_c exists for the onset of the instability in the uniform alignment and is given in [4.4].

The threshold voltage diverges at a critical frequency, and at higher frequencies the space-charge distribution is constant in time. Electrostatic torques cause d to oscillate in the x–y plane with the driving frequency. This is called the "dielectric regime" and also has a threshold voltage.

Threshold voltages for the conduction regime at low frequency are generally on the order of 8–10 V (rms). The threshold voltages for the dielectric regime are usually much higher, on the order of 100 V (rms). Above the threshold both modes produce a periodic spatial modulation of the director orientation with a spatial frequency on the order of the liquid-crystal thickness l. At higher voltage in the "conduction regime", the structure of the conduction regime breaks up into an incoherent turbulance known as "dynamic scattering". With the same geometry as above, but with $\varepsilon_a > 0$ the same processes occur, but above the Freedericksz transitions [4.12]. Many other geometries can be used to give interesting kinds of hydrodynamic phenomena. Most practical DSM cells need homeotropic (not homogeneous) alignment because of lower light scattering in the off state.

In the absence of an electric field, hydrodynamic effects may still take place through "back flow" processes mentioned earlier [4.13, 14]. We see from (4.15, 16) that in the absence of any fluid flow, an angular velocity $\boldsymbol{\Omega}$ gives contributions to the net viscous force proportional to α_2 and α_3. Since in (4.16) the inertial forces are small, this contribution must be balanced by the other viscous forces through hydrodynamic flow. The gradients of this flow then appear in (4.17) and modify the viscous torque opposing the original director rotation. The ability of the fluid to flow offers additional freedom for the liquid crystal to respond to electrostatic torques, and in general reduces the overall resistance to director rotation. It will also affect the geometry of the director rotation, and can cause certain characteristic phenomena in some kinds of devices.

4.2.4 Optical Characteristics of LCD

The optical characteristics of liquid-crystal devices come about through the physical anisotropy of the molecules and the ordered nature of the mesophase. As a consequence of the macroscopic optical anisotropy, the transmission of light through the liquid crystal is controlled by an applied external field through variations in the director. In most devices the important optical parameter is the refractive index. This has one value n_o (ordinary index) for light polarized perpendicular to the director and another value n_e (extraordinary index) for light polarized parallel to the director. The optical birefringence $n_e - n_o$ may be as high as 0.3 in some nematics, which is much larger than the birefringence of almost all natural optically uniaxial crystals. This high birefringence combined with the ability to achieve large director reorientations defines the unique role played by liquid crystals in electro-optic applications. By way of contrast, the most efficient electro-optic crystals have maximum electric-field-induced index changes on the order of 10^{-4}.

There are three principal ways in which director reorientation can be used to control light transmission. The first is through altering the polarization vector of light passing through the birefringent material. In systems using this phenomena the liquid crystal is placed between optical polarizers arranged so that one director orientation allows originally polarized light to pass through the second polarizer, while another director orientation gives light that does not pass through the second polarizer. The on and off states of the device are thereby defined. Either of the two states is produced with the application of an external field and the other state occurs without the field. This principle is used in various forms of field-effect devices. The second phenomena used to control light transmission is scattering produced by hydrodynamic structures or turbulance which results in extensive and random refraction of light passing through the liquid crystal. This phenomenon is dynamic scattering. It is also possible to use the liquid crystal to impart a uniform orientation to the molecules of dissolved dichroic dye. Such dyes have an absorption which depends on the relative direction between the dye molecular axes and the local optical electric field. Then, by altering the orientation of the liquid crystal with respect to the incoming light, the overall absorption may be changed. No polarizers are necessary.

4.3 Large-Scale Displays

One of the most enticing goals of past and current work with liquid crystals has been the development of low-cost area displays capable of replacing cathode-ray tubes for many applications. The essential simplicity of LCD and their planar format has encouraged many efforts to develop large scale display panels. The most significant problem in such technology is that of addressing information to appropriate parts of the display. For small displays (for example, watches with four 7-segment numerals) it is possible to have a separate electrical contact to each display element. This becomes impractical for displays having more than about 50 elements or for two dimensional arrays requiring electrical crossovers.

This section reviews various solutions that have been developed for this problem. One of them, multiplexing, has characteristics tailored to liquid-crystal-device physics and will be discussed in detail.

4.3.1 Addressing Techniques for Large-Scale Displays

Multiplexing

The simplest addressing system is to form the array of display elements as a rectangular matrix as shown in Fig. 4.3. On one surface all the terminals of the elements in a given row are connected to a common bus which runs out to the edge of the array. Each row has its proper bus. On the opposite surface of the liquid crystal, the element terminals in a given column are connected to a common bus. Thus, every display element has one terminal connected to a row bus and the other to a column bus. Addressing is accomplished by applying voltages to the row and column busses with the appropriate magnitude and polarity such that the voltage difference across selected elements is above threshold and gives an optical response, while the voltage across the unselected elements is insufficient to generate a response.

In a typical display system, vertical scanning is achieved by applying pulses in sequence to the horizontal busses $x_1, x_2, \ldots$. Simultaneously, with each scanning pulse, the information to be displayed is applied as a set of pulses to

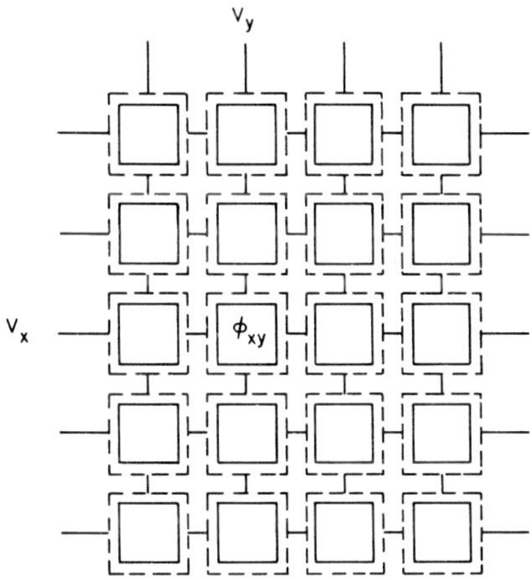

Fig. 4.3. Multiplex array of display elements

all the vertical busses $y_1, y_2, \ldots$. Crosstalk is eliminated by using the polarity of the vertical pulses to code the information [4.15]. To produce a response, the scanning and information pulses have opposite polarity and apply a maximum voltage across the liquid crystal.

The limitation in number of lines scanned comes from the fact that information pulses are applied to all the display elements during the scanning of each line. Since the display elements respond to the rms voltage of the applied signal (see Sect. 4.2) they all cooperate to build up a background response that reduces the contrast.

As we will see in the section on multiplexing theory, there is a serious tradeoff between the number of elements to be multiplexed and both the display contrast and the response rate [4.16–17]. This tradeoff occurs because the display elements must be operated close to their rms threshold voltages in order to prevent cross talk. In order to have good display characteristics, the optical response must be a strong function of the rms excitation voltage. This may be achieved in two ways: First by choosing a liquid-crystal operating mode that has a sharp dependence of director orientation on rms voltage, and secondly, by choosing an optical system that has a sharp dependence on director orientation. At present it seems that certain dynamic-scattering systems [4.18] can give sharp threshold, and certain twisted-nematic systems [4.19] give a strengthened dependence of optical response on director orientation.

With the best current technology, the optical contrast, viewing-angle range, and response speed of liquid-crystal displays are adequate for general display purposes. However, any performance degradation in these areas is a considerable price to pay for the economies of multiplexing. Nevertheless, progress in this area is rapid and practical displays incorporating multiplexing are becoming important commercial items.

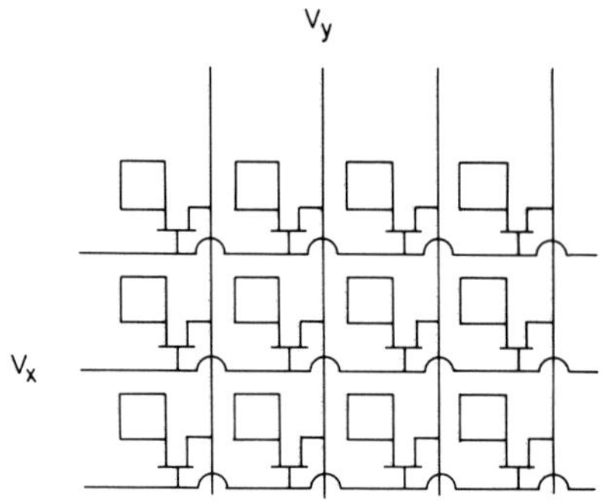

Fig. 4.4. Multiplexed array with switching transistor at each display element

Switching-Device Arrays

The problems inherent in multiplexing liquid devices can be eliminated by incorporating an electronic switch into each device. This switch prevents voltage from being applied to nonaddressed devices in an array. The operating voltage need not remain near the threshold voltage but can be chosen to optimize the optical characteristics of the liquid-crystal device itself. The most attractive switch for this purpose appears to be a MOS transistor. Figure 4.4 shows an array of display devices incorporating such transistors.

As in the case of multiplexed displays, the individual display elements are connected in an array. In this structure, all the connections are on the same surface. The horizontal busses $x_1, x_2, \ldots$ are connected not to the display electrodes but to the gates of MOS transistors. The voltage applied to the gate of a transistor controls the resistance between the other two terminals. One of these terminals is connected to a display-element electrode while the other is connected to one of the vertical busses $y_1, y_2, \ldots$. When a voltage is applied to a horizontal bus x_1, all the transistors connected to this bus conduct, and the voltage applied to the vertical busses is connected to the displays of that line of the display. When the voltage is removed from the transistor gates, the source-drain resistance becomes very high and isolates the display elements from the vertical busses. Charge is stored capacitively on the display elements until the next addressing cycle. Since each display element is isolated from the signal busses except when the elements are being addressed, the contrast and response-time limitations of multiplexed systems do not apply to displays using switching-device arrays. Consequently, the size of the display is limited only by the ability to fabricate arrays of suitable transistor switches. Furthermore, a variable voltage may be applied to provide for gray scale. Finally, charge storage on the display element throughout the display cycle reduces the required voltage. Taking advantage of these characteristics, liquid-crystal displays with speed and resolution comparable to commercial television and operating at around 10–20 V have been demonstrated [4.20].

The principal technological problem for displays using switching devices is not with the liquid-crystal devices but with the economical fabrication of large transistor arrays in which each device functions properly. The technology of integrated circuits is developing rapidly, but the problems of achieving economic yields for chip sizes on the order of inches are formidable.

The earliest reported work in this area was based on thin-film MIS transistors, in which the semiconductor was deposited in polycrystalline form as a thin film on an inert substrate such as glass [4.21]. This material may be potentially of very low cost, since it uses low-cost substrates and processes suitable for continuous mass production. However, the technology of thin-film devices is not nearly as developed as that of silicon devices, so the manufacturability and economics of displays using such devices is quite speculative.

Recent work using single-crystal silicon MOS devices has resulted in the real-time television system referred to above. This system takes advantage of the highly developed integrated-circuit technology, and may expect to benefit from the current rapid progress in this area. Since it starts with a crystalline silicon wafer, it may be difficult to produce low-cost large-area device arrays with this approach.

Photoconductive Light Valves

Displays using this technology abandon electronic addressing in favor of an optical system that modifies the transmission of the liquid crystal in accordance with the spatial and temporal variation of an externally generated image projected onto the device. The liquid crystal then controls the transmission of a much more intense light beam which can be projected onto a screen or other viewing device. The system may be thought of as an image amplifier or transformer.

Figure 4.5 shows the structure of a liquid-crystal photoconductive light-valve device. As in other devices, the liquid crystal is sandwiched between transparent conductors. However, one surface is coated with a photoconductive material whose resistance drops when exposed to light of suitable

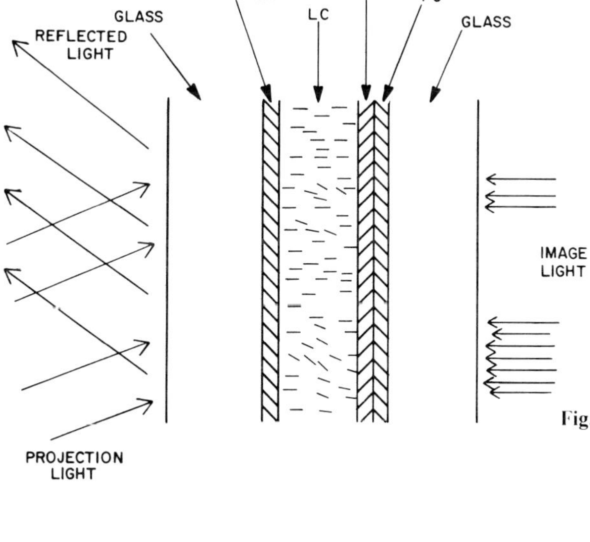

Fig. 4.5. Liquid-crystal light valve

wavelengths and intensity. An image projected onto this material produces a conductivity pattern of corresponding form. Electrical conduction through this pattern into the liquid crystal causes dynamic scattering which then modulates the intensity of a second light source. The liquid-crystal layer sits in the image plane of a projection system that uses the second light source to form an amplified and larger image corresponding to the one produced on the photoconductor.

A principal technical problem is to find a photoconductor sensitive to a weak image source, and yet one that is not affected by the much more intense projection light. One approach is to use a photoconductive material that is sensitive only to uv light and to use light of longer wavelength for the projection system [4.22]. A more versatile approach is to expose the photoconductor only to the image light by using an opaque barrier between it and the liquid crystal [4.23–24].

Such systems, using dielectric mirrors and high efficiency photoconductors, have been demonstrated with real-time television [4.25]. In principle, all of the liquid-crystal electro-optical phenomena may be used with these devices.

Light-valve systems of the type described above lack the flat-panel structure of other liquid-crystal displays, since they require optical systems both for imaging and for projection. They may be, however, more compact and lightweight than cathode-ray tubes of comparable size, especially for screen sizes larger than those commonly used on television. More important, they can project more light on the screen. A crucial factor for the utility of each system is the optical efficiency.

Thermally Addressed Light Valves

Unlike the other devices described previously in this section, the present system does not use the electro-optic properties of nematic liquid crystals, but rather the fact that certain smectic materials or nematic–cholesteric mixtures are stable in either of two textures: a clear, uniformly aligned configuration and a highly nonuniform configuration that scatters light similarly to dynamic scattering. A layer of such materials may have different regions in one or the other of the two states, and thus, it can store an optical image that can be viewed directly or by projection.

The scattering state is formed when the liquid crystal is heated into the isotropic phase and then allowed to return to the liquid-crystal phase. This is carried out by scanning the liquid-crystal layer with a laser beam of intensity sufficient to locally heat regions into the isotropic phase. When an infrared laser is used, this process may be assisted by coating one of the surfaces with a material that transmits visible light yet strongly absorbs the laser infrared radiation. Those regions heated by the laser beam cool to the scattering texture, while the rest of the liquid crystal remains in the clear state. To erase all or part of the display, an ac electric field is applied to the liquid crystal. This causes a transition from the scattering to the isotropic phase.

Various display systems using thermal addressing have been reported [4.26–29]. Operating speed is limited by the thermal time constant of the liquid crystal and its surrounding material, and by the need for sequential scanning to reduce overall power demands. As a result, such systems have not achieved television performance, but appear suitable for applications such as the display of complex graphics for computer terminals.

4.3.2 Multiplexing Theory

In this section we analyze the operation of a multiplexed array of liquid-crystal devices. For simplicity we consider only the field-effect devices discussed in Sect. 4.2.2, but the results are applicable also to dynamic-scattering devices. Here, we restrict ourselves to linear operation.

We consider an x–y array of field-effect devices as illustrated in Fig. 4.3. The response of each device as a function of time is represented by $\phi_{xy}(t)$, where xy indicates the x row and the y column, and ϕ indicates the amount of director rotation. A given row x is addressed by applying a voltage pulse V_x to the x row bus, otherwise all the row buses are at ground potential. The voltages applied simultaneously to the column electrodes, $\pm V_y$ convey the information to be displayed. We use (4.10) to describe the response of the element (x, y) to the applied voltages.

The array under consideration has N rows; therefore, there are N separate signals applied to element (x, y). The first, finishing at time t_0, is the difference between the row-addressing voltage V_x and the signal voltage $\pm V_y$. The others, finishing at time intervals Δ, are signal voltages $\pm V_y$ representing information to be displayed on subsequent rows. In this analysis we regard ϕ_{xy} as an average director rotation and neglect the variation of ϕ with z.

Using then (4.10), we get

$$\begin{aligned}
&\phi_{xy}(t_0)=\phi_{xy}(t_0-\Delta)\,\mathrm{e}^{-\Delta/\tau_0}\\
&\phi_{xy}(t_0+\Delta)=\phi_{xy}(t_0)\,\mathrm{e}^{-\Delta/\tau_1}=\phi_{xy}(t_0-\Delta)\,\mathrm{e}^{-\Delta(1/\tau_0+1/\tau_1)}\\
&\vdots\\
&\phi_{xy}[t_0+(N-1)\Delta]=\phi_{xy}[t_0+(N-2)\Delta]\,\mathrm{e}^{-\Delta/\tau_1}\\
&\qquad\qquad=\phi_{xy}(t_0-\Delta)\,\mathrm{e}^{-\Delta[1/\tau_0+(N-1)/\tau_1]}\\
&\qquad\qquad=\phi_{xy}(t_0-\Delta)\,\mathrm{e}^{-N\Delta/\tau_f}
\end{aligned}\tag{4.19}$$

$$\tau_0^{-1}=\frac{\varepsilon_a}{\gamma_1 l^2}[V_c^2-(V_x\pm V_y)^2]$$

$$\tau_1^{-1}=\frac{\varepsilon_a}{\gamma_1 l^2}(V_c^2-V_y^2)$$

$$V_c^2=\frac{K_{11}\pi^2}{\varepsilon_a l^2}.$$

It is evident that coding the information with the polarity of V_y prevents crosstalk, since the polarity of V_y only enters when V_x is also present, i.e., when the given row i is being addressed. Furthermore, the response at the end of the cycle has time constant

$$\frac{1}{\tau_0}+\frac{N-1}{\tau_1}=\frac{\varepsilon_a}{\gamma_1 l^2}[NV_c^2-(N-1)V_y^2-(V_x\pm V_y)^2] \tag{4.20}$$

which is exactly equal to that of an equivalent device responding over the same time interval $N\Delta$ to a voltage equal to the rms voltage applied to the display element of the array:

$$V_{\text{rms}}^2=\frac{1}{N\Delta}[\Delta(V_x\pm V_y)^2+(N-1)\Delta V_y^2]\,. \tag{4.21}$$

Thus, the response of a field-effect liquid-crystal device to a complex waveform is equivalent to the response to a continuous waveform of magnitude equal to the rms value of the complex waveform. Though this is shown here only for field-effect devices, it appears [4.17] to be a general characteristic of dynamic-scattering devices as well.

The operation cycle represented by (4.19) will be repeated continuously. Since they are solutions to the linearized equations, the limit of many cycles will be

$$\begin{aligned} &\phi_{xy}\to 0,\ V_{\text{rms}}<V_c\\ &\phi\to\infty,\ V_{\text{rms}}>V_c\,. \end{aligned} \tag{4.22}$$

This would suggest that the contrast ratio of a multiplexed array is infinite, or at least very large. This is manifestly not true. We will see that the conditions of multiplexing put severe limits on the available contrast ratio and also on the response rate. The problem with (4.19) is that they neglect the nonlinear terms which control the saturation behavior of the director response.

While a nonlinear analysis of multiplexing has not yet been presented, it is usually assumed [4.17] that the device response will be proportional to the rms voltage applied over one frame cycle. Using (4.21), we define the quantity

$$\begin{aligned} R^2&=\frac{\overline{V_{\text{on}}^2}}{\overline{V_{\text{off}}^2}}=\frac{(N-1)V_y^2+(V_x+V_y)^2}{(N-1)V_y^2+(V_x-V_y)^2}=R^2(\gamma)\\ \gamma&=\frac{V_y}{V_x}\,. \end{aligned} \tag{4.23}$$

It is assumed that the ratio of director rotation between the on and off states will be just R. The optical response of the device is related to R according to the optical-transfer function of the device.

To maximize the contrast ratio at a given number of lines, N, one maximizes $R(\gamma)$, and finds

$$\eta = 1/\sqrt{N} \tag{4.24}$$

$$R^2(1/\sqrt{N}) = \frac{N-1+(\sqrt{N}+1)^2}{N-1+(\sqrt{N}-1)^2}. \tag{4.25}$$

We now write the response rate for the display.

$$\tau_{F,\mathrm{on}}^{-1} = \frac{\varepsilon_a V_c^2}{\gamma_1 l^2 \beta^2}\left[\beta^2 - \frac{N-1}{N}\eta^2 - \frac{1}{N^2}(1+\eta)^2\right] \tag{4.26a}$$

$$\tau_{F,\mathrm{off}}^{-1} = \frac{\varepsilon_a V_c^2}{\gamma_1 l^2 \beta^2}\left[\beta^2 - \frac{N-1}{N}\eta^2 - \frac{1}{N^2}(1-\eta)^2\right] \tag{4.26b}$$

$$\beta = V_c/V_x. \tag{4.26c}$$

We chose an operating point by setting the condition

$$\tau_{F,\mathrm{on}}^{-1} = -\tau_{F,\mathrm{off}}^{-1}. \tag{4.27}$$

Using (4.24, 26) this establishes the relationship

$$\beta = \left(\frac{2}{N}\right)^2. \tag{4.28}$$

The response rate is then the same for the on and off states:

$$|\tau_F^{-1}| = \frac{\varepsilon_a V_c^2}{\gamma_1 l^2}\frac{1}{\sqrt{N}}. \tag{4.29}$$

From this model we can draw the following conclusions about multiplexed operation of liquid-crystal displays:

1) The contrast ratio decreases with increasing N. Figure 4.6 shows the contrast ratio determined from (4.25) on the assumption that the optical-transfer function is that of a DAP device (see Sect. 4.2.2)

$$I(\phi) \sim \phi^4, \tag{4.30}$$

hence the optical-contrast ratio is

$$C_R = R^4. \tag{4.31}$$

It is clear that the contrast ratio decreases rapidly as the number of scanned lines increases. On a display, this result would also cause a sharp reduction in the viewing-angle range.

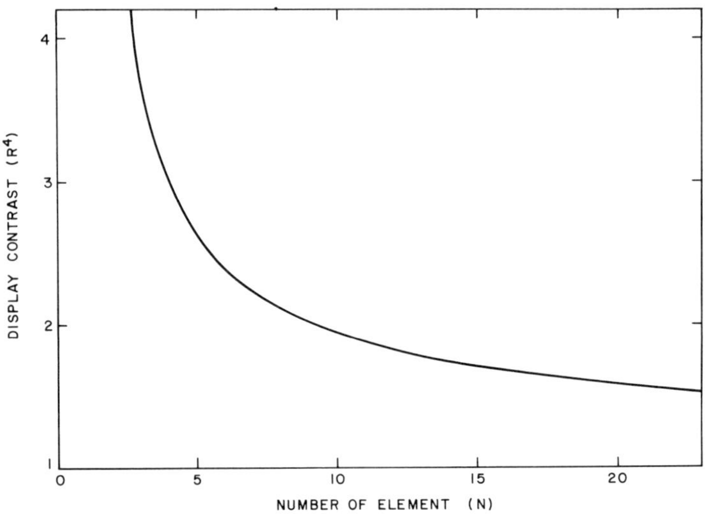

Fig. 4.6. Contrast of multiplexed display vs number of display elements

2) The operating voltage increases with increasing N. Equations (4.23, 24, 26c, 28) show that the operating voltages are proportional to $\sqrt{N}$. A display whose elements have a threshold voltage of 2 V will require 6.3 V for a 10-line display and 20 V for 100 lines.

3) The response rate of the display decreases with increasing N. Equation (4.29) shows that the response rate is proportional to $(\sqrt{N})^{-1}$. This means that a multiplexed display with many lines will appear visually sluggish.

Physically, this slowing of the response results from the fact that with increasing N the display elements are operated with the rms voltage approaching V_c, the threshold voltage. In Sect. 4.2.2 we saw that the time constant of a liquid-crystal device becomes infinite at the Freedericksz threshold.

These conclusions have been derived from a specific model for multiplexed display operation, but experience with both dynamic-scattering and field-effect displays demonstrates that they are generally applicable. Very large (400 lines) multiplexed displays using dynamic scattering have been demonstrated [4.18], and such displays have response times on the order of seconds. Large scale (128 effective lines) twisted-nematic displays have been demonstrated with high contrast ratios, but at the sacrifice of viewing-angle range [4.19].

4.4 Device Performance

To achieve maximum display performance, consideration must be given to the liquid crystal and its material parameters, how they are utilized in the display package and energized by the external circuit. In other words, the liquid crystal should be considered from a systems standpoint. The liquid-crystal parameters have already been introduced; it will be shown in the following sections how these physical properties may be utilized by control of the boundary conditions and the electrical drive to give optimum display performance.

The container for the liquid crystal, in its idealization, should be totally hermetic to protect the liquid and surfaces from the environment. Otherwise, it has no effect upon the liquid, except for determining the boundary conditions which govern the orientation of the liquid, particularly in its off conditions. The anchoring of the orientation to the display electrode surfaces also controls the return of the orientation to its original state when the energizing signal is removed. (In the case of the cholesteric-nematic transition, [4.30] the restoring force derives both from the twist elastic energy [4.31] and nucleation process [4.32].)

The anchoring at the surface [4.33] has two characteristics: the alignment angle and the strength of the attachment. The anchoring of the liquid and of defects such as disclination walls [4.34] is a mutual property of the liquid crysal and the surface. As will be seen, the attachment angle may be adjusted by different techniques to yield particular effects.

4.4.1 Dynamic Scattering

In dynamic scattering [4.35], the display exhibits a threshold voltage [4.36] for domains [4.37], the orderly vertical flow made visible because of the alignment of the optically anisotropic fluid in the direction of flow [4.38]. With increasing voltage, the flow becomes turbulent [4.39], and the display is able to scatter incoming light from rapidly moving regions [4.40]. Further increases of voltage result in changes into distinct flow patterns which have been considered as separate phases [4.41].

The unactivated state may be either the uniform planar (sometimes referred to as "homogeneous") orientation, or the perpendicular (referred to as "homeotropic") orientation[1]. The former may be achieved by the same techniques used to produce the planar orientation in field-effect displays, and will be discussed in that section. Perpendicular orientation is usually achieved by the use of surfactants, either dissolved in the liquid [4.42, 43] or directly produced on the electrode surface using chemical means [4.44]. A particular problem with surfactants dissolved in the liquid is the tendency for nonuniform distribution to occur because of chromatographic effects. This results in nonuniform alignment. The same chromatographic effects can also produce nonuniform distribution of the ionic doping agent, with the result that there is a variation of scattering intensity with temperature. This is due to the change in a critical frequency f_c which is approximately the ionic-relaxation frequency. If the operating frequency exceeds f_c, no dynamic scattering occurs (Sect. 4.2.3).

1 The terms "homogeneous" and "homeotropic" should be abandoned in favor of "planar" and "perpendicular", respectively, with appropriate modifiers such as tilted, uniform, etc.

Procedures are available to eliminate both these variations [4.45] by applying the dopants to the electrode surfaces from solution.

The viewability of dynamic scattering has good angular characteristics but the contrast for viewing by ambient light is not high. The use of mirrored backgrounds improves the contrast but at the expense of some information confusion and viewing-angle restrictions. Illumination of the display from behind through inclined blackened vanes, with the viewer seeing as background the sides of the vanes gives excellent contrast and good angular viewability [4.46].

The two boundary conditions can lead to different kinetic effects. When the alignment is perpendicular, dynamic scattering is preceded by a Freedericksz transition to a planar orientation. This results in a slower rise time than for the alignment which is initially planar. The perpendicular boundary conditions are also susceptible to certain operating defects caused by a combination of field and flow when their effects are additive. This can lead to (temporary) conversion of the boundary conditions to nonuniform planar orientation. This produces a different contrast, a different relaxation time, and a different off-state transmission from the unaffected areas [4.47]. The production of secondary hydrodynamic structures in the bulk due to changes in flow pattern results in a higher contrast and faster off-time response if the wall orientation is perpendicular, but producing the opposite effect if the orientation is planar [4.48].

Dynamic scattering may be observed either under ac or dc excitation. There are some differences between the mechanism for the generation of flow between ac and dc [4.49], and difference in optical characteristics when the ac frequency falls below the reciprocal of the ionic transit time [4.50], $t_{tr}^{-1} = \mu V/l^2$ where μ is the ionic mobility, V the rms voltage, and l the inter-electrode spacing.

The use of dc for dynamic-scattering (and the related storage effect [4.51]) facilitates its use with photoconductor electrodes [4.22] in image processing systems and in direct driven video displays [4.25]. One drawback of using dc along with ionic doping agents is that the operating life is found to be inversely proportional to the charge passed through the display [4.52]. That life is, however, much longer than that expected for a 100% efficient (faradic) electrode reaction. The deterioration of alignment and change of chemical composition have both been observed [4.53]. In order to explain the life, it has been suggested that the liquid-crystal material itself may be able to undergo almost reversible electronation and de-electronation reactions without undergoing chemical change [4.54]. Use of redox dopants which act as electron acceptors at the cathode and which return the electrons to the anode at voltages below which the liquid crystal becomes involved, thereby protecting the latter, has resulted in a dramatic increase in operating life [4.55]. The investigation of such processes is still an active subject [4.56]. Under "true" ac, with frequency greater than the inverse transit time, operating life is satisfactory.

The widespread use of dynamic scattering displays probably has been precluded not because of operating-life problems but rather because of the power requirements. First, operation at voltages below $12\,V_{rms}$ is poor, and second, the conductance necessary to keep the cutoff frequency sufficiently high at lower temperature (0 °C) leads to current drains at the upper temperatures (60 °C) which are of the same magnitude as for LEDs. This is a result of the variation of the carrier mobility with temperature [4.57]. This variation of conductance with temperature also requires that multiplexing schemes using two frequencies [4.58] be temperature compensated to keep the low frequency below f_c and the high frequency above f_c.

Another drawback to dynamic scattering is that at low temperatures, the spatial variation of refraction caused by the turbulent flow does not vary sufficiently quickly (if it occurs at all) compared with the temporal response of the eye, so that the "scattering" will not appear to have sufficient contrast.

Efforts to improve the dynamic-scattering display materials include reducing the threshold by increasing the magnitude of the negative dielectric anisotropy [4.59].

4.4.2 Twisted-Nematic Field-Effect Display

The commercial market for liquid-crystal displays consists almost entirely of the twisted-nematic type [4.60]. Time keeping is at present (1979) the major application in the form of watches and clocks, both with direct-driven displays. The multiplexed calculator use is next, followed by direct-driven displays for portable electronic equipment. The power requirements are extremely modest, a few microwatts per square centimeter, and at voltages that are compatible with the outputs of integrated circuit decoder-drivers. This success was achieved in spite of a contrast viewing-angle distribution that is not uniform, either for displays viewed by reflected or transmitted light [4.61], and with a brightness reduced by at least 50% by the absorption of the necessary polarizers. Some light-gathering schemes have improved the brightness of displays illuminated by ambient light [4.62], and as will be seen, the angular viewability can be made quite satisfactory.

There are two different modes of operation of the twisted-nematic display, each of which requires a different display construction and different operating parameters: direct drive and multiplexed. The basic structure of the device is not dissimilar for the two types, and is illustrated in Fig. 4.7. The alignment of the liquid crystal in the electrode plane is made to be parallel or perpendicular to the polarizer axis. For watches, it is conventional to place the front polarizer axis vertical, so as to be parallel to the axis of polarizing sunglasses; calculators are usually made with the polarizer axis 45° to the vertical in order to give the midplane alignment a vertical axis of symmetry. The front and rear electrode surfaces are set to produce an alignment which differs by 90°; the liquid-crystal alignment, anchored to the surfaces therefore undergoes a 90° twist, either clockwise or counterclockwise. This creates a midplane alignment axis of $\pm 45°$, depending on the sense of the twist. The twist degeneracy may be removed by

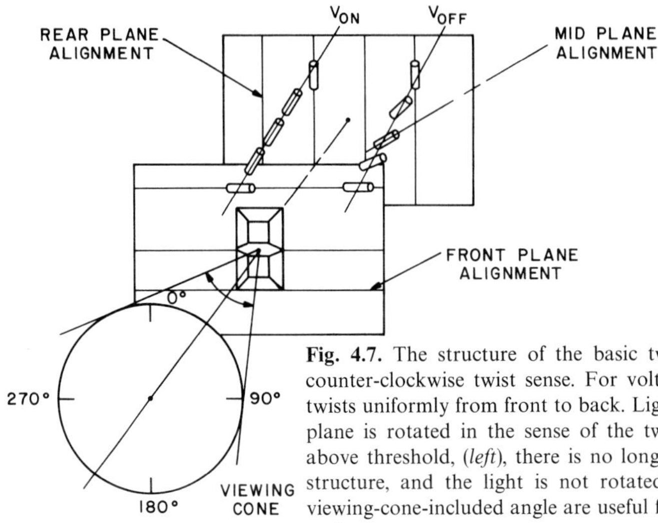

Fig. 4.7. The structure of the basic twisted-nematic device with a counter-clockwise twist sense. For voltage off (*right*), the alignment twists uniformly from front to back. Light orthogonal to an electrode plane is rotated in the sense of the twist. With voltage sufficiently above threshold, (*left*), there is no longer any but a residual twisted structure, and the light is not rotated. The azimuthal angles and viewing-cone-included angle are useful for characterizing the contrast performance

the inclusion of a small amount of optically active material into the liquid to impart the desired twist sense to the nematic by making it "slightly" cholesteric [4.63], with a cholesteric pitch larger than the interelectrode spacing. Because of the twisted structure and the optical properties of the liquid, light polarized orthogonal to the alignment entering the front (back) electrode emerges rotated by the twist angle from the back (front) polarizer. With applied voltage sufficiently above threshold, the twist is converted to a "perpendicular" alignment, so that the polarized light is no longer rotated. Therefore if the structure originally passed polarized light without extinction with the voltage off, the application of voltage causes extinction and visa versa. This is of course an idealization; the actual optical performance is much more complex, as will be seen.

The alignment can be made parallel to the electrode surfaces (planar orientation) or given a tilt with respect to the electrode plane(s). The device will have a threshold only when the alignment is planar. For direct drive, there is no need of a threshold, and the boundary conditions may be adjusted to tilt the average orientation and thus improve the contrast and view angle [4.64]. The pretilt behaves somewhat like a voltage bias; its magnitude should not be great enough to reduce contrast within a desired viewing cone. The tilt bias has been accomplished by preparing the surfaces with a combination of rubbing and placing a surfactant in the liquid [4.65], or by vacuum evaporation of inorganic materials at angles near glancing incidence, up to about 15° ("low angle" evaporation) [4.66]. In the latter method, the liquid crystal assumes an alignment tilted out of the plane at an angle which depends on the exact angle of evaporation [4.67], the specific chemical composition of the liquid and the evaporant [4.68], and the temperature [4.69]. For alkoxy Schiff bases on a silicon monoxide film evaporated at 15° to the plane, the alignment angle is approximately 40° at room temperature [4.70]. The tilt angle may be modified by either simultaneous or sequential evaporation at low angles and high angles (greater than 20° but less than 45°), the two evaporations being made at right angles to each other [4.71].

If both electrodes are prepared with a tilt, the directions of the tilts are compatible with only one twist sense; when a liquid of the "wrong" twist sense is used, an unusual Freedericksz transition may be observed [4.72]. Where there is a finite tilt angle, the twist angle may exceed 90° and remain stable [4.73].

For the case of a multiplexed display, the threshold conditions are necessary in order for the select voltage to produce a satisfactory "on" appearance, but for the application of the nonselect voltage, no observable response should be obtained within the desired viewing cone. This requires a close-to-planar orientation with just enough pretilt to remove the degeneracy of the two possible tilt directions [4.74]. Such tilt degeneracy may be observed while the display is activated, appearing as blotchy regions of two different contrasts when observation is made along the axis of symmetry, but at off-normal incidence. Viewing from the other direction with respect to the normal axis reverses the blotch pattern contrast. If the twist degeneracy is not eliminated, the "off" state will exhibit two sets of regions of opposite twist, with equal contrast separated by disclination walls. When activated, the two regions will show differences in contrast when viewed off the normal: the azimuthal directions for reversal of contrast are the two degenerate midplane axes which differ by 90°. It is possible to have both tilt and twist degeneracy simultaneously in planar displays.

The electro-optics of the twisted display are complex due to the asymmetric birefringent liquid structure. As there are no two rays which pass through the display that trace out identical optical paths, the optical density and response time of the display varies nonuniformly and sometimes surprisingly with both the direction of the incoming light and the position of the observer. For an idealized planar display, the optical properties can be related to the ratio of the operating voltage to the Freedericksz threshold [4.75]. The latter is a property of the fluid, independent of the interelectrode spacing which can only be measured with boundary conditions of strong anchoring and planar alignment [4.76]. For a real display, an effective electro-optic threshold may be defined; depending on the tilt angle [4.77] and the strength of the anchoring [4.78], it is a joint property of the liquid and the package. Two displays for which the only difference is the true threshold of the liquid-crystal material will exhibit nearly identical view-angle-symmetry properties if measured with respect to their effective thresholds. For materials of large dielectric anisotropy, the contribution of field distortion [4.79] may be reduced by defining the effective threshold as the voltage for 50 percent of full "on".

For the same material, in a display viewed head-on, the higher contrast at a given voltage will be obtained the higher the initial tilt angle. There is however a limit on how much tilt is acceptable, since the inactivated areas within the

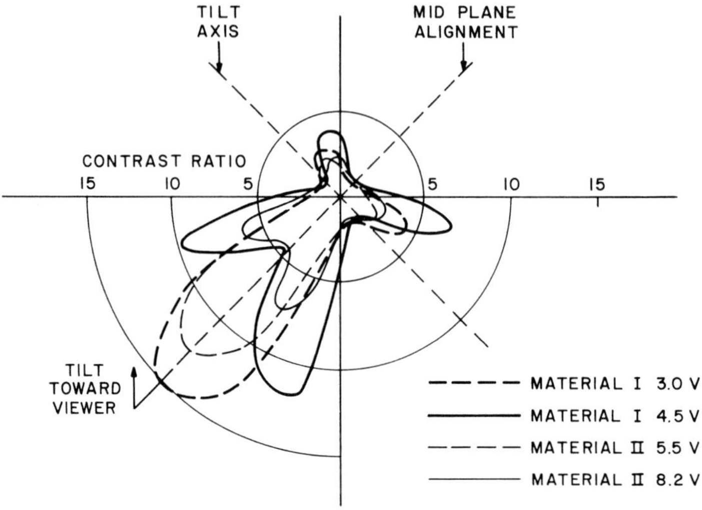

Fig. 4.8. The viewing symmetry as influenced by the alignment symmetries, at an operating to effective threshold ratio of 2 and 3. The thresholds of materials I and II are 1.5 and 2.8 V, respectively. The tilt angles, alignment symmetries and interelectrode spacing are identical. The display is viewed by reflected light, with uniform illumination. The angles are defined in Fig. 4.7

desired viewing cone must remain "off". Figure 4.8 shows the symmetry of the contrast ratio for two displays with identical tilt angles and interelectrode spacing, but whose materials have different thresholds. When operated at voltages in the same ratio to their respective effective thresholds, the symmetries of the contrast are virtually indistinguishable. The absolute contrast difference, however, is quite evident. This may be understood by realizing that for other than head-on viewing, because of the symmetry of the activated twisted-liquid structure, for almost all voltages and light propagation directions, the path of the light rays passing through the display are never orthogonal to the director. With increasing voltage, the structure is converted to one which has more perpendicular structure and less twist, but there is always a residual twist and tilt near the electrode surface, and at the surface, a region which retains the unaltered orientation. The width of this region, called the coherence length is given by $lV_{\text{th}}/\pi V$ [4.80]. Over this length, the alignment imparted by the walls changes to the alignment required by the field. Therefore, no ray passing through the display at any angle can remain linearly polarized and thus be extinguished by the analyzer. This is why a liquid-crystal display cannot achieve the contrast ratios shown by polarizers themselves. It is, therefore, important to reduce the influence of the coherence region, by making the optical path through that region as short as possible. This can be accomplished by making the interelectrode spacing small, and by reducing the optical anisotropy of the fluid. The lower limit to this improvement is the requirement that the effective optical spacing exceed a few wavelengths of light

[4.81]. A tradeoff with small spacing and low optical anisotropy is the possibility of pronounced color tints in the inactive areas under off-axis viewing. The influence of the interelectrode spacing is shown in Fig. 4.9. Note that increasing the operating voltage does not necessarily improve the extinction.

A measure of the optical thickness of a display may be made using the property called the retardation, given by $\Delta n l \cos^2 \Theta/\lambda$ where Δn is the optical anisotropy, Θ the average tilt angle (with respect to the plane), and λ the wavelength. Two displays with the same retardation may be considered to have the same effective optical spacing. Such displays will exhibit identical interference colors at normal incidence when placed between crossed polarizers, 45° to the alignment angle. Displays with the same effective optical spacing, the same midplane-symmetry-axis direction, the same tilt direction and operated at the same multiple of the effective threshold voltage will exhibit the same viewing symmetry and contrast, as long as the voltage is a few times the effective threshold and the viewing angles are not extreme. Figure 4.10 shows this result for three displays with 5°, 20°, and 40° tilt angles. The contrast is measured around a 60° included-angle cone, with uniform illumination [4.82] and the identical reflector material.

The reflector can influence the viewing characteristics, depending on how the incoming light is scattered back. The surface of the reflector can be considered the source of the illumination. A peculiar property of reflective displays is the appearance of a shadow on the activated display segment for some angles of illumination which gives the display the appearance of being "on" at viewing angles for which it is really "off" by transmission [4.83]. This broadening of the viewing cone can be significant if the parallax can be reduced by decreasing the distance between the planes of the liquid crystal and the reflector, and by selecting reflector textures that keep the display-edge contour sharp.

The dynamic performance of displays with respect to the average tilt angle may be briefly summarized. For the same physical spacing, the display with the tilted alignment will have faster rise time but a slower decay time than a display with planar boundary conditions [4.84]. For nearly planar alignments, the response time also depends on the duty cycle. A segment turned back on after being off for a time not long enough to allow complete relaxation of the alignment will turn on faster than a segment that has been off longer than the alignment-relaxation time [4.85].

There is another field-effect system, in which the liquid crystal is used as an aligning medium for dissolved dye molecules: the guest–host effect [4.86]. If the optical absorption depends on the relative orientation of the dye molecular axis with respect to the electric vector of the light, changing the orientation of the dye with respect to the incoming illumination will change the perceived color [4.87]. Most of the pleochroic dyes used for this phenomenon have a maximum absorption when the electric vector is parallel to the long molecular axis of the dye, which is also found to be approximately parallel to the liquid-crystal

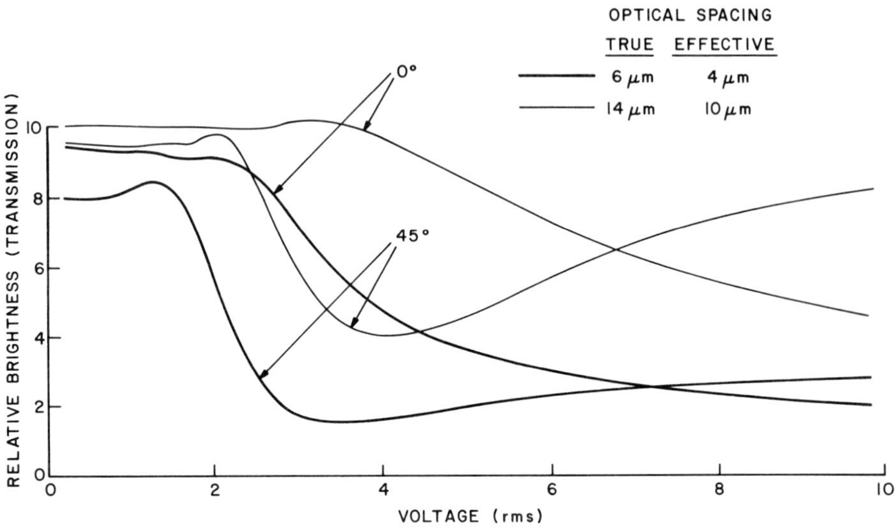

Fig. 4.9. The influence of the optical spacing in contrast is shown for two azimuthal angles around a 60° included angle viewing cone. The displays both have an average alignment tilt angle of 40°. The true threshold of the material is 2.3 V. The angles are defined in Fig. 4.7

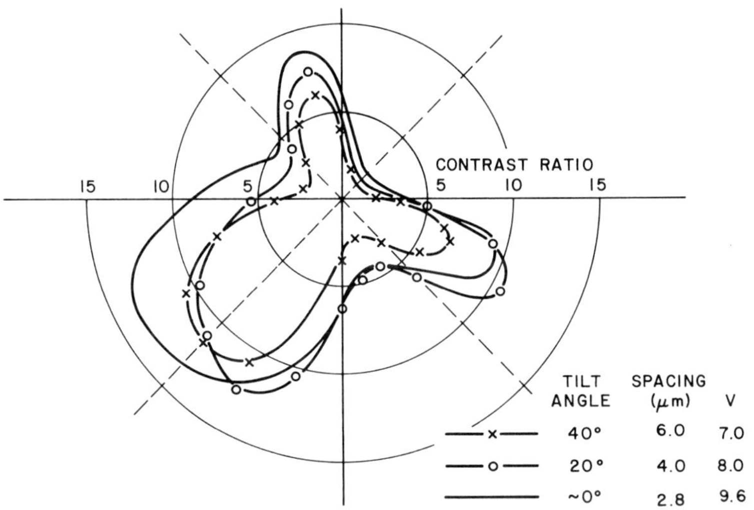

Fig. 4.10. Contrast as a function of azimuth around a 60° included angle viewing cone, for three displays with the same effective spacing, 4 μm, and at 3 times the effective threshold voltage. The displays have the same tilt and twist symmetry axes but the average tilts are different. The liquid-crystal material is the same. Angles are the same as Fig. 4.7

director. (The alignment of the dye may also be described by an order parameter [4.88].) The symmetry of rod-like dye molecules therefore requires that for maximum contrast without polarizers, it is best to have the dye axis perpendicular to the electric vector (and the electrode surfaces) for the nonabsorbing state and parallel but not necessarily uniformly ordered for the other. Various methods have been developed to obtain these results [4.89]. The most promising is the combination of dye with the cholesteric-nematic transition. Once suitable dyes are available to obtain (almost) black on white capability with photo- and electrochemical stability, the dye systems should present a challenge to the twisted-nematic display. This is because they can be multiplexed, there is no polarizer brightness loss, the viewability is symmetric about the normal, being determined by the order parameter of the dye, the dichroic ratio and the optical anisotropy of the fluid.

Field-effect displays operate as dielectric devices, but the presence of finite conductance due to residual ions [4.90] may lead to rectification if the operating frequency is lower than the inverse transit time of the carriers. The rectification is a result of asymmetric polarity-dependent electrosorption when the electrodes are electrochemically dissimilar. Degradation because of oxidation–reduction reactions has been observed [4.91]. Another defect is caused by hydrodynamic instabilities which can cause permanent contrast loss if the disclinations generated by the instabilities anchor to the electrode surfaces. The combined effects of asymmetric back flow [4.92] and electrochemical changes can lead to accumulation of degradation products at certain edges of display segments, determined by the symmetry of the tilt and twist. The electrochemical reactions which occur in typical materials used in displays are the subject for continuing investigation.

4.5 Manufacturing Technology

The methods for manufacturing liquid-crystal displays are varied, but basically two packaging techniques may be used; glass frit and polymer ("plastic") seal. A totally hermetic cell may be produced by using glass frit to seal the front and rear faces together, except for the opening which is soldered closed after the liquid is introduced by vacuum filling. The solder bond is usually made to a metal film sputtered to the glass. In order to withstand the temperatures required for frit sealing, aligning surfaces must be inorganic. The usual method is to produce an alignment by low-angle evaporation on one or both electrodes; the most commonly used material is silicon monoxide. Hermetic packages allow the use of materials which are sensitive to the ambient, such as Schiff bases or esters, to be used successfully under conditions of high humidity. The polarizing material, however, is still subject to deterioration unless it is protected from moisture. It is possible to produce the almost planar alignment required for multiplexing by using techniques of combined low- and high-angle evaporation, either simultaneous or sequential, or planar evaporation followed by rubbing, but the usual method of using an organic aligning film is compatible only with the low-temperature polymer seal. The polymer seal may be used in conjunction with slope evaporation, but more often with a polymer film such as polyvinyl alcohol applied to the electrode surfaces by dipping from solution. After drying, the film is "rubbed" to impart the alignment. The alignment produced by this technique is parallel to the rubbing direction, with a small tilt out of the plane parallel to the vector direction of the rubbing stroke.

The seal itself may be thermosetting, such as epoxy, or thermoplastic, applied in a solvent to the perimeter of the electrode surface(s) and allowed to dry before the application of the alignment film. Then excess liquid crystal is placed between the electrode surfaces, and the seal is accomplished by heating the parts under modest pressure.

Many liquid-crystal compounds such as esters and Schiff bases may suffer deterioration of alignment in contact with the alkali which (eventually) diffuses out of soft glass [4.93]. This is usually prevented by incorporating a blocking layer of a material such as silica, produced by sputtering or by chemical vapor transport.

4.6 Final Comments

Liquid-crystal display devices of the twisted-nematic type have become standard technology for low-power applications under ambient illumination. Various new developments hold promise for substantial improvements in the usefulness of liquid crystals. Guest–host dye systems offer the possibility of increased brightness and angular viewability. Improvements in material will extend the capabilities of multiplexed liquid crystals for portable electronic products. Microprocessor techniques [4.94] can extend the multiplexing voltage requirements in specialized displays, such as for oscilloscope use [4.95]. Electrical addressing [4.96] of high-information-density storage systems offers the potential for replacing cathode-ray tubes in data terminals with flat panels.

References

4.1 P.G.de Gennes: *The Physics of Liquid Crystals* (Oxford University Press, London 1974)
4.2 E.B.Priestly, P.J.Wojtowicz, P.Sheng (eds.): *Introduction to Liquid Crystals* (Plenum Press, New York 1974)
4.3 S.Chandrasaekar: *Liquid Crystals* (Cambridge University Press, London 1977)
4.4 M.J.Stephen, J.R.Straley: Rev. Mod. Phys. **46**, 617 (1974)
4.5 D.J.Channin, D.E.Carlson: Appl. Phys. Lett. **28**, 300 (1976)
4.6 D.J.Channin: Soc. Information Display 1976 Symp., Dig. Tech. Papers, Vol. VII, p. 38
4.7 E.Dubois-Violette, P.G.de Gennes, O.Parodi: J. Phys. (Paris) **32**, 305 (1971)
4.8 P.H.Penz, G.W.Ford: Phys. Rev. **6**, 414, 1676 (1972)
4.9 I.W.Smith, Y.Galerne, S.T.Lagerwall, E.Dubois-Violette, G.Duraul: J. Phys. (Paris) **36**, C-1–237 (1975)
4.10 G.Labrunie, J.Robert: J. Appl. Phys. **44**, 4869 (1973)
4.11 M.Schiekel, K.Fahrenschon, H.Gruler: Appl. Phys. **7**, 99 (1975)
4.12 M.I.Barnik, L.M.Blinov, M.F.Grebenkin, A.N.Trufanov: Mol. Cryst. Liq. Cryst. **37**, 47 (1976)

4.13 D. Berreman: Appl. Phys. Lett. **25**, 12 (1974); J. Appl. Phys. **46**, 3746 (1975)
4.14 C.Z. Van Doorn: J. Appl. Phys. **46**, 3738 (1975)
4.15 A.R. Kmetz: In *Nonemissive Electro-Optic Displays*, ed. by A.R. Kmetz, F.K. von Willison (Plenum Press, New York 1976)
4.16 A. Sobel: IEEE Trans. ED-**21**, 146 (1974)
4.17 P.M. Alt, P. Pleshko: IEEE Trans. ED-**21**, 146 (1974)
4.18 K. Ono, E. Matani, E. Kaneko, M. Sato: Soc. Information Display 1976 Symp., Dig. Tech. Papers, Vol. VII, p. 34
4.19 M. Godcianski: J. Appl. Phys. **48**, 1426 (1977)
4.20 L.T. Lipton, M.H. Meyer, D.O. Massetti: Soc. Information Display 1975 Symp., Dig. Tech. Papers, Vol. VI
4.21 T.P. Brody, J.A. Asars, G.D. Dixon: IEEE Trans. ED-**20**, 995 (1973)
4.22 J.D. Margerum, J. Nimoy, S.Y. Wong: Appl. Phys. Lett. **17**, 51 (1970)
4.23 D.L. White, M. Feldman: Electron. Lett. **6**, 837 (1970)
4.24 W.E. Haas, G.A. Dir: Appl. Phys. Lett. **29**, 325 (1976)
4.25 W.P. Bleha, J. Grinberg, A.D. Jacobson: Soc. Information Display 1973 Symp., Dig. Tech. Papers, Vol. VII, p. 34
4.26 G.N. Tagler, F.J. Kahn: J. Appl. Phys. **45**, 4330 (1974)
4.27 A. Sasaki, K. Kurahashi, T. Takayi: J. Appl. Phys. **45**, 4356 (1974)
4.28 F.J. Kahn: Appl. Phys. Lett. **22**, 111 (1973)
4.29 P.E. Cladis, S. Torzu: J. Appl. Phys. **46**, 584 (1975)
4.30 J.J. Wysocki, J.J. Adams, W. Haas: Phys. Rev. Lett. **20**, 1024 (1968)
4.31 W. Grubel: Appl. Phys. Lett. **25**, 5 (1974)
4.32 G.H. Heilmeier, L.A. Zanoni, J.E. Goldmacher: In *Liquid Crystals and Ordered Fluids*, ed. by J.F. Johnson, R.S. Porter (Plenum Press, New York 1970) p. 215
4.33 E. Guyon, W. Urbach: In *Nonemmissive Electrooptic Displays*, ed. by A.R. Kmetz, F.K. von Willisen (Plenum Press, New York 1976) p. 121
4.34 J. Nehring: Phys. Rev. A**7**, 1737 (1973)
4.35 G.H. Heilmeier, L.A. Zanoni, L.A. Barton: Appl. Phys. Lett. **13**, 46 (1968)
4.36 W. Helfrich: J. Chem. Phys. **51**, 4092 (1969)
4.37 R. Williams: J. Chem. Phys. **39**, 384 (1963)
4.38 W. Helfrich: J. Chem. Phys. **50**, 100 (1969)
4.39 P.A. Penz: Phys. Rev. Lett. **24**, 1405 (1970)
4.40 E. Jakeman, P.N. Pusey: Phys. Lett. **44**A, 456 (1973)
4.41 S. Kai, K. Hirakawa: Solid State Commun. **18**, 1573 (1976)
4.42 I. Haller, H.A. Huggins: U.S. Patent # 3,656,838 (1972)
4.43 W. Haas, J. Adams, J.B. Flannery: Phys. Rev. Lett. **25**, 269 (1970)
S. Matsumoto, M. Kawamoto, N. Kaneko: Appl. Phys. Lett. **27**, 269 (1975)
4.44 F.J. Kahn: Appl. Phys. Lett. **22**, 386 (1973)
S. Matsumoto, D. Nakagowa, N. Kaneko, K. Mizunoya: Appl. Phys. Lett. **29**, 67 (1976)
G. Porte: J. Phys. (Paris) **37**, 1245 (1976)
4.45 H. Sorkin, R.I. Klein: U.S. Patent # 3,698,449 (1972)
4.46 A. Sussman: U.S. Patent # 4,021,945 (1977)
4.47 P.M. Alt, M.J. Freiser: J. Appl. Phys. **45**, 3237 (1974)
J. Nehring, M.S. Petty: Phys. Lett. **40**A, 307 (1972)
4.48 A. Sussman: Appl. Phys. Lett. **21**, 269 (1972)
4.49 R.J. Turnbull: J. Phys. D**6**, 1745 (1973)
F. Gaspard, R. Herino, F. Mondon: Chem. Phys. Lett. **25**, 449 (1974)
H.S. Lim, J.D. Margarum: J. Electrochem. Soc. **123**, 837 (1976)
4.50 D. Meyerhofer, A. Sussman: Appl. Phys. Lett. **20**, 337 (1972)
4.51 G.H. Heilmeier, J. Goldmacher: Proc. IEEE **57**, 34 (1969)
4.52 A. Sussman: Appl. Phys. Lett. **21**, 126 (1972)
4.53 F. Nakano, M. Sato: Jpn. J. Appl. Phys. **15**, 1937 (1976)
B. Gosse, J.P. Gosse: J. Appl. Electrochem. **6**, 515 (1976)
4.54 S. Barret, F. Gaspard, R. Herino, F. Mondon: J. Appl. Phys. **47**, 2375–2378 (1976)

4.55 Y.Onishi, M.Ozutsumi: Appl. Phys. Lett. **24**, 213 (1974)
A.I.Baise, I.Teucher, M.M.Labes: Appl. Phys. Lett. **21**, 142 (1972)
4.56 H.S.Lim, J.D.Margerum: Appl. Phys. Lett. **28**, 478 (1976)
4.57 A.Sussman: Mol. Cryst. Liq. Cryst. **14**, 182 (1970)
4.58 P.J.Wild, J.Nehring: Appl. Phys. Lett. **19**, 335 (1971)
C.R.Stein, R.A.Kashnow: Appl. Phys. Lett. **19**, 343 (1971)
4.59 Y.Hori, M.Fukai: J. Electrochem. Soc. **124**, 1752 (1977)
W.H.DeJeu, J.van der Veen: Phys. Lett. A**44**, 277 (1973)
4.60 M.Schadt, W.Helfrich: Appl. Phys. Lett. **18**, 127 (1971)
4.61 D.Meyerhofer: J. Appl. Phys. **48**, 1179 (1977)
G.Bauer, F.Windscheid, D.W.Berreman: Appl. Phys. **8**, 101 (1975)
4.62 M.Bechtler, H.Kruger: Electronics **50**, 113 (1977)
4.63 A.Sussman: IEEE Trans. PHP-**8**, 24 (1972)
4.64 D.Meyerhofer: Phys. Lett. **51** A, 407 (1975)
4.65 W.Urbach, M.Boix, E.Guyon: Appl. Phys. Lett. **25**, 479 (1974)
4.66 J.L.Janning: Appl. Phys. Lett. **21**, 173 (1972)
4.67 M.Yamashita, Y.Amemiya: Jpn. J. Appl. Phys. **17**, 2087 (1978)
4.68 W.A.Crossland, J.H.Morrisy, B.Needham: J. Phys. D**9**, 2001 (1976)
4.69 F.J.Kahn: Mol. Cryst. Liq. Cryst. **38**, 109 (1977)
A.Toda, H.Mada, S.Kobayashi: Jpn. J. Appl. Phys. **17**, 261 (1978)
4.70 D.Meyerhofer: Private communication
4.71 D.Meyerhofer: Soc. Information Display 1976 Biennial Display Conf. **76 CH1124-7ED**, p. 48
D.Meyerhofer, A.Sussman: U.S. Patent # 3,967,883 (1976)
E.P.Raynes, D.K.Rowell, I.A.Shanks: Mol. Cryst. **34**L, 105 (1976)
M.R.Johnson, P.A.Penz: Proc. SID **18**, 21 (1977)
4.72 G.Porte, J.P.Jadot: J. Phys. (Paris) **39**, 213 (1978)
4.73 A.Miyaji, M.Yamaguchi, A.Toda, H.Mada, S.Kobayashi: IEEE Trans. ED-**24**, 811 (1977)
4.74 K.Toriyama, T.Ishibashi: In Ref. 4.4, p. 146
4.75 D.W.Berreman: J. Opt. Soc. Am. **63**, 1374 (1973)
4.76 W.H.DeJeu, W.A.P.Claassen, A.M.J.Spruijt: Mol. Cryst. Liq. Cryst. **37**, 269 (1976)
4.77 T.Motooka, A.Fukuhara, K.Suzuki: Appl. Phys. Lett. **34**, 305 (1979)
4.78 J.Nehring, A.R.Kmetz, T.J.Scheffer: J. Appl. Phys. **47**, 850 (1976)
4.79 H.Gruler, L.Cheung: J. Appl. Phys. **46**, 5097 (1975)
H.Deuling: Mol. Cryst. Liq. Cryst. **27**, 81 (1974)
4.80 P.Sheng: In Ref. 4.2, p. 114
4.81 D.W.Berreman: U.S. Patent # 4,033,671 (1977)
4.82 G.G.Barna: Proc. SID 1976 Biennial Display Conf. **76 CH1124-7ED**, p. 29
4.83 P.A.Penz: Soc. Information Display Intern. Symp., Dig. Tech. Papers, Vol. IX (Lewis Winner, Coral Gables 1978) p. 68
4.84 M.Schadt, F.Muller: IEEE Trans. ED-**25**, 1125 (1978)
4.85 A.Sussman: Unpublished
4.86 G.H.Heilmeier, L.A.Zanoni: Appl. Phys. Lett. **13**, 91 (1968)
4.87 A.Bloom, E.B.Priestly: Proc. SID **18**, 39 (1977)
4.88 D.L.White, G.N.Taylor: J. Appl. Phys. **45**, 4718 (1974)
4.89 T.J.Scheffer: In Ref. 12, p. 66
4.90 A.Sussman: J. Appl. Phys. **49**, 1131 (1978)
4.91 A.Sussman: J. Electrochem. Soc. **126**, 85 (1979)
4.92 D.W.Berreman, A.Sussman: "Net Lateral Flow During a Twist-Cell Cycle", in preparation
4.93 L.Goodman, F.DiGeronimo: 5th Intern. Liq. Cryst. Conf., Stockholm (1974)
4.94 M.G.Clark, I.A.Shanks, N.J.Patterson: SID Intern. Symp., Dig. Tech. Papers, Vol. X, (Lewis Winner, Coral Gables 1979) p. 110
4.95 I.A.Shanks, P.A.Holland: In Ref. 93, p. 112
4.96 D.Coates, W.A.Crossland, J.H.Morris, B.Needham: J. Phys. D. **11**, 2205 (1978)

5. Electrochromic Displays Based on WO_3

B. W. Faughnan and R. S. Crandall

With 9 Figures

Electrochromic displays based on thin films of WO_3 as the active layer are reviewed with emphasis on the basic mechanisms of the electrochromic effect. These include the concept of double injection of electrons and protons into the WO_3 film, resulting in the formation of a colored tungsten bronze H_xWO_3. It is suggested that the broad optical absorption band which peaks at $\lambda = 0.95\,\mu m$ arises from an electron transfer between tungsten ions.

A detailed theory of the dynamics of coloring and bleaching is presented. In the coloring mode, the WO_3-electrolyte interface and the x-dependence of the chemical potential of H_xWO_3 play a critical role. Speed limitations due to diffusion of electrons or protons in the WO_3 film are shown to be unimportant for practical devices. During bleaching, the positive and negative charges must be separated and exit from opposite sides of the film. The limitation on the bleaching speed is the space charge limited current flow of protons through the bleached portion of the film.

Various methods of preparing WO_3 films are discussed. Methods of measuring the water content of films and its relationship with film reproducibility and device performance are emphasized.

Device performance, starting with the simplest WO_3-film/proton-electrolyte system, is considered. Equations for coloring and bleaching speeds are developed. More complicated devices including liquid electrolytes in which positive ions other than protons are used and all solid state devices are described.

Finally, the critical problems of device lifetime and degradation mechanisms are reviewed.

5.1 Background Information

5.1.1 Electrochromic Displays

An electrochromic material has the property of changing color when a voltage is applied across the material, or alternately, a current is passed through it. This color change should be reversible when the voltage is removed or when the polarity of the voltage or current is reversed. Known electrochromic materials are both inorganic and organic, and the color change may be from clear or transparent to colored, or from one color to another color. The last case is characteristic of organic materials.

An electrochromic display device utilizes one of the above materials usually in the form of a thin film, to build a passive display. The characteristics of electrochromic displays are: low-voltage operation (1.5 V down to approximately 0.4 V), low power requirements [somewhere between liquid-crystal displays (LCD) and light-emitting diode (LED) displays], storage of the display without power dissipation, potentially low cost and simplicity in building, and possibility of an aesthetically pleasing display with good contrast and wide viewing angle.

The last characteristic probably has provided the motivation for work in this area; that is, the desire to obtain a display which looks better than an LCD display, whose device properties it most closely resembles. One area of keen interest has been small displays suitable for watches or calculators, but somewhat larger displays suitable for automotive or aircraft use remain an attractive possibility.

5.1.2 Historical Outline

One of the first discussions of electrochromism was by *Platt* [5.1] who coined the term. However, a field-induced increase in optical density goes back to *Franz* and *Keldysh* [5.2], who explained the shift of optical absorption caused by an electric field. The Franz–Keldysh effect depends on the shift of an optical absorption band caused by an electric field, requires high voltage and usually results in a small optical density (OD) change. This is to be contrasted with the low-voltage effect discussed in this review. This low-voltage electrochromic (EC) effect occurs only when charge is injected and can result in large optical-absorption changes.

Perhaps the earliest example of this EC effect, as well as the simplest, is the creation of F centers in an alkali halide crystal when a voltage is applied between two metal electrodes attached to opposite faces of the crystal, which is heated to about 700 °C [5.3]. Briefly, the effect occurs because electrons injected from the negative pointed electrode are trapped at anion vacancies thereby forming F centers. Charge neutrality and current continuity is maintained by anion vacancy motion originating from the positive electrode. This implies a net mass transport of anions to the positive electrode where halogen gas is evolved. If the electrode potentials are reversed, hole centers are produced accompanied by the release of alkali metal at the negative electrode.

This simple example illustrates most of the requirements for an electrochromic material. First of all, it must have a color center, or some optical absorption in the visible. A second feature is the presence of mixed conduction; i.e., electronic and ionic conduction. This is necessary since charge neutrality must be preserved. These requirements severely restrict the range of available materials. In the example above, the crystal must be heated to 700 °C to raise its ionic conductivity sufficiently to produce a measurable current and, hence, optical density change. This renders it impractical for most device applications.

Another example is the electrocoloration of transition-metal-doped $SrTiO_3$ [5.4]. In a typical case the $SrTiO_3$ is doped with Fe and Mo which can exist in $SrTiO_3$ in a variety of oxidation states, some of which have optical absorption bands. In this case, a color front moves into the crystal from each electrode. Iron is oxidized starting from the positive electrode and Mo is reduced at the negative electrode. Oxygen vacancies move toward the cathode to maintain charge neutrality. It appears that three types of carriers are involved in different parts of the crystal, electronic conduction in the Mo^{5+} rich region, hole conduction in the Fe^{4+} rich region, and ionic (vacancy) conduction in the central uncolored region. Again, the crystal must be heated to approximately 200 °C to obtain a reasonable conductivity for the least conducting carrier. Thus, practical devices have not been developed.

Clearly, for good electrochromic performance in a solid at room temperature, the activation energy for the slowest carrier must be low, 0.4 eV or less. Since the slowest carrier is usually the ionic carrier, this places the search for good EC materials within the class of fast-ion (or superionic) conductors [5.5].

Probably the first mention of color change in a material having such qualities was in 1951 by *Brimm* et al. [5.6] who were studying sodium tungsten bronze (Na_xWO_3). They found that when the tungsten bronze was used as the electrode in an electrochemical cell it changed color depending whether it was used as the anode or cathode. However, this phenomena does not appear to have been exploited as a display device.

It was not until 1969 that an actual electrochromic display device was reported on by *Deb* [5.7]. He described an effect in amorphous films of WO_3 much like the F center coloration described above. Fortunately, the coloration took place at room temperature. In analogy with the alkali halides, he ascribed the EC effect to injection of electrons and their subsequent trapping at oxygen vacancies to form F centers [5.8]. He correctly recognized that the H_2O in the films was necessary for the device operation. The current explanation for the effect is that the electric field dissociates H_2O in the film and protons migrate in the field to the cathode where they compensate the injected electrons [5.9]. This region with a high density of electrons and protons is converted into hydrogen tungsten bronze H_xWO_3 $(0<x<1)$. Because the H^+ is required for charge compensation the device speed is governed by the proton mobility. Fortunately the proton mobility is anomalously high in WO_3, making it color quickly enough for room temperature operation [5.10, 11].

The WO_3-based device structure that is under current study and holds the promise of a useful device is based on double injection of electrons and protons or other ions into the film [5.12]. This structure has subsecond switching speeds and is thus useful in watch applications or other alphanumeric displays [5.12–14].

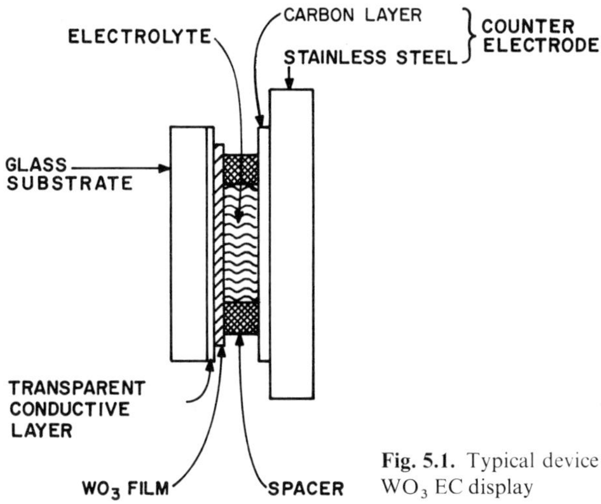

Fig. 5.1. Typical device structure of a liquid electrolyte WO_3 EC display

5.1.3 Nature of EC in WO_3

Amorphous films of tungsten or molybdenum trioxide are superior EC device materials for two reasons. Ion mobilities are high in these materials and electron injection produces a strong optical absorption in the visible. A high ion mobility is required because coloration entails injection of positive ions, and bleaching entails extraction of positive ions [5.12]. Ions such as Li^+, Na^+, Ag^+ have been used in WO_3 EC devices [5.15–18], however, with some sacrifice in device speed. In this review we will refer to protons as the ions used for electron charge compensation even though other ions could equally well be utilized without changing the general nature of the effects.

To simplify the ensuing discussion we represent what is a common EC device structure in Fig. 5.1. To color the WO_3 film, a battery is connected between the transparent conductive electrode (hereafter referred to as ITO for indium tin oxide which is commonly used for this purpose) and the carbon film. With a negative voltage on the ITO electrode, electrons are injected from the ITO electrode and protons from the acid electrolyte into the WO_3 film. This continues until the WO_3 is converted into the tungsten bronze H_xWO_3. A value of $x \sim 0.5$ appears to be the upper limit for the amount of hydrogen that can be injected into the films [5.12]. Of course the coloration process can be stopped at lower values of x. Under open circuit conditions the color remains in the film for some time [5.16]. Thus, one has a long-term memory device. To bleach the film, the polarity is reversed so that electrons leave at the anode and protons at the cathode; i.e., electrons and protons leave at the same electrode from which they enter the film. Current flows until the entire film is restored to its original WO_3 (colorless) state. It is convenient to think of the color and bleach process as the charging and discharging of a battery.

The injected ion does not appear to play an essential role in color-center formation; it enters the crystal merely for charge compensation. It has been proposed [5.12] that the color arises when an electron trapped on a W^{5+} ion makes an optically induced transition to a W^{6+} ion. The question as to whether the electron actually distorts the lattice around the tungsten ion to further lower its energy by forming a bound polaron [5.19–20] is still unclear. The original models that an F-center-like defect is formed or that a band-to-band transition produces the optical absorption are inconsistent with the experimental data [5.21]. The broad Gaussian-shaped absorption band peaking at $\sim 1\,\mu m$ is best explained by the transfer of an electron from a W^{5+} to W^{6+}, as will be discussed in more detail elsewhere in this review.

5.2 Principles of Electrochromism in WO_3

5.2.1 Basic Phenomenon of the EC Effect in WO_3

In general terms the electrochromic phenomenon in WO_3 can be expressed by the chemical formula

$$WO_3(\mathrm{CLEAR}) + xe^- + xM^+ \xrightleftharpoons{\mathrm{VOLTAGE}} M_xWO_3(\mathrm{BLUE}) \qquad (5.1)$$

where M^+ is a metal ion or a proton. This reaction goes to the left or right depending on the sign of the voltage V. Written as above, the EC phenomenon is a simple redox reaction. However, this simple explanation hides much that makes electrochromism in WO_3 unique and interesting. Forming M_xWO_3 is like doping a substance, whereas in a redox reaction a new substance is grown at the anode or cathode and the volume of material depends on the time that the process continues. In M_xWO_3 growth, the volume is fixed equal to the WO_3 volume. During growth the value of x increases. M_xWO_3 is not a single phase but rather a continuous phase, whereas in a normal redox growth the material is a single phase. The fact that M_xWO_3 forms a continuous phase has profound effects on the dynamics of coloration of the WO_3 [5.22]. The bleach process in a redox reaction is governed by surface phenomena at the electrolyte–material interface whereas in M_xWO_3 it is the solid-state properties of WO_3 that determine bleaching [5.23].

The blue color is associated with the formation of the amorphous M_xWO_3. To our knowledge films made in different ways and with different metal ions all have essentially the same color or optical absorption band. This is in contrast to the single-crystal bronzes that have different colors depending on the value of x. However, in this case, the color is determined by both the reflectivity and absorption and it is the change in reflectivity with x that gives the apparent color change in single-crystal bronzes [5.24].

These general physical principles of electrochromism give a rough idea of how the device works but only hint at the complexity of its operation which requires concepts of solid-state physics, chemistry, and electrochemistry. In what follows we shall discuss each aspect of the device operation in some detail to see how it affects the overall performance and operation of the display.

5.2.2 Dynamics of Coloring and Bleaching

Coloring and bleaching (charge injection and extraction) are not symmetric phenomena. Coloring is governed mainly by the properties of the interface between the WO_3 and the proton-injecting contact, whereas bleaching is determined by proton transport in the WO_3. The electron motion is so fast compared with proton motion that the proton motion determines the device speed. Theories of both coloring [5.25] and bleaching [5.23] have been developed that permit one to choose the best operating conditions for a particular device. These theories were developed to explain experiments on H_xWO_3 devices but are general enough to apply to other proposed device configurations [5.16].

We shall consider the coloration process first. To visualize the operation we refer to Fig. 5.2 where a schematic diagram of the coloration process for a device with the sandwich structure of Fig. 5.1 is shown. The transparent ITO electrode makes an electron injecting contact to the WO_3. The electrolyte is usually of low pH and can readily supply protons to the WO_3. The counterelectrode can be carbon, H_xWO_3, metal, or a composite material. In the following discussion we assume that there is a negligible voltage drop at the electrodes.

Because WO_3 and the counterelectrode are dissimilar materials there will be an emf between them proportional to their chemical potential difference [5.26]. The size and magnitude of this emf depend on the nature of the counterelectrode as well as whether the WO_3 film is colored or bleached. As the film colors (WO_3 is converted to H_xWO_3) its chemical potential changes [5.22]. It changes

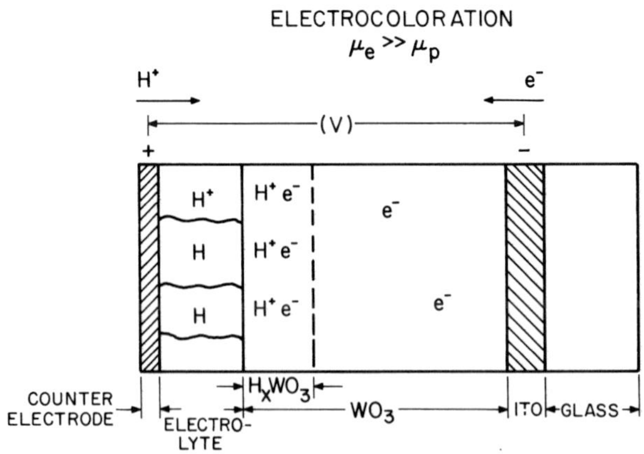

Fig. 5.2. Schematic diagram of the electrocoloration process in WO_3

in such a way as to make the H_xWO_3 film more positive with respect to the counterelectrode. This can have a profound effect on device operation because this emf is directed opposite to the applied voltage in the coloring mode and in the same direction as the applied voltage in the bleaching mode.

The dynamics of the coloration process is complicated because there are potentially many different mechanisms which can limit the coloring current. These are

1) transport of the electrons and protons through the bulk of the WO_3 film;
2) a barrier at the electron-injecting contact;
3) a barrier at the proton-injecting contact (electrolyte-WO_3 interface);
4) a barrier at the counterelectrode; and
5) charge transport in the electrolyte.

Since the ITO makes an ohmic contact to WO_3, 2) is unimportant under all conditions. By choosing a low-resistance electrolyte 5) can be made unimportant except at the highest current levels. Mechanism 4) can be similarly eliminated by a suitable choice of counterelectrodes. Mechanism 3), we find, dominates the coloration under most experimental conditions [Ref. 5.26, p. 111]. The conditions under which 1), 4), and 5) affect the coloration will be discussed when appropriate.

The situation in which current flow is limited by thermal activation over a barrier is well known and leads to the result that the current depends exponentially on the applied voltage [Ref. 5.26, p. 862]. The ideas involved can be applied to the electrolyte–WO_3 interface to derive a current voltage curve. However, a novel feature is present in these WO_3-based EC devices. As coloration proceeds and x increases, an emf appears across the interface that opposes current flow. This emf appears because, as the hydrogen concentration in H_xWO_3 increases, the chemical potential increases [5.22]. This back emf, V_r, is just equal to the chemical potential change caused by coloring. If the film is colored to the maximum x value of ~ 0.5, V_r can be as high as ~ 0.7 V. Therefore, if the applied voltage were less than this value, the current flow would cease at an x value less than 0.5. The current flow ceases because V_r has increased until it just equals the applied voltage.

From this we see the importance of using as high a voltage as is practical for coloring. Because V_r increases as the film colors, the current will decrease. These ideas have been used to derive an expression for the time dependence of the coloring current [5.25] that explains the often observed result that this current is inversely proportional to the square root of the current. This particular result is obtained if proton diffusion in the film is fast enough to be neglected. Expressions for the current when diffusion cannot be ignored are derived in the appendix. However, if diffusion in the film limits device speed then generally the device will be too slow to be of any practical value.

Neglecting diffusion, the coloring current (A.6) is

$$J_c = J_0 \exp(-x/x_1) \exp(V_a/2RT) \left(\frac{1-x}{x}\right) \tag{5.2}$$

where J_0 is the exchange current defined in Appendix A, x_1 a constant the order of 0.1, R the gas constant, T the absolute temperature, and V_a the applied voltage. The conditions for the validity of (5.2) are discussed in the appendix. Even though (5.2) does not give the time dependence of the current, it does show its behavior with applied voltage and proton density. At a given x value J_c increases exponentially with increasing applied voltage. This is a direct consequence of the thermally activated proton transport across the barrier. As the film colors and x increases J_c decreases. This is caused by the increase in the back emf.

To find the functional dependence of x on time, t, we equate J_c to the proton current in the film. This is done in the appendix for the two situations in which diffusion in the film is either faster or slower than the time scale of the measurement. The characteristic time for diffusion is

$$\tau_D = l^2/4D_p \tag{5.3}$$

where l is the film thickness and D_p is the proton diffusion coefficient. In amorphous films of WO_3 we have found D_p varying from 2.5×10^{-7} to 2.5×10^{-9} cm^2/s [5.23]. In recrystallized films it could be as low as the single-crystal value of 1×10^{-12} cm^2/s [5.27]. Since typical film thicknesses vary between 0.1 and 1.0 μm, we expect τ_D in amorphous films to vary between 10^{-4} and 10^{-1} s. Useful devices will have τ_D in the low end of this range. Thus the time scale of coloring, 0.1–1 s, will normally be much longer than τ_D. In this case the time dependence of the coloring current, for $x/x_1 \ll 1$, is given by (A.9) as

$$J_c = J_0 \exp(V_a/4RT)\sqrt{\frac{\tau_0}{t}}\,. \tag{5.4}$$

The constant τ_0 is defined in Appendix A. This expression exhibits the commonly observed inverse dependence of the current on the square root of time.

Others have interpreted this time dependence as due to diffusion either in the electrolyte or in the film [5.14, 28]. We believe these interpretations are incorrect as the diffusion coefficients required to explain the observed results are orders of magnitude smaller than the diffusion coefficients measured in the film [5.11, 23] or the electrolyte [Ref. 5.26, p. 471]. Furthermore, (5.2) shows a strong dependence on applied voltage which would not be expected if only diffusion were important. Equation (5.4) has been verified for proton- [5.25] as well as lithium-ion- [5.16] based devices.

For times short compared with τ_D, diffusion in the film will indeed be important in determining coloring speed. Nevertheless, because the barrier at the electrolyte–WO_3 interface is still important, the expression for the coloring current will be modified from the standard diffusion expression [Ref. 5.26, Chap. 4]. There is an additional time dependence because the surface proton concentration x_s is time dependent. It is useful to define τ_s as the characteristic time for the surface concentration to increase to its maximum value ($x_s \approx 0.5$). *For* $\tau_s < t < \tau_D$, x_s is constant and the coloring current has the standard form [Ref. 5.26, Chap. 4]

$$J_c = e\varrho x_s \sqrt{D_p/\pi t} \tag{5.5}$$

where ϱ is the tungsten atom density and e the electron charge. This expression shows the usual $t^{-1/2}$ dependence for a diffusion limited process and furthermore it is independent of voltage.

For the short time regime $t < \tau_s$ where ($x_s \ll 0.5$) a new time dependence appears. For small x, (A.9, 10) can be combined to give

$$J_c = \exp(V_a/4RT) \sqrt{J_0 e\varrho}(D_p/\pi t)^{1/4} . \tag{5.6}$$

This expression shows that J_c depends not only on the diffusion coefficient but also has a strong dependence on voltage. The weak $t^{-1/4}$ time dependence arises because the increase in x_s opposes the decrease of current due to diffusion. As shown by (A.11), $x_s \propto t^{-1/4}$.

Because (5.4, 5) have the same time dependence, the observation that $J_c \propto t^{-1/2}$ is not sufficient to characterize the transport as diffusion limited. One must also show that $J_c \propto t^{-1/4}$ at short times. Furthermore, measured values of D_p range between 2.5×10^{-7} and 2.5×10^{-9} cm/s [5.11, 23]. Substituting these values in (5.5) gives currents that are orders of magnitude higher than observed currents. Thus it is unlikely that a diffusion-limited process has been observed in amorphous WO_3.

Bleaching is accomplished by changing the sign of the applied voltage so that protons move from the H_xWO_3 into the electrolyte and electrons into the ITO electrode. In contrast to the coloring, bleaching is not limited by proton transport over the barrier at the electrolyte–WO_3 interface, the main reason being that the back emf contributes to current flow because it is in the same direction as the applied voltage.

During the bleach cycle, electrons do not enter the film from the electrolyte nor do protons enter it from the ITO. Therefore, in a region adjacent to the electrolyte interface, the current in the H_xWO_3 must flow as a pure space-charge-limited (SCL) current of protons. Similarly at the ITO interface there is a SCL electron current. However, because of the large ratio of electron to proton mobility, electron transport will not limit the current flow and can be neglected. Therefore, current flow will be determined by the SCL proton current.

The situation inside the film a short time after the beginning of bleaching is shown in Fig. 5.3. Protons have moved out of region III leaving an electron SCL current. However, this region is so thin that the voltage drop is negligible. Region II contains the neutral low resistance H_xWO_3. In region I, however, electrons have left, leaving behind a proton space charge. Because region II has

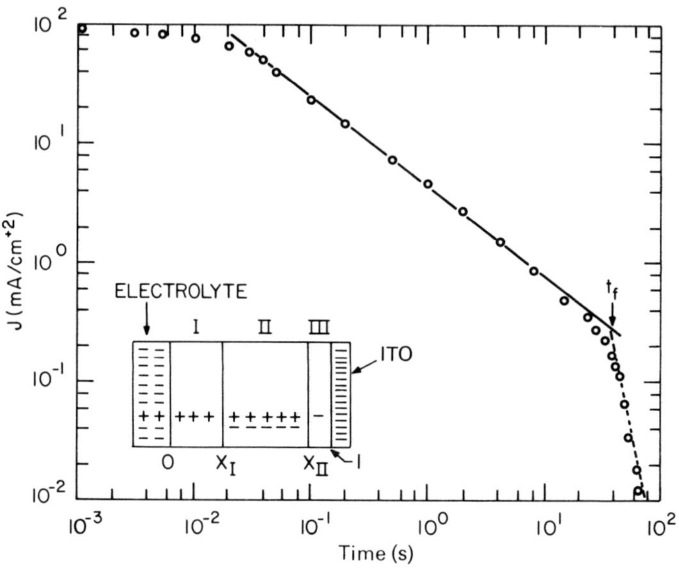

Fig. 5.3. Bleaching curve of a WO_3 electrochromic film

a low resistance compared with region I, there will be little voltage drop across it. In what follows, therefore, we assume that the applied voltage appears across region I.

Region II is the source of protons that feeds region I. The thickness of region II shrinks with time as protons and electrons leave at the II–I and II–III interface, respectively. Therefore, region I grows. The SCL proton current [5.23] J_B in region I is

$$J_B = K\varepsilon_0 \mu_p V_a^2 / x_I^3(t) \tag{5.7}$$

where K is the relative dielectric constant, ε_0 the permittivity of free space, μ_p the proton mobility, and x_I the thickness of region I. J_B decreases with time as x_I increases. The time dependence of x_I can be obtained by realizing that the charge extracted from region II in unit time is just J_B. This leads to the result that $J_B \propto t^{-3/4}$, which has been verified experimentally [5.16, 23]. Similarly the time (t_B) to bleach the sample, that is, the time for x_I to become equal to the sample thickness (l), is given by [5.23]

$$t_B = \frac{ex\varrho l^4}{4\mu_p V_a^2 K\varepsilon_0}\,. \tag{5.8}$$

This shows the strong dependence of the bleaching time on the thickness of the film. Also, the lower the mobility the longer the time to bleach. A higher coloration density (high x) also implies a longer bleach time.

SYMBOL	MATER-IAL	TREATMENT	THICKNESS	(OD) MAX	OD/μm	E(eV) PEAK	ΔE(eV) HWHM
———	SINGLE CRYSTAL	COLORED IN HCl:Zn 5 MIN., T=45°C	~.25mm CRYSTAL .27μm COLOR LAYER	3.0	5.6	0.90	1.0
·········	EVAP FILM	H.T. IN AIR, 500°C, 1-2 MIN. COLORED ELECTROLYTICALLY	0.5μm	1.45	2.9	0.95	0.6
- - - - -	EVAP FILM	AS EVAPORATED COLORED HCl:Zn 5MIN., 45°C	0.1μm	0.85	8.5	1.46	1.0

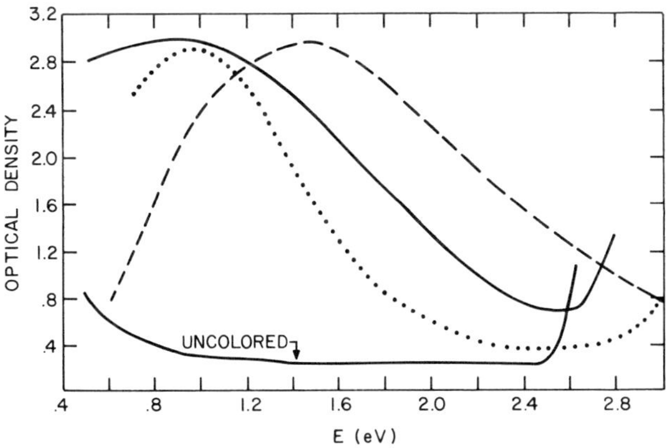

Fig. 5.4. Optical absorption in amorphous and crystalline WO_3

5.2.3 Optical Properties

Experiment

The blue color [5.8] of electrochromic WO_3 observed in both crystalline and amorphous films is due to a broad absorption band which peaks between 0.90 eV and 1.46 eV depending on film properties [5.11, 19]. Also, its full width at half maximum (FWHM) varies between 1.1 eV and 2.0 eV. Figure 5.4 shows typical optical-absorption curves.

The optical absorption in amorphous WO_3 peaks at approximately 1.46 eV and the width of the absorption broadens as the density of coloration increases to its maximum, corresponding to $x \sim 0.5$. The amorphous WO_3 curve shown in Fig. 5.4 is for a thin film (~ 0.1 μm) colored to a high optical density and, therefore, is broad.

When amorphous WO_3 is crystallized by heating in air [5.8], or deposited in a polycrystalline form, its absorption peak shifts to lower energy. This is shown in Fig. 5.4 for a film which has been crystallized by heating in air at 500 °C for 2 min. This shifts its absorption peak to 0.95 eV. The coloration density is not as great for this film as for the amorphous film and so its absorption is not as broad; e.g., 1.2 eV instead of 2.0 eV for FWHM.

In the same figure, we show the optical absorption arising from a single-crystal platelet. This crystal was colored by immersing in heated HCl in the presence of Zn [5.29] pellets. Analysis by SIMS (secondary ion mass spectroscopy) [5.30] shows that it was colored to a density corresponding to $x=0.5$ in a surface layer 0.27 μm deep. The depth of this surface layer increases with increased time of immersion in the HCl:Zn. Interestingly, the properties of the single-crystal absorption are very similar to the crystallized film and to the amorphous film except for the shift in the peak by approximately 0.5 eV. All of these curves can be fit quite well to a Gaussian shape.

The optical properties described above are consistent with a trapped or localized center, and the various optical properties can be interrelated through Smakula's equation which was originally proposed for color centers in alkali halides. Smakula's equation [5.32] is

$$Nf = 8.7\times 10^{6}\frac{n}{(n^2+2)^2}(k_{\max}l)\,W_{1/2} \tag{5.9}$$

where N is the number of color centers formed per cm^3, n the index of refraction of the film, k_{max} the absorption coefficient at the band peak in cm^{-1}, $W_{1/2}$ the full width at half maximum in eV, and f the oscillator strength.

The oscillator strength is defined so that a fully allowed electric-dipole transition involving a single electron can have a maximum value of unity. Putting the numbers appropriate for WO_3 into the above equation we find $f\sim 0.1$. For comparison, the oscillator strengths [5.32] for F-center transitions in alkali halides typically lie between 0.5 and 0.6.

Model for the Optical Transition

The optical absorption in amorphous EC WO_3 arises from a transition of an electron from one W^{6+} trapping site to an adjacent W^{6+} site. According to this model electrons injected from the cathode during the coloration process are trapped at transition-metal sites while protons (or other positively charged metal ions) injected from the anode remain in interstitial positions in their charged state. We proposed [5.1] a particular model in terms of an intervalence charge transition which can be written as shown below:

$$W^{5+}(A)+W^{6+}(B) \overset{\text{light}}{\rightleftharpoons} W^{6+}(A)+W^{5+}(B)\,. \tag{5.10}$$

This model can explain, at least qualitatively, the features of the optical absorption including the magnitude of the oscillator strength. An alternative model based on small-polaron theory has been proposed [5.19–20]. This model is similar to the intervalence-transition model except that it postulates a self-trapping term which lowers the energy of the site with the trapped electron relative to other W^{6+} sites.

A systematic study of the optical properties of crystallized WO_3 films, including the shift in the energy peak upon crystallization has been made [5.19]. This shift, plus the asymmetry of the high-energy side of the optical absorption in amorphous films has led the authors to postulate a disorder term in their expression for the energy peak. In a disordered system the W^{6+} sites do not all have equivalent energy. The injected electrons will be trapped primarily in those sites of lowest energy. Hence, optical transitions will be made with a range of energies. This explains the asymmetry in the optical-absorption width as well as the shift to lower energy for crystalline films. The latter effect occurs because the disorder term disappears for crystalline films.

These authors [5.19] also claim that the optical absorption in crystalline films is of the Drude or free-electron type but we feel that this is probably not correct for the following reasons: 1) Shape of the optical absorption band is Gaussian for both amorphous and crystalline films. This is consistent with localized centers but free-electron absorption should be Lorentzian in shape. 2) The shift to lower energy by 0.5 eV is explainable by the loss of the disorder term as explained above if the absorption is due to localized centers in both cases. If the mechanism of the absorption changes, then this significance to the energy shift is lost. 3) The intensity of the transition is approximately the same as the film changes from amorphous to crystalline. This would be fortuitous if the absorption mechanism changed from local to nonlocal. 4) The interpretation of the results [5.21] of crystallization of mixed-oxide WO_3/MoO_3 films to be explained below is consistent only with the notion that the electrons remain localized upon crystallization. 5) The small-polaron concept could provide the mechanism of localization in the crystalline state.

The optical properties [5.21] of mixed-oxide WO_3/MoO_3 EC films lend support to the above model. Briefly, the optical absorption in these mixed-oxide films occurs at a higher energy than in either WO_3 or MoO_3 films. There is a systematic variation of absorption peak as the concentration of one constituent is varied. The density of coloration also affects the absorption-peak position. These systematics can be explained if it is assumed that the trapped electrons are localized and that Mo^{6+} is an electron trap which lies 0.7 eV below W^{6+}. When a mixed-oxide film is crystallized, its absorption peak shifts by the same amount as in a pure WO_3 films. This supports the idea that the electrons remain localized in the crystalline state but that the disorder term disappears.

5.3 Material Preparation and Characterization

5.3.1 Preparation

Tungsten trioxide used for electrochromic devices is usually in the form of amorphous films. They range in thickness from 0.3 to 1.0 μm. The films are most commonly thermally evaporated. However, other methods are also used, such as electron-beam evaporation, reactive and non-reactive sputtering, and spray-on technique.

In thermal evaporation, various boats may be used to hold the WO_3 powder, e.g., tantalum, platinum, or Al_2O_3 coated molybdenum. They all give equivalent results. The WO_3 is usually evaporated onto glass or quartz substrates, heated to 80–120 °C. Substrate temperature higher than 270 °C produces blue rather than clear films [5.33]. Boat temperatures in the range 1000–1300 °C yield transparent films whereas higher temperatures produce blue films [5.33]. Usually thermal evaporation is carried out in a diffusion-pump vacuum system with a base pressure of 10^{-6} Torr prior to evaporation and a pressure of 10^{-5} Torr during evaporation. Variation of the partial pressure of O_2 by means of a controlled leak during evaporation has no measurable effect on the WO_3 films. Evaporation has also been carried out in better vacuums [5.33]. We found, using a vac-ion system with an order of magnitude lower base and evaporation pressures that the films had different properties from those prepared in a diffusion-pump system. Nevertheless, reproducible films could be made in both systems. Films made under higher vacuum were poorer electrochromically (slower to color and bleach) and also contained less water, as indicated by a decrease in the infrared water-absorption bands.

The presence of water in evaporated WO_3 films appears to be the principal variable affecting their performance [5.9, 34]. Variability in water content can lead to nonreproducible films even within one laboratory. The methods for measuring the amount of water in the WO_3 films will be discussed in the next section. However, we have monitored the amount of water vapor present in the vacuum system during WO_3 evaporation. We find that the partial pressure of the water vapor is the primary contributor to the total pressure. For evaporation carried out under high pressure (10^{-5} Torr during evaporation) the water content in the evaporated film reaches a maximum value of about 50%. That is, the film can be represented by the chemical formula $(H_2O)_{0.5}WO_3$. For evaporation carried out under better vacuum, the amount of water present in the film is less, although some water is always found in evaporated films.

Sputtered WO_3 films have been used by a number of workers [5.29, 33, 35]. Reactive sputtering from a tungsten target and sputtering from a WO_3 target with or without a small amount of O_2 in the sputtered films do not have any water incorporated in them [5.33]. Furthermore, they have poor electrochromic properties. We have found that they only color in a thin layer (<1000 Å) near the surface. High sputtering rates can lead to microcrystalline rather than amorphous films, presumably due to substrate heating [5.35]. Annealing in air between 350 and 400 °C leads to crystallized films [5.6]. In fact, we find that twenty minutes at 400 °C is sufficient to produce large crystallites (>2000 Å) which give an x-ray diffraction pattern indistinguishable from single-crystal WO_3. At higher temperatures a correspondingly shorter time is required to produce crystallization.

Films 2–5 μm thick can be prepared by spraying an atomized solution of metatungstic acid ($H_6W_{12}O_{39} \cdot 15 \cdot 5\,H_2O$) onto heated quartz substrates [5.9]. This technique permits a systematic variation of the water content of the films. *Hoppmann* and *Salje* have colored single-crystal platelets of WO_3 by proton injection using a silver-paste contact on one face and a proton electrolyte on the other [5.36a].

5.3.2 Film Characterization

In this section, we will discuss the various physical and chemical methods used to characterize WO_3 electrochromic films. Optical and transport measurements which can also be used to infer properties of the films will be discussed in separate sections.

x-ray diffraction spectra are used to show whether the WO_3 film is crystalline or amorphous. Evaporated films deposited on substrates below 100 °C are amorphous. Sputtered films can be either amorphous or polycrystalline depending on sputtering conditions. *Dautremont-Smith* et al. [5.18], find their films sputtered from an oxide target polycrystalline with crystallite size between 200 and 800 Å. *Shiojiri* et al. [5.36b] studied the crystal structure of evaporated films of WO_3. x-ray patterns of grown films showed the typical halo of the amorphous structure. Annealing of the film at 350 °C for two hours produced crystallization.

Stoichiometry of deposited WO_3 films has been determined in a variety of ways and is found to range from WO_2 up to WO_3 [5.9, 33, 37]. Water content of WO_3 films has been measured by a number of workers. Their results are summarized in Table 5.1. The measurement techniques include: differential scanning calorimetry [5.38]; differential thermal analysis [5.9, 35]; differential thermogravimetric analysis [5.35]; and nuclear reaction for detection of hydrogen [5.20]. Another technique not listed in the table, which can detect both hydrogen and oxygen, is SIMS [5.30, 31]. Water content in WO_3 films has also been measured by the frequency-shift technique [5.39]. In this method a WO_3 film is deposited on a quartz substrate which is also the resonator of a crystal evaporation thickness monitor. The amount of water absorbed by the film upon letting air into the vacuum system after evaporation can be

Table 5.1. Some Estimates of H_2O content and stoichiometry of $WO_{3-y} \cdot x(H_2O)$

Measurement method	Film preparation	x	y	Reference
DSA	Evaporation	0.5	–	[5.38]
DSA	Reactive sputtering	0	–	[5.38]
DTA, TGA	Evaporation	0.4→1.0		[5.34]
Nuclear reaction	Evaporation	$0.1 < x < 0.25$	0.3	[5.20, 33]
Nuclear reaction	Sputtered	– Wide Range –		[5.20, 33]
TG, DTA	Spraying (metatungstic acid)	$0.03 < x < 0.23$	–	[5.9]

determined by the change in the quartz-crystal frequency due to the added weight of water. However, this method will not detect the amount of water incorporated during evaporation which can be considerable. The infrared absorption bands due to water have been used by some workers to show the presence of water [5.40]. However, it is hard to obtain a quantitative result since the width of these bands depends on the type of bonding between the water and WO_3. Two different kinds of infrared bands have been reported [5.40]; one associated with water incorporated during evaporation, the other with water introduced from the atmosphere after deposition. The latter water appears to be more tightly bound.

5.4 Device Technology

5.4.1 Introductary Remarks

As already noted, WO_3 EC display devices use either a liquid or solid electrolyte. In either case, the basic device geometry is the same and is shown for a reflective display in Fig. 5.1. Light enters from the transparent substrate (glass) side, passes through the WO_3, and is reflected by either the electrolyte layer or the counter-electrode. Transmissive cells have been constructed by making both the electrolyte and the counterelectrode transparent. Alternatively, the active counterelectrode may be confined to the edge of the display and a transparent back substrate (i.e., glass) used to complete the cell.

Many variations are found in actual devices, and we now discuss each element of the device in detail. The first two elements are most commonly ITO coated glass, since this is cheap and readily available. In some cases, a thin ($\sim$50–100 Å) semitransparent evaporated gold layer is used for the conductive layer instead of ITO. If a segmented display is desired, such as for a numeric display, the ITO is etched photolithographically [5.41]. The ITO layer should have both high transparency and high conductivity. The latter is required since an EC display is essentially a current device and the resistance of the leads to the individual display segments can limit the performance of the device. Sputtered ITO layers with good transmission in the visible (90% for 20 $\Omega/\square$ resistance) are available commercially, representing state-of-the-art in transparent conductive coatings.

In liquid-electrolyte cells, the edges of the ITO layer may come into direct contact with a liquid-acid electrolyte. In this case, the layer must be able to withstand the acid environment and this may influence the choice of conductive layer. Another solution is to protect the edges with an evaporated layer of a transparent insulator which is not attacked by acid. This requires an extra evaporation through a mask in register with the ITO segments.

The WO_3 layer is normally deposited on top of the transparent conductive coating. For a display device, the WO_3 may be evaporated through a mask in the form of numeric segments. The WO_3-layer thickness is in the range 500–5000 Å, with perhaps 2000–4000 Å being typical. Thicker films can be colored to a higher ultimate contrast but take a longer time to bleach. Since practical EC display devices are quite acceptable operating at a reflective contrast ratio of 5:1, a thickness of a few thousand Angstroms combines adequate coloration with fairly rapid bleaching.

5.4.2 Liquid–Solid Devices

The earliest liquid electrolyte consists of a mixture of water, H_2SO_4, glycerol, and a white pigment powder [5.13]. Apart from lifetime problems, which will be discussed later, this electrolyte works very well and is the standard with which other electrolytes, including solids, must be compared. The electrolyte takes the form of a paste whose consistency is adjustable by varying the relative proportions of the constituents. The H_2SO_4 provides a large reservoir of protons for injection into the WO_3 film. It additionally assures that the electrolyte has a high enough conductivity so that it does not limit current flow through the device. The glycerol provides a low volatility solvent. The white pigment renders the electrolyte opaque and provides a white diffuse background for optimal viewing of the colored WO_3. TiO_2 and other oxides such as ZnO_2 can be used for the pigment.

Liquid electrolytes other than aqueous H_2SO_4 and glycerol have been used [5.34]. Because water in the electrolyte is deleterious (lifetime limiting) [5.34, 42] to the WO_3 layer, attempts have been made to completely eliminate water. Investigators have used various aprotic electrolytes with Li^+ or Na^+ as the conducting ion [5.16, 34, 43]. This introduces several new problems. If Li^+ or Na^+ are to be injected into WO_3, the coloring and bleaching dynamics may be different. The evidence to date is that Li^+ can enter WO_3 quite readily although there may be some problems of Li-metal precipitation at high Li^+ concentrations [5.16]. Also, bleaching is slower than with protons because of the lower Li^+ mobility [5.16]. A further problem is that the ionic conductivity of Li^+ (or Na^+) in aprotic electrolytes is considerably lower than protons in water and the speed of coloring and bleaching is considerably reduced [5.43]. Since the switching speed even for the proton system is typically 1/2 s, this raises a serious problem for a practical device. Nevertheless, much work has been done on the Li^+–aprotic-solvent systems and this is a serious contender for a practical commercial device. A typical combination is $LiClO_4$ electrolyte in solvents such as ethylene glycol, acetonitrile, acetone, propylene carbonate, or dimethylformamide. Care must be taken to remove all water from the nonaqueous solvents. Not much work has been reported on using Na^+ with liquid electrolytes, although Na^+ injection has been studied in connection with solid electrolytes [5.18].

Ideally the counterelectrode should be nonpolarizing. Since positive ions are leaving the electrolyte at the WO_3–electrolyte interface, ions must pass freely across the electrolyte–counterelectrode interface. Most metal electrodes which have this property for electrons are unsuitable, that is, they are polarizing for ionic current flow. With such electrodes, some current will flow initially but it will decay as the electrolyte near the metal becomes depleted of positive ions. Of course, if a high enough voltage is applied across this electrode a new electrode reaction produces current flow, usually accompanied by the evolution of gas. This is not a satisfactory solution since high voltages can also appear across other parts of the device and the evolution of gas may seriously degrade the device. An ideal nonpolarizable counterelectrode would be H_xWO_3 for the proton system and the corresponding alkali-metal bronze for Li^+ and Na^+ ions. However, when a WO_3 film is colored and used as a counterelectrode, it gradually loses its color, leading to a short device lifetime.

A common electrode consists of graphite painted on stainless steel. The graphite forms a porous structure with a large surface area. The operation of this electrode is not well understood. However, we have found that the electrode is much better after some use than when it is fresh. A fresh electrode is completely polarizable in acid electrolyte. In any case, a voltage of 0.2–0.4 V must be maintained across this electrode during the coloring cycle. This necessitates the use of higher voltages with this electrode than with a tungsten bronze counterelectrode.

5.4.3 All–Solid State Devices

In the all-solid-state EC device, the liquid electrolyte is replaced by a solid. These devices can be divided into two distinct classes: those that depend on water and those that do not. The first class may be dubbed the "damp" electrolyte as opposed to the "wet" electrolytes discussed earlier. Early devices belong to this category. One of the first consisted of a SiO_x layer evaporated over the WO_3 layer, and a gold electrode evaporated over the SiO_x [5.7]. The role played by water in these sandwich devices was not immediately obvious.

Nevertheless, it was realized that water played a role in the electric-field coloration of WO_3 in the planar geometry with metal electrodes [5.8]. Specifically, no coloration occurred in a vacuum of 10^{-5} Torr. In this geometry, the WO_3 is in direct contact with the ambient and potentials on the order of a kilovolt are used to produce coloration. Under these circumstances, water on the surface of WO_3 can be dissociated in the high fields.

The dominant role of water in the solid-state sandwich cell structure ITO/WO_3/SiO_x/Au was studied by *Hughes* and *Lloyd* [5.44]. Recent work has clarified the situation [5.39, 45]. A promising new approach using insulating evaporated layers of ZrO_2 or TaO_5 has obtained switching speeds of a few seconds [5.39]. In this structure weight changes during coloring and bleaching have been measured indicating absorption and desorption of H_2O between the cells and their surroundings. Mass-spectrometer measurements showed that O_2 left the cell during coloration and H_2 left during bleaching. Also, coloration does not occur until a threshold voltage is reached across the insulating layer. The threshold voltage corresponds to that of the electrolysis of water. The insulating layer contains water that can be dissociated by the electric field, thereby providing the protons necessary for coloration of the WO_3 film. This explains why higher coloring voltages are always found for the solid-state cells compared with liquid-electrolyte cells.

A similar dependence on humidity was found for a solid-state cell using LiF for the solid electrolyte [5.45]. In addition, switching speeds as fast as liquid cells (0.3 s) could be obtained. Proton transport through the insulator was believed to be the coloration mechanism. A new solid-state device has been reported which depends on trapped water [5.46]. It is unaffected by ambient conditions including a vacuum of 10^{-6} Torr, because the water is tightly bound. This device which has the structure In_2O_3/a-WO_3/Cr_2O_3/Au has several unusual features. The Cr_2O_3 layer also undergoes a color change along with the WO_3. Under open-circuit conditions the color bleaches rapidly due to bimolecular recombination at the WO_3–CrO_3 interface. This is a disadvantage for EC-display applications. On the other hand, the device has a good switching lifetime.

Solid-state devices which do not depend on water for their operation have also been constructed. These cells all utilize the transport of Na^+ ions through a superionic conductor. In one case [5.18], a thin ($\sim$1 mm) ceramic plate of β–Al_2O_3 is the solid electrolyte. The WO_3 is then evaporated on the β–Al_2O_3 and a transparent conductive coating is sputtered on top of the WO_3. Na_xWO_3 is used as a counterelectrode. Coloring and bleaching are slow, requiring high voltages compared with liquid electrolyte cells. The WO_3 films are polycrystalline and it is believed that coloring and bleaching are aided by rapid diffusion of Na^+ ions down grain boundaries.

Other solid-state devices (5.37, 47] make use of the superionic-conductor electrolytes such as $Na_3Zr_2Si_2PO_{12}$, $Na_{1+x}Zr_2Si_xP_{3-x}O_{12}$, and $Na_5YSi_4O_{12}$ with deposited films of Na_xWO_3 as the counterelectrode. Good contrast and response times less than 0.5 s have been observed for voltages of 3–5 V.

All of these true solid-state devices are comparatively new and very little is known about device lifetime and the ultimate limitations on switching speeds. In general they use higher voltages and are slower than liquid-cell devices. Their significance must await further development.

5.5 Device Performance

5.5.1 General Considerations

Device structures have been discussed in some detail in the previous section. The physical principles involved in the electrocoloration of WO_3 films have also been explained insofar as they are understood. In this section we will discuss device performance and interpret it in terms of the above principles.

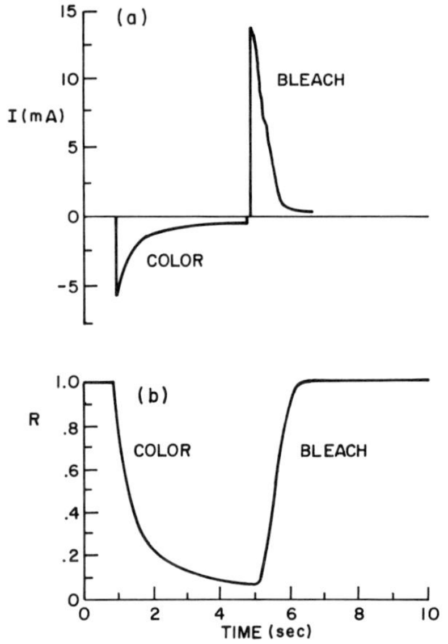

Fig. 5.5a, b. Current and optical reflectance during the color bleach cycle (constant applied voltage pulses) of a WO_3 EC device

EC display devices can be operated either at constant voltage or constant current or some combination of the two. Constant voltage is the simplest from the device point of view as well as conceptually. Therefore, we will treat that case first. In this section we consider only the liquid-cell structure. All the phenomena discussed here should apply equally well to solid-state devices. However, there may be additional complications caused by the more complex processes arising in the solid electrolyte.

The current vs time curve for coloring and bleaching of a typical liquid electrolyte cell is shown in Fig. 5.5. According to the theory presented in Sect. 5.2.2, the coloring current decays as $t^{-1/2}$ until either the coloring voltage is removed or the film becomes completely colored. During the period following the coloring pulse the color is stored in the WO_3 film. Upon applying a bleaching voltage, the current decays as $t^{-3/4}$ until all the color leaves the film at which point the current drops rapidly to zero. The charge removed during bleaching will normally equal the charge injected during coloration. Factors such as WO_3 film properties, electrolyte composition and conductivity, nature of the counterelectrode, and magnitude of the applied voltage all can affect the nature of the coloring and bleaching. Therefore, the ideal behavior is not always observed. Nevertheless the qualitative shape of the curves is always the same and can be described empirically as $I \propto t^{-n}$ where n varies between 0.5 and 1.0.

The optical absorption in the film is related to the injected charge by

$$I/I_0 = \mathrm{e}^{-kl} = \mathrm{e}^{-\gamma \Delta Q} \tag{5.11}$$

$$\mathrm{OD} = \log I_0/I = \gamma \Delta Q \tag{5.12}$$

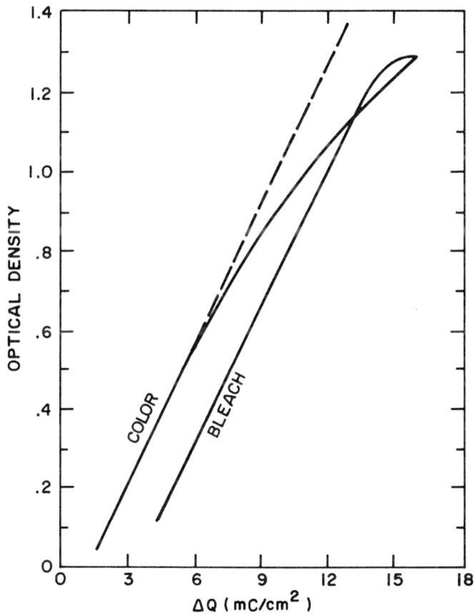

Fig. 5.6. Coloration density at $\lambda = 0.633$ μm vs injected charge for a WO_3 EC device

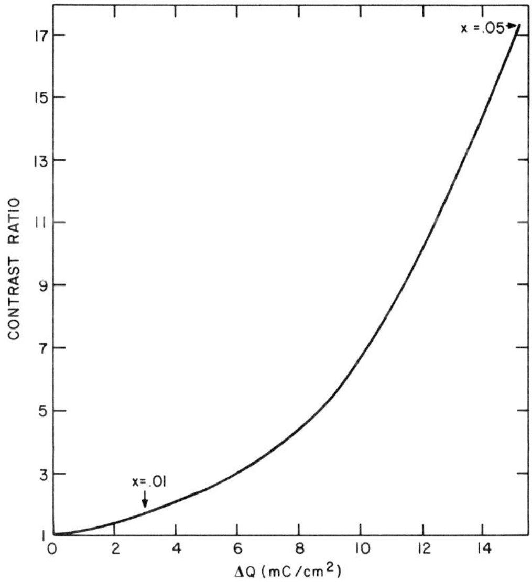

Fig. 5.7. Contrast ratio at $\lambda = 0.55$ μm vs injected charge for a WO_3 EC device

where I_0 is the intensity of light entering film, I the intensity of light transmitted through the film, k the absorption coefficient, γ the constant, and ΔQ the injected charge.

If the film coloration is measured in reflection with a white diffuse background, the contrast ratio (CR) is defined as

$$\mathrm{CR} = R_0/R_x \tag{5.13}$$

where R_0 is the intensity of light reflected in uncolored state, and R_x the intensity of light reflected in colored state.

The contrast ratio can be written in terms of the optical-absorption coefficient of the film,

$$\mathrm{CR} = \mathrm{e}^{2kl}. \tag{5.14}$$

The 2 occurs in the exponent to account for two light passes through WO_3 in a reflective display. If the light is incident from a range of angles the measured CR will also depend on viewing angle and will be higher since, on the average, light travels a distance greater than twice the film thickness.

Figure 5.6 shows a plot of OD vs ΔQ for a WO_3EC device. The curve is linear for small ΔQ but eventually shows some saturation at high coloration levels. This could have several causes: 1) At high coloration not every electron injected leads to a color center. For example, electrons might pass right through the film and combine with protons to form H atoms and then H_2 gas. 2) At high coloration density the addition of each electron produces a smaller increment of optical density either through a decrease in the oscillator strength per electron, or through an overall broadening of the optical-absorption curve. All these mechanisms are expected but the relative contribution of each has not been studied in detail. From the slope of the straight line in Fig. 5.6 the oscillator strength can be determined. The same data is presented in a different form in Fig. 5.7 where the visible contrast ratio for a reflective device is plotted vs the charge injected into the WO_3 film.

5.5.2 Switching Speed

The time to color and bleach can be determined from the theoretical considerations discussed in Sect. 5.2.2. The coloring time (t_c) is defined as the time required to inject a given charge Q_c. This charge can be found by integrating (5.4). Inverting the resulting expression gives:

$$t_c = Q_c^2 \exp(-V_a/2RT)(4J_0\tau_0)^{-1}. \tag{5.15}$$

For example, according to Fig. 5.7, a CR of 5:1 requires a Q_c of about $0.0085\,\mathrm{C/cm^2}$. Assuming $V_a \approx 0.7\,\mathrm{V}$, $J_0 \approx 1\times10^{-1}\,\mathrm{A/cm^2}$, $l \approx 0.5\,\mu\mathrm{m}$, and $\tau_0 \approx 2\times10^4\,\mathrm{s}$, we find that $t_c \approx 0.25$. Using the same parameters in (5.8), with $\mu_p = 1\times10^{-6}\,\mathrm{cm^2/Vs}$, we find the bleach time is $t_B = 40\,\mathrm{s}$. This is considerably longer than the coloring time. Nevertheless, because of the strong dependence of t_B and l, this time can be reduced considerably. However, there is a limit to how much the bleaching time can be reduced if a high contrast ratio is desired. The reason is that x, l, and CR cannot be chosen independently. These parameters are related by

$$\ln(\mathrm{CR}) = 2\beta x l \tag{5.16}$$

where $\beta = k/x \approx 3\times10^5$. This expression can be used to eliminate l in (5.8) for the bleaching time to give

$$t_B = e\varrho(\mu_p V_a^2 K\varepsilon_0 64\beta^4)^{-1}[\ln(\mathrm{CR})]^4 x^{-3}. \tag{5.17}$$

This expression shows that to attain the fastest speed for a given contrast ratio the film should color to a high x value. However, for successful long term device operation it may not be advisable to color to arbitrarily high x.

The reason is that above $x \sim 0.28$ a new electrode reaction at the H_xWO_3–electrolyte interface becomes possible [5.22]. This is the dissolution of the H_xWO_3 by spontaneous generation of hydrogen via the reaction

$$H_{0.28} + \delta WO_3 \rightleftharpoons \frac{\delta}{2} H_2 + H_{0.28}WO_3. \tag{5.18}$$

Another difficulty with coloration to high x is that once the film has colored to its maximum value then further electrons flowing into the film will pass through it and combine with protons in the electrolyte to form H^0. This phenomenon is observed experimentally by the failure of the coloring current to increase linearly with injected charge at high coloration densities.

For example, choosing $x=0.28$, $CR=5$, $\mu=1\times10^{-6}\,cm^2/V\,s$, using (5.17), we find a bleach time of 0.35 s. For this case $l\simeq 0.1\,\mu m$. This emphasizes the severe restriction on film thickness for fast high-contrast operation.

5.5.3 Constant–Current Operation

When constant-voltage operation is used the switching time is slow because of the drop in both coloring and bleaching current with time. Since it is the charge injected into the film which determines the contrast ratio, faster switching speed will be obtained if charge is injected at constant current. The problem here is that the total voltage will rise as the current is kept constant with time. Since high voltages can lead to unwanted electrochemical side reactions, care must be taken to limit the maximum voltage used, and also to be sure that the extra voltage does not occur at the electrodes where it can do the most damage. With these precautions in mind, higher switching speeds can be obtained with constant current at the expense of an increase in circuit complexity.

5.5.4 Power Requirements

An electrochromic display requires more power to operate than a liquid-crystal display but less than a light-emitting-diode display. One advantage of electrochromics in this regard is their storage capability. Once a display segment has been colored, it will remain colored for some time so that power is required only to change the colored state of a given segment. This memory effect is not indefinite, however. For proton-based WO_3 ECs using liquid electrolytes, significant color fading will occur in the time span of minutes to hours although Li-based ECs have a considerably longer memory storage [5.16]. This loss of color is ascribed to the presence of O_2 gas in the electrolyte, so that if O_2 is removed from the electrolyte, longer storage times can be obtained.

It is instructive to calculate the power required to operate an EC display. Consider a small ($0.01\,cm^2$) numeric segment to be used for a watch display. Assume that on the average, four segments must change each minute to display the new time (seconds are not displayed). Therefore, the total area $=0.04\,cm^2$; assume 0.8 V for coloring and bleaching. To achieve a contrast ratio of 4:1 requires approximately $4\,mC/cm^2$. Power: $VI=V\,\Delta Q/60$ where ΔQ is the charge injected every minute to change numbers. Therefore, power $=0.8\times4\times0.04/60=32/15\sim2\,\mu W$ per pulse or $4\,\mu W$ including coloring and bleaching. If a voltage of 1.5 V were used the power would be $8\,\mu W$. If seconds were displayed it would become $\sim 500\,\mu W$. Eight microwatts can be supplied by a standard watch battery although $500\,\mu W$ cannot.

5.5.5 Device Lifetime and Degradation Mechanism

The principal obstacle to commercialization of EC displays appears to be device lifetime. When this problem is solved there will undoubtedly be a market for electrochromic watch displays, as well as many large-area displays currently supplied by LCDs and LEDs.

By now it is well recognized that water is detrimental to WO_3 electrochromic films. The problem is that most WO_3 films dissolve in water and to a lesser extent in acidic environments. The extent to which this dissolution is dependent on film preparation and properties has not been fully investigated. Single-crystal WO_3 is virtually insoluble in water. This and other experimental findings have led *Arnoldussen* to postulate that the normal EC WO_3 is more akin to a molecular solid of metatungstate ions ($H_2W_{12}O_{39}$) than to a true amorphous solid [5.42]. In any case, normally evaporated and otherwise untreated WO_3 films will dissolve in H_2O at the rate of approximately 25 Å per day [5.34]. This poses a shelf-life problem for these devices. This has led to a search for aprotic electrolytes such as Li, Na, and K salts in organic solvents such as ethylene glycol, acetonitrile, acetone, sulfolane, formic acid, acetic acid, etc. The dissolution rate of WO_3 is found to be less severe in these nonaqueous solvents. However, the kinetics of coloring and bleaching is much slower and most are not suitable for a practical display device for this reason.

Devices based on Li^+ ions in aqueous solvents have reasonable switching speeds even though the mobility of Li^+ in Li_xWO_3 appears to be an order of magnitude slower than H^+ in H_xWO_3 [5.16]. Lithium has some further advantages; Li_xWO_3 is much more stable against color fading, especially when dissolved oxygen is present in the electrolyte. Li_xWO_3 is also thermodynamically more stable than H_xWO_3. Nevertheless, *Mohapatra* and *Wagner* [5.48] have studied $Li_xWO_3/LiClO_4$ in propylene carbonate/Li cells and found the long coloration/bleaching cycle to be a serious drawback.

The dissolution of WO_3 in aqueous solution is actually a strong function of pH. This is shown in Fig. 5.8 where dissolution rate is measured as a function of pH. In this experiment, the change in thickness of a WO_3 film with time is monitored as the change in optical path length (interference fringes) of a laser transmitted though the electrolyte and reflected from the WO_3. The etch rate varies over four orders of magnitude as the pH varies between 0 and 10. Even at $pH \approx 0$ the etch rate is 10 µm/yr or approximately 300 Å per day. This is still too high for a useful device. It is also an order of magnitude higher than the etch rate found by *Randin* [5.34]. He found, under storage conditions in sealed capsules for an electrolyte of glycerin–H_2SO_4 (10:1) at 50 °C, a dissolution rate of 20–25 Å per day. The difference we believe is that our measurements were performed under flowing-electrolyte conditions, giving the true etch rate for a

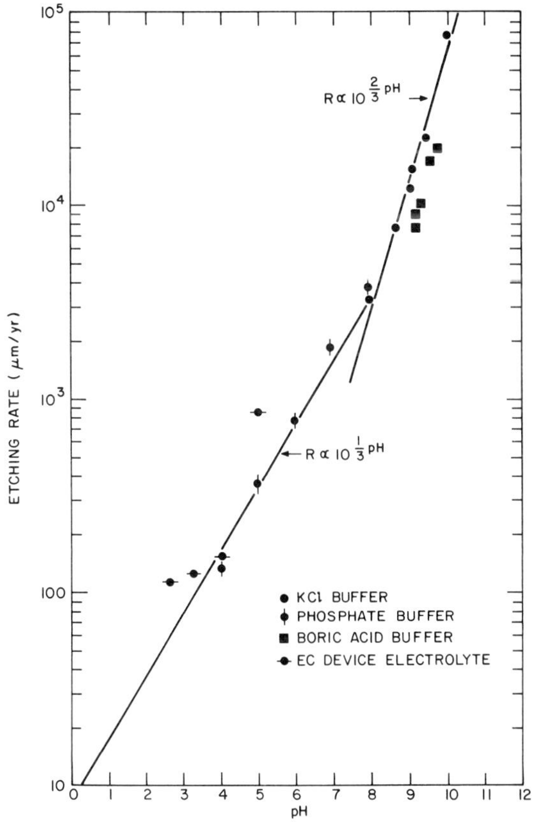

Fig. 5.8. Etch rate (R) vs pH for WO_3 films immersed in different buffered solutions

given pH. Under Randin's static conditions fresh etchant will not have access to the surface as time passes. His experiments cover a period of 60 d, whereas ours covered a period of less than 1 h. Randin's electrolyte was also saturated with WO_3 which may slow the dissolution rate of WO_3. To summarize, the dissolution rate of WO_3 in aqueous electrolyte is a strong function of the pH of the electrolyte, the true chemical conditions of the WO_3–electrolyte interface, and the properties of the WO_3 film itself.

In addition to the shelf-life problem discussed above, there are additional lifetime-limiting mechanisms when the cells are cycled. Typically, the higher the contrast ratio to which a cell is cycled, the fewer the number of cycles before device failure occurs. Experimental results for a number of cells are shown in Fig. 5.9. When one life tests a large number of cells, failure can occur at a point below the straight line of Fig. 5.9, but no cell of a given type will be above this line. Although the mechanism is unknown it seems to represent a fundamental lifetime limitation. Of course, differently prepared films or different (improved) electrolytes may lead to a longer lifetime, but we conjecture that the dependence of lifetime and contrast ratio would persist. From Fig. 5.9 it is seen

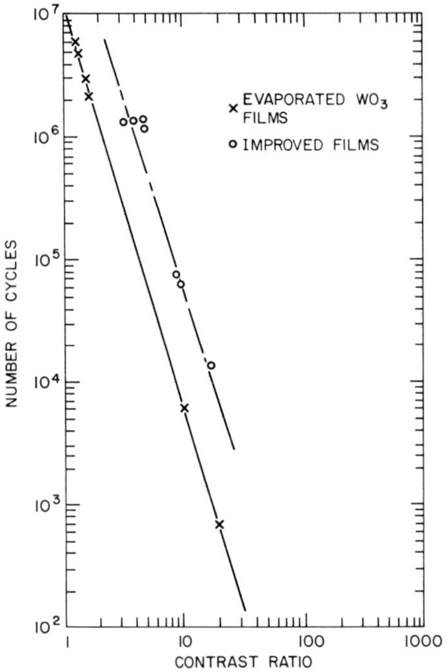

Fig. 5.9. Lifetime (number of color-bleach cycles) vs visible contrast ratio for a series of WO_3 EC cells. Complete color-bleach cycle is 2 s

that if a lifetime of at least 10^6 cycles is desired the contrast ratio must be in the range 2:1 to 4:1.

If a cell which has been cycled for several hundred thousand cycles is taken apart and the WO_3 film examined under an optical microscope, it is found to have a granular texture, and does not adhere to the glass substrate as well as the original "as evaporated" film. The most usual form of degradation under cycling conditions is the gradual reduction of the charge, and hence color, which is injected into the film with each cycle. But other problems may arise, such as incomplete bleaching or the separation of part of the WO_3 film from the glass substrate. Frequently these effects are first observed at the edges of display segments. *Knowles* [5.49] has found a decay of the injected charge of a factor of two after 100 h of cycling for a WO_3/lithium perchlorate–propylene carbonate/Li_xWO_3 cell. This degradation can be partially removed by flooding the WO_3 film with uv light while a bleaching voltage is applied across the cell.

A further degradation can occur if the electrolyte attacks the transparent conductive layer on the glass substrate. In this case the transparent conductive layer must be protected from direct exposure to the acid or a transparent conductive layer must be chosen which is etch resistant. For example, SnO_2 is more etch resistant than Sb doped In_2O_3 although the latter can be made more conducting. H_2 gas can be generated in an EC cell if voltages across the electrodes are too high, or if the WO_3 film is colored beyond $x \simeq 0.28$. This will obviously have a deleterious effect on a sealed cell.

5.6 Conclusions

The properties of WO_3 films and their incorporation into electrochromic display devices have been extensively investigated at a number of laboratories. The basic electrochromic phenomena in WO_3 have turned out to be exceedingly complex. Their proper understanding requires a knowledge of solid-state physics, including amorphous materials, chemistry, and electrochemistry. There are still gaps in our understanding of electrochromism in WO_3, especially in the material characterization of the films, and the transport mechanism of positive ions in WO_3. Nevertheless, an overall understanding of the basic effect has been obtained and this knowledge is very useful for the proper design of an EC display device.

The chief limitation of EC display devices utilizing WO_3 films is the lifetime, both on the shelf and under cycling conditions. Some progress has been made in understanding these problems and cells with cycling lifetimes in the range of 10^6–10^7 have been constructed.

The work on WO_3 electrochromics has stimulated the development of other electrochromic display systems. For example, an inorganic system based on IrO_2 has been developed [5.50]. This system colors anodically rather than cathodically as with WO_3. An interesting feature of this system is that the film can be grown in the supporting electrolyte. If, after many cycles, the thickness of the film is reduced, it can be regrown in situ. This may circumvent the film-dissolution problem which plagues the WO_3 films.

Another recent all-solid-state cell has the structure SnO_2/Phosphotungstic acid/Au or Cu, where the chemical formula for phosphotungstic acid is $H_3PO_4 \cdot (WO_3)_{12} \cdot nH_2O$ [5.51]. This system has switching speeds and optical properties similar to WO_3-based devices. However, the authors hope it will not have the long-term stability problems of the electrolyte-based WO_3 cell.

A number of organic systems have been discovered. The organic display based on the viologen dyes [5.52] was developed quite early. More recently a number of new organic systems are under study. These include lutetium diphthalocyanine compounds [5.53] and 5-OH-triaryl-pyrazoline and tetrathiaflavalene compounds [5.54]. The organic systems have the advantage of considerable flexibility in the synthesis of particular molecules to optimize certain electrochromic properties such as optical-absorption wavelength and intensity. However, they tend to work by switching from one color to another color which may or may not be an advantage. This depends on the application. These materials (with the exception of the viologen system) are too new to have much data on lifetime limitations.

The all-solid-state cells based on WO_3 are also too new to have firm data on lifetime. Most suffer from slow response for coloring and bleaching and these problems must be solved first before lifetime data is accumulated.

To summarize, no electrochromic display device has yet achieved widespread commercial application. The WO_3/aqueous-electrolyte systems are the best understood in terms of basic mechanisms and lifetime-degradation effects. Problems with the above systems have motivated researchers to consider first aprotic electrolytes to circumvent the film-dissolution problem and then the all-solid-state WO_3/electrolyte system. Which of these systems, if any, will win out cannot be predicted at this time. In addition, new electrochromic display systems have been developed which hold considerable promise.

Appendix A

A theory for the time dependence of the coloring current has been developed under the assumption that diffusion in the WO_3 film can be ignored [5.25]. In what follows we will extend this theory to cover the situation where diffusion in the film affects coloring.

The coloring speed is influenced by proton transport across the electrolyte–WO_3 interface because of a potential barrier there. The current across a symmetric barrier is expressed by [Ref. 5.25, Eq. (4)], which is

$$J_B = \frac{J_e x_m}{A^{1/2} x_s}\left[1 - A\left(\frac{x_s}{x_m}\right)^2\right] \tag{A.1}$$

where J_e is the exchange current density, x_m the maximum x value (x_m depends on the applied voltage) and x_s the surface proton concentration at the WO_3–electrolyte interface. The function A is given by

$$A = \frac{(1-x_m)^2}{(1-x_s)^2} \exp\left[\frac{-0.53}{kT} e(x_m - x_s)\right]. \tag{A.2}$$

In steady state the current across the barrier is equal to the current in the film J_F. The form of J_F depends on the time scale of measurements relative to the diffusion time τ_D.

For times long compared with τ_D the protons are uniformly distributed in the film. Thus

$$J_F = e\varrho l \frac{dx}{dt} \tag{A.3}$$

which when equated to (A.2) gives the result quoted in [5.25].

The above derivation, however, is restricted to low voltages such that $x_m < 0.5$. In device operation high voltage is necessary to increase the coloring speed. The general expression for the barrier current for high applied voltage using Eqs. (1, 2) of Ref. [5.25] is

$$J_B = J_e \exp\left(\frac{V_a}{2RT}\right) \exp\frac{+\Delta\mu(x)}{2RT} \tag{A.4}$$

where $\Delta\mu(x)$ is the change in chemical potential upon coloration. It is [5.25]

$$\Delta\mu(x)=2RT\ln\left[\left(\frac{1-x_0}{1-x}\right)\frac{x}{x_0}\right]+0.53(x-x_0) \tag{A.5}$$

where x_0 is the proton density in the film before application of the applied voltage V_a. Substituting $\Delta\mu(x)$ into (A.4) gives

$$J_B=J_0\exp\left(\frac{V_a}{2RT}\right)\exp(-x/x_1)\left(\frac{1-x}{x}\right) \tag{A.6}$$

where

$$J_0=J_e e\frac{0.53x_0}{RT}\left(\frac{x_0}{1-x_0}\right) \tag{A.7}$$

redefines the exchange current in terms of x_0. $x_1=2RT/0.53$. Equating (A.3) into (A.6) gives a differential equation for x. Its solution for small x is

$$x=e\exp(V_a/4RT)\sqrt{t/\tau_0} \tag{A.8}$$

where $\tau_0=e\varrho l/2J_0$. The coloring current is then found by substituting (A.8) into (A.6) to give

$$J_c=J_0\exp(V_a/4RT)\sqrt{\tau_0/t}\,. \tag{A.9}$$

If we are interested in the time domain $t<\tau_D$ where diffusion in the film is important then the current in the film is given by the solution of the one-dimensional diffusion equation [Ref. 5.26, Chap. 4]. Thus

$$J_F=e\varrho x_s\sqrt{D_p/\pi t}\,. \tag{A.10}$$

The unknown surface concentration (x_s) in this expression can be found by equating (A.10) to (A.6). The resulting trancendental equation can be readily solved for small x_s. The x appearing in (A.6) is to be replaced by x_s. Thus

$$x_s^2=J_0\exp(V_a/2RT)(\pi t/D_p)^{1/2}/e\varrho\,. \tag{A.11}$$

This expression shows that x_s increases slowly with time. This is reflected in the current which is found by substituting (A.11) into (A.10). The result is

$$J_c=\sqrt{J_0 e\varrho}\exp(V_a/4RT)(D_p/\pi t)^{1/4}\,. \tag{A.12}$$

As the surface proton concentration increases toward the maximum value (A.11) is no longer valid. In this regime, x_s is replaced by a constant in (A.9) to give $J_c\propto t^{-1/2}$.

References

5.1 J.R.Platt: J. Chem. Phys. **34**, 862 (1961)
5.2 W.Franz: Z. Naturforsch. **13**A, 44 (1958)
L.V.Keldysh: Zh. Eksp. Teor. Fiz. **34**, 1138 (1958)
5.3 J.H.Shulman, W.D.Compton: *Color Centers in Solids* (Pergamon Press, New York 1963) p. 38
5.4 J.Blanc, D.L.Staebler: Phys. Rev. B**4**, 3548 (1971)
5.5 S.Geller (ed.): *Solid Electrolytes*, Topics Appl. Phys. 21 (Springer, Berlin, Heidelberg, New York 1977)
M.B.Salamon (ed.): *Physics of Superionic Conductors*, Topics Current Phys. 15 (Springer, Berlin, Heidelberg, New York 1975)
5.6 E.O.Brimm, J.C.Brantley, J.H.Lorenz, M.H.Jellinck: J. Am. Chem. Soc. **73**, 5427 (1951)
5.7 S.K.Deb: Appl. Opt. Suppl. **3**, 193 (1969)
5.8 S.K.Deb: Philos. Mag. **27**, 801 (1973)
5.9 R.Hurditch: Electron. Lett. **11**, 142 (1975)
5.10 P.G.Dickens, D.J.Murphy, T.K.Halstead: J. Solid State Chem. **6**, 370 (1973)
5.11 K.Nishimura: Solid State Comm. **20**, 523 (1976)
5.12 B.W.Faughnan, R.S.Crandall, P.M.Heyman: RCA Rev. **36**, 177 (1975)
5.13 J.H.McGee, W.E.Kramer, H.N.Hersh: SID. Tech. Digest **50** (1975)
R.D.Giglia: SID. Tech. Digest, 52 (1975)
5.14 I.F.Chang: IEEE Trans. ED-**22**, 749 (1975)
I.F.Chang, B.L.Gilbert, T.I.Sun: J. Electrochem. Soc. **122**, 955 (1975)
5.15 H.N.Hersh, W.E.Kramer, J.H.McGee: Appl. Phys. Lett. **27**, 646 (1975)
5.16 S.K.Mohapatra: J. Electrochem. Soc. **125**, 284 (1978)
5.17 M.Green, D.Richman: Thin Solid Films **24**, 545 (1975)
5.18 W.C.Dautremont-Smith, M.Green, K.S.Kang: Electrochem. Acta **22**, 751 (1977)
5.19 O.F.Shirmer, V.W.Hwer, G.Baur, G.Brandt: J. Electrochem. Soc.: Solid-State Sci. Tech. **124**, 749 (1977)
5.20 P.Gerard, A.Deneuville, G.Hollinger, T.M.Duc: J. Appl. Phys. **48**, 4252 (1977)
5.21 B.W.Faughnan, R.S.Crandall: Appl. Phys. Lett. **31**, 834 (1977)
5.22 R.S.Crandall, P.J.Wojtowicz, B.W.Faughnan: Solid State Commun. **18**, 1409 (1976)
5.23 B.W.Faughnan, R.S.Crandall, M.H.Lampert: Appl. Phys. Lett. **27**, 275 (1975)
5.24 B.W.Brown, E.Banks: J. Am. Chem. Soc. **76**, 963 (1954)
5.25 R.S.Crandall, B.W.Faughnan: Appl. Phys. Lett. **28**, 95 (1976)
5.26 J.O.M.Bockris, A.K.N.Reddy: *Modern Electrochemistry* (Plenum Press, New York 1973)
5.27 M.Stanley Whittingham, Robert A.Huggins: "Fast Ion Transport in Solids", ed. by W.Van Gool, Proc. of the NATO Sponsored Adv. Study Inst. on Fast Ion Transport in Solids, Belgrate, Italy, September 5–15, 1972 (North-Holland/American Elsevier 1973) p. 645
5.28 M.Green, W.C.Smith, J.A.Weiner: Thin Solid Films **38**, 89 (1976)
5.29 P.G.Dickens, J.H.Moore, D.J.Neild: J. Solid State Chem. **7**, 241 (1973)
5.30 A.Benninghoven: Appl. Phys. **1**, 3 (1973)
5.31 In Ref. 5.26
5.32 D.L.Dexter: In *Solid State Physics*, Vol. 6, ed. by F.Seitz, D.T.Turnbull (Academic Press, New York 1958) p. 353
5.33 A.Deneuville, P.Gerard: J. Electron. Mat. **7**, 559 (1978)
5.34 J.P.Randin: J. Electron. Mat. **7**, 47 (1978)
5.35 E.K.Sichel, J.I.Gittleman, J.Zelez: Appl. Phys. Lett. **31**, 109 (1977)
5.36a G.Hoppmann, E.Salje: Phys. Status Solidi (a) **37**, K187 (1976)
5.36b M.Shiojiri, T.Miuano, C.Kaito: Jpn. J. Appl. Phys. **17**, 567 (1978)
5.37 G.G.Barna: J. Electron. Mat. **8**, 153 (1979)
5.38 H.R.Zeller, H.U.Beyeler: Appl. Phys. **13**, 231 (1977)
5.39 Y.Hajimoto, M.Matsushima, S.Ogura: Program of the 20th Electronic Materials Conf., Santa Barbara, CA, June 28–30, 1978, p. 12

5.40 V.Wittwer, P.Schlotter: Program of the 20th Electronic Materials Conf., Santa Barbara, CA, June 28–30, 1978, p. 13
5.41 E.Spiller, R.Feder: In *x-Ray Optics*, ed. by H.J.Queisser, Topics Appl. Phys. **22** (Springer, Berlin, Heidelberg, New York 1977)
5.42 T.C.Arnoldussen: Program of the 20th Electronic Materials Conf., Santa Barbara, CA (1978) p. 19
5.43 T.B.Reddy, A.Battistelli: Program of the 19th Electronic Materials Conf., Cornell University, Ithaca, NY (1977) p. 35
5.44 A.J.Hughes, P.Lloyd: Program of the 19th Electronic Materials Conf., Cornell University, Ithaca, NY (1977) p. 44
5.45 H.J.Stocker, S.Singh, L.G.Van Uitert, G.Zydzik: Program of the 20th Electronic Materials Conf., Santa Barbara, CA (1978) p. 12
5.46 I.Shimizu, M.Shizukuishi, E.Inoue: Jpn. J. Appl. Phys. (to be published)
5.47 M.Bayard: Program of the 20th Electronic Materials Conf., Santa Barbara, CA (1978) p. 12
5.48 S.K.Mohapatra, S.Wagner: Program of the 20th Electronic Materials Conf., Santa Barbara, CA (1978) p. 13
5.49 T.J.Knowles: Appl. Phys. Lett. **31**, 817 (1977)
5.50 S.Gottesfeld, S.D.McIntyre, G.Beni, J.L.Shay: Appl. Phys. Lett. **33**, 208 (1978)
5.51 B.Tell, S.Wagner: Appl. Phys. Lett. **33**, 837 (1978)
5.52 C.J.Schoot, J.J.Ponjee, H.T.van Dam, R.A.van Doorn, P.T.Bolwijn: Appl. Phys. Lett. **23**, 64 (1973)
5.53 M.M.Nicholson, R.V.Galiardi: Tech. Report, U.S. NTIS AD Rep. Govt. Rep. Announcement Index (U.S.) 1977, 77 (15)
5.54 F.B.Kaufman, E.M.Engler: Program of the 20th Electronic Materials Conf., Santa Barbara, CA (1978) p. 21

6. Electrophoretic Displays

A. L. Dalisa

With 16 Figures

The choice of a display technology best suited for a particular application depends upon many parameters. Some of these parameters are determined by the overall system requirements such as operating voltage, power consumption, size, cost, etc. Others are determined by the environment in which the system will be used, such as ambient light levels, operating temperature range, etc. Finally, and perhaps the most important requirements of the display are those related to the prospective viewer, such as brightness, contrast, angular viewing range, colors, as well as other subjective qualities which tend to minimize the visual accommodation required by the user. In most cases, the major interaction and sometimes the only interaction of the user with a system is via the display; hence the visual impact and subjective appearance of the display must be recognized as a critical part of a successful product.

A display technology with a distinctive and pleasant appearance as well as many desirable system and environmental characteristics is under development and is known as the *e*lectro*p*horetic *i*mage *d*isplay (EPID). This chapter will describe the present state of the art in this technology, and indicate trends in its future development.

Fig. 6.1. Several prototype EPID devices including a numeric display, a miniature status indicator, a bar graph and an annunciator panel

EPID is a nonemissive display concept based upon the transport of charged pigment particles in a colloidal suspension. The charged pigment particles are transported by means of an applied electric field; this phenomenon is commonly known as electrophoresis. The first use of electrophoresis for recording or displaying information was the establishment of liquid development processes for electrophotographic applications [6.1, 2].

The use of an electrophoretic process as a reversible display technique was first reported by *Evans* [6.3]. In 1973, *Ota* et al. [6.4] described the implementation and preliminary characteristics of an electrophoretic display device which has become known as EPID. Based upon this initial work and that of others [6.5–7] the EPID concept has emerged as a promising new flat-panel nonemissive display technique. Figure 6.1 shows several prototype EPID devices, including a numeric display, a miniature four-element status indicator, a bar graph, and an annunciator panel, which have been fabricated and show strong commercial promise. A particular application which would take advantage of EPID's flat-panel format, nonemissive nature, and distinctive appearance is that of a page display which would have a comfortable, easy-to-read appearance similar to ink on paper, and yet be addressed and erased electronically.

6.1 Principles of Operation

6.1.1 General Characteristics

A simple EPID cell consists of a thin layer ($\approx 50\,\mu m$) of a colloidal suspension sandwiched between the transparent electrode surfaces of two glass plates. The colloidal suspension consists of pigment particles dispersed in a dyed non-aqueous suspending liquid of contrasting color. The pigment particles are submicron in size and electrically charged to the same polarity. Figure 6.2 shows a cross section of a simple cell, in which the particles are assumed to be charged negatively. Information can be displayed on this cell, by segmenting one of the transparent electrodes using standard photolithographic techniques. If one electrode segment is then connected to a voltage source of positive polarity, the other segment connected to a voltage source of negative polarity, the pigment will be driven to opposite sides of the cell as shown in the figure. In the region of the cell in which the pigment has been packed on the front electrode, the color of the pigment will be seen by the observer. In the region of the cell in which the pigment is on the rear electrode, only the portion of the ambient room light that is transmitted by the dye and scattered by the pigment is observed.

There are three principal device parameters that determine the contrast of an EPID device [6.7]. The first such parameter is the composition of the suspension. Consider a suspension that consists of clear fluid and dispersed

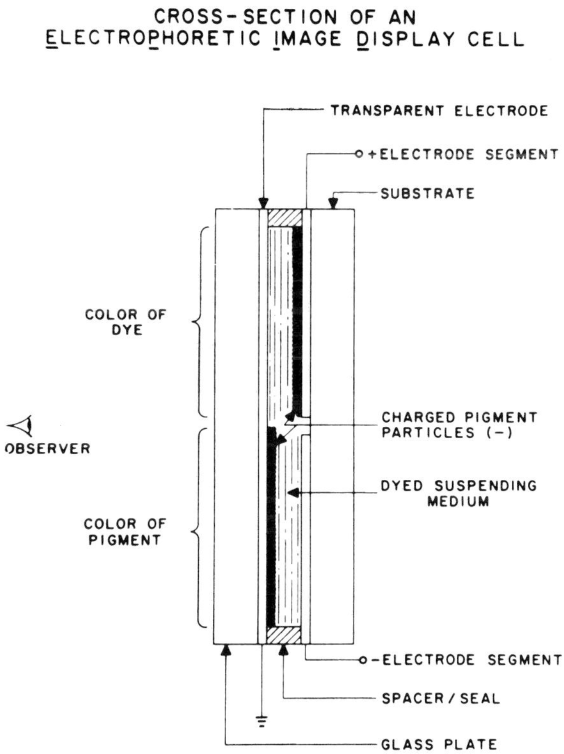

Fig. 6.2.
Cross section of a simplified EPID cell

pigment. Without dye there would, of course, be no contrast. As dye is added to the suspension, pigment on the rear electrode starts being obscured and the contrast increases. At a certain concentration of dye the contrast will peak and then begin to decrease as the dye concentration continues to increase. The decrease starts when the opacity of the dye starts reducing the brightness of the ON state, i.e., pigment on the viewers side of cell, more than it improves the darkness of the OFF state, i.e., pigment on the rear side of the cell. Figure 6.3 shows the dependence of the contrast ratio on the concentrations of dye and pigment for three operating voltages.

The second principal device parameter that determines the contrast is the cell thickness. Figure 6.4 shows that for a given suspension composition and applied electric field, as the cell is increased in thickness, the contrast would first increase and then level off when essentially all the light that enters the cell and scatters off the pigment on the rear electrode is absorbed in the layer of dyed liquid. However, by decreasing the dye concentration and further increasing the cell thickness, higher contrast would be available since the brightness of the pigment on the front electrode would improve.

The third device parameter that determines the contrast is the applied electric field. Consider an EPID cell in the ON state. The brightness of this cell

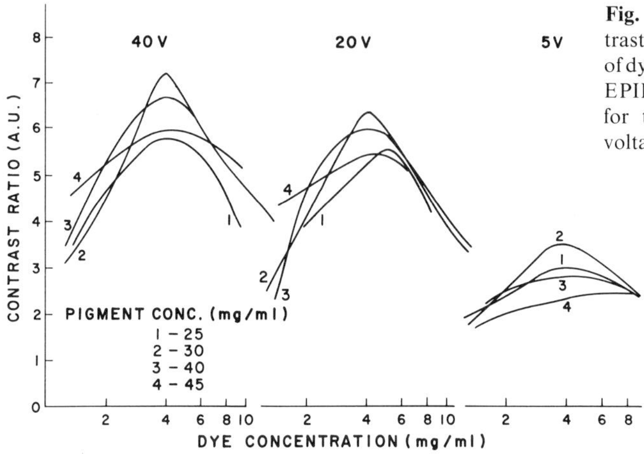

Fig. 6.3. Dependence of contrast on the concentrations of dye and pigment in a typical EPID device (50 µm thick), for three different operating voltages

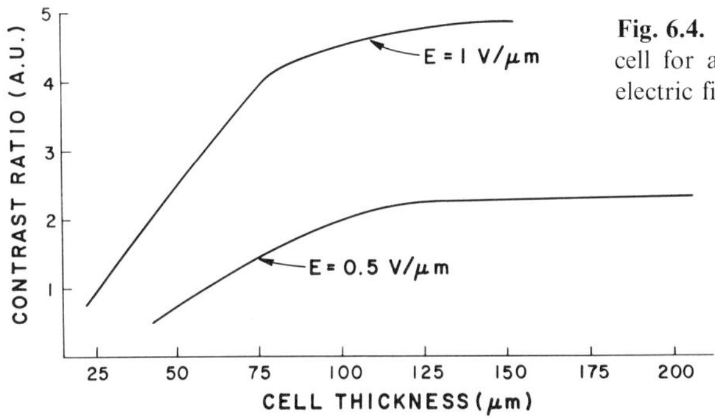

Fig. 6.4. Contrast vs thickness of EPID cell for a typical suspension at applied electric fields of 1 V/µm and 0.5 V/µm

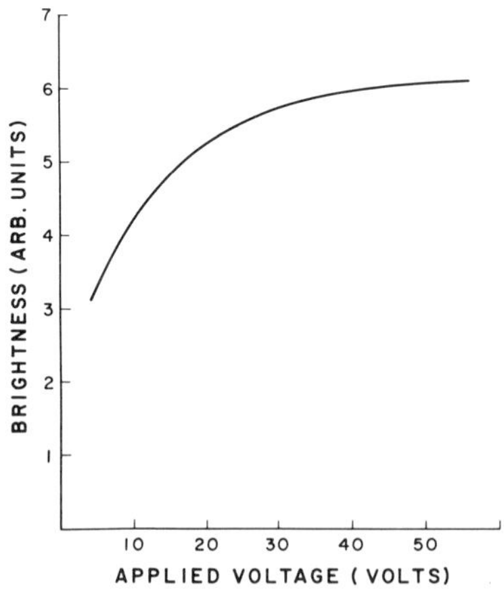

Fig. 6.5. Brightness of the ON state of EPID device vs applied voltage

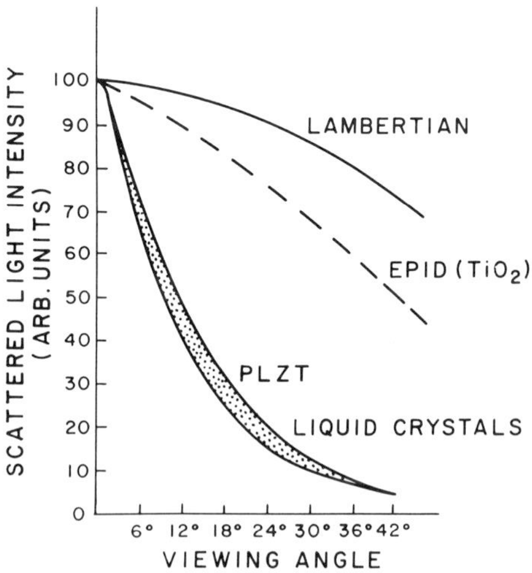

Fig. 6.6. Contrast Merit Factor for several non-emissive display media

for a given suspension composition and cell thickness is dependent upon the density of light scatterers (pigment particles) on the electrode. The effect of the applied electric field which packs the pigment layer is shown in Fig. 6.5.

A distinctive feature of EPID device is their high contrast over a wide range of viewing angles. Figure 6.6 shows a comparison of the contrast merit factor [6.10] for several display media. An EPID device with a suspension of white pigment and black dye greatly surpasses the performance of the other media and approaches that of black ink on white paper as a nonemissive display.

Based upon these operating principles, several general characteristics of EPID devices can be expected. By proper selection of the pigment and dye, a variety of color combinations are possible, including black and white. Due to the nonemissive nature of the display, the near Lambertian scattering of the pigment, and the strong absorption of the dyed liquid, EPID devices exhibit excellent contrast over a wide range of viewing angles and ambient-light levels. If the polarities of the applied voltages are reversed, the position of the pigment and hence the color tone can be reversed. Since the pigment is transported through a fluid, the inherent speed of EPID is in the tens of milliseconds range, with applied voltages in the range of 25–75 V. Although the voltages are moderately high, the devices basically respond to the electric field, and since the suspension fluids are highly resistive the currents are rather low, e.g., $0.1\,\mu A/cm^2$. In addition, when the applied voltage is removed, the pigment remains on the electrodes, giving this device inherent memory. Hence, these devices operate with very low power concumption. The materials in the suspensions as well as the device structure are inexpensive and hence EPID devices are capable of providing large area displays.

6.1.2 Electrical Addressing of EPID Devices

To electrically address a display device, each display element can be individually connected to and activated by the electronic drive circuitry. However, for devices with even a modest number of display elements, such as multiple-digit numerics or an alphanumeric display, it is crucial for reasons of cost and reliability to be able to reduce the number of electrical interconnections and drivers. Since the basic electrophoretic mechanism of EPID devices does not exhibit a voltage threshold in its optical response, it has been necessary to investigate techniques to permit the writing of selected display elements without partially activating any other display elements, i.e., writing without half-select problems.

The first attempts to matrix address electrophoretic displays employed suspensions that exhibit a voltage threshold for the removal of the pigment layer from an electrode surface [6.11]. The mechanism of this threshold is not well understood, but is is known to depend upon many factors, such as the shape, size, and distribution of the pigment; the molecular structure and charging effectiveness of the stabilizers; the applied electric field; the Hamaker

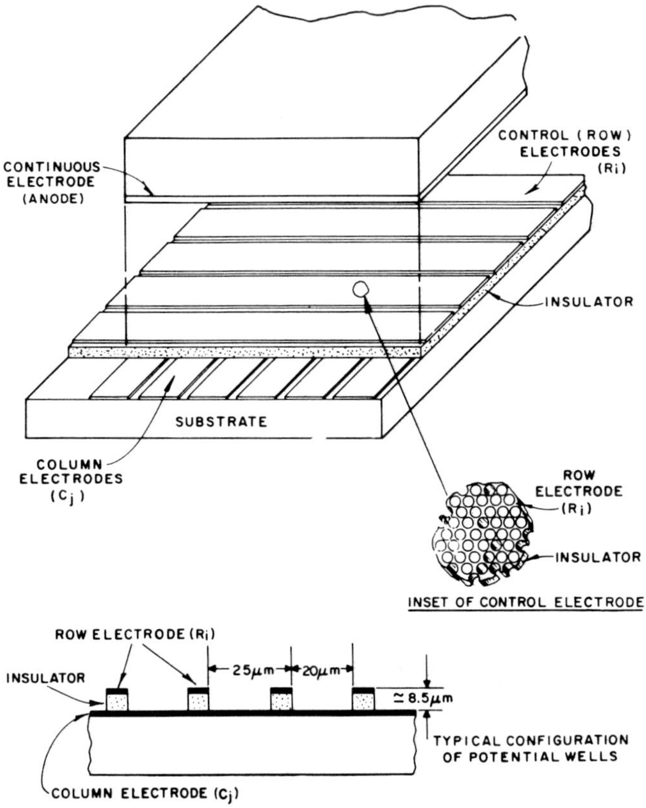

Fig. 6.7. Schematic of a control-electrode EPID device

constants[1] of the liquid, pigment, and electrodes; the conductivities and dielectric constants of the liquid and pigment; etc. Some preliminary performance characteristics of this technique will be discussed in Sect. 6.3.3.

Recently an alternative technique based upon the use of a control electrode has been demonstrated [6.9], and one embodiment of this technique is shown in Fig. 6.7. One side of the cell is a continuous transparent electrode which shall be called the anode. On the other side of the cell the transparent electrode is patterned into a set of isolated electrode strips called the column electrodes. An insulator covers these column electrodes, and an electrode layer on top of the insulator is divided into a second set of isolated electrode strips called the row electrodes, which are oriented orthogonal to the column electrodes. The row electrodes are patterned into a dense array of holes or grid, beneath which the exposed insulator has been removed forming a multiplicity of physical and potential wells as shown in the figure. By application of the appropriate voltages between the row, column, and anode electrodes, the pigment can be

1 A discussion of the origin and significance of the Hamaker constant can be found in [6.12].

retained in or released from any display element without half-select problems. Preliminary results of prototype control electrode devices will be described in Sect. 6.3.3.

6.2 Device Technology

6.2.1 Colloidal-Suspension Technology

The colloidal suspension is the active and most crucial part of an EPID device. It consists of a suspending liquid, a soluble dye, stabilizing agents, and the submicron pigment particles. The composition of the suspension determines to a large degree the lifetime, optical properties, and response times of the device. In an ideal suspension the pigment would neither settle nor float in the suspending liquid, i.e., it would be gravitationally stable. Secondly, the individually dispersed pigment particles would remain separate and not bunch together or agglomerate under all operating conditions; this requirement is called colloidal stability. In addition, the interaction between the fluid and the pigment, i.e., hydrodynamic effects, must be controlled to maintain a uniform pigment-layer thickness. Finally, the constituents of the suspension must be chemically compatible with each other and with other materials present in the EPID device.

Some of the general considerations in the selection of suitable suspension constituents are discussed below. A major consideration in the selection of a suspending liquid is that it contribute to high electrophoretic mobility. The electrophoretic mobility [6.13] of a particle in a suspending liquid is given by

$$\mu=\varepsilon\zeta/6\pi\eta \tag{6.1}$$

where ε is the dielectric constant, ζ the zeta potential, and η the viscosity. Hence, liquids with the highest ε/η ratio should be chosen for fast display operation. This choice must be consistent with the other requirements on the suspending liquid, such as: ability to function over a wide temperature range, low toxicity, excellent chemical stability, proper specific gravity, chemical inertness, and high resistivity ($>10^{12}\,\Omega\,\mathrm{cm}$).

Selection of the dye in the suspension is based upon the following requirements: solubility in the suspending liquid, chemical stability, chemical compatibility with suspension constituents, and high optical density at the portion of the optical spectrum reflected by the pigment.

The particles used in EPID devices can be either organic or inorganic pigments. The general requirements on the selction of the pigments are: acceptable optical characteristics (e.g., scattering power, color, opacity, etc.), insolubility in the liquids, little swelling or softening, ability to be easily and stably charged, chemical stability, and a specific gravity that can be matched by suitable suspending liquids.

The stabilizing agents used in EPID suspensions are of critical importance, but their interactions with the pigment surface, e.g., the charging mechanisms, are very complex and in general poorly understood. The charging mechanism for one type of EPID suspension has been determined and reported by *Fitzhenry* [6.14]. In general, effective stabilizing materials for a given pigment are determined by empirical testing guided by the chemical structures and expected chemical interactions of the pigment surface and stabilizer. Examples of typical EPID suspensions have been described in [6.4, 7].

The most direct method for achieving gravitational stability is to insure that the pigment and suspending liquid have equal specific gravities. This can generally be accomplished by using a suspending liquid that is composed of two liquids, one of relatively high specific gravity and one of lower specific gravity. The two are mixed such that the resulting specific gravity matches that of the particles. Although this matching can in general be achieved at only one temperature (e.g., room temperature), the slight mismatch at other temperatures has not proved to be a problem. The use of a dense inorganic pigment, e.g., TiO_2, has been reported on by *Ota* et al. [6.11a]. *Ota* has obtained promising results with suspensions with bare TiO_2 as well as demonstrated techniques for microencapsulation of this pigment to reduce its effective specific gravity.

Colloidal suspensions are inherently unstable systems, since the large surface free energy of a suspension is reduced when the particles bunch together, i.e., agglomerate. To stabilize a colloidal system against agglomeration a repulsive force between particles must be developed to counteract the attractive London dispersion force. The repulsive force can be provided by charging the pigment particles to produce electrostatic repulsion.

Deryaguin and *Landau* [6.15], and later *Verwey* and *Overbeek* [6.13] calculated the potential energy for two charged colloid particles and established a quantitative theory of electrostatic stabilization of colloid systems, referred to as the DLVO theory.

In nonaqueous systems, such as a typical EPID suspension, electrostatic stabilization is of limited effectiveness [6.16]. An alternative mechanism called steric repulsion is of more importance in stabilizing such suspensions. Colloidal particles can be prevented from agglomerating by the presence of polymer molecules (50–200 Å) on their surfaces. Steric repulsion is not well understood, and only some general principles which govern this type of stabilization have been developed. The origin of the repulsive potential energy in steric stabilization is the interpenetration and/or compression of the polymer chains. Hence this type of repulsion only appears when the particle separation approaches twice the stabilizer chain length. It is possible to combine the electrostatic and steric stabilization mechanisms to greatly enhance colloidal stability.

The colloid stability requirements on a suspension for an EPID device are extremely strenuous. Stability as defined in standard colloid chemistry is generally measured by observing the time duration before noticeable separation of the solid and liquid phases of a suspension occurs in a test tube

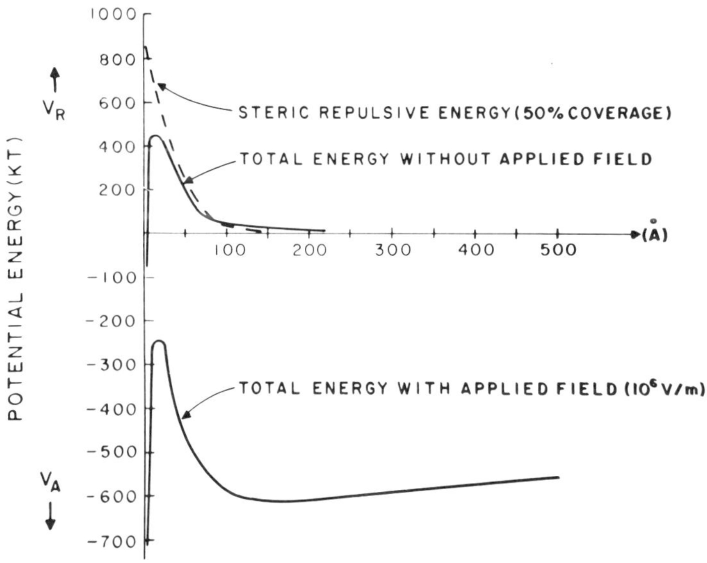

Fig. 6.8. Calculation of total potential energy for two charged pigment particles, assuming they are covered with adsorbed polymer with a 50 Å chain length under no applied electric field and with an applied electric field of 1 V/μm

because of agglomeration. When no separation is observed for some arbitrary period (e.g., 2 days) the suspension is considered to be stable. Suspensions which do not separate for many months in such a quiescent test will be severely agglomerated in several switching operations in an EPID cell. The difficulty of achieving stability in an EPID cell can be appreciated by understanding that the optical contrast of the device is achieved by compressing the pigment onto the electrode. Under the applied electric field, particles are forced into close proximity, and the probability of agglomeration is greatly increased. In addition, the particle concentration on the electrode is increased by as much as an order of magnitude over the concentration in the bulk suspension, and it has been found that the stability of a suspension decreases rapidly as the particle concentration increases [6.17]. However, even though the stability requirements are strenuous, EPID suspensions have been developed that can sustain more than 10^8 switching operations and still continue to operate satisfactorily. Figure 6.8 shows the calculated potential energy for an EPID suspension using a combination of electrostatic and steric stabilizers while compressed on an electrode surface with an electric field of 10^6 V/m [6.18]. A potential energy barrier to agglomeration even under these strenuous conditions is evident in the figure.

The application of a dc electric field to a liquid of low conductivity can produce macroscopic motion of the medium. This hydrodynamic motion is related to the motion of ions which form the space charge in the suspension. A condition of *e*lectro*h*ydro*d*ynamic (EHD) instability can be established which causes the fluid to break up into a regular pattern of convection cells. The pigment acts as tracers for this fluid pattern and appears as loosely bound clusters whose size and pattern are closely related to the amount of background charge in the suspension [6.19].

In addition, the fluid is also set in motion by the electrophoretic movement of the pigment during switching. This fluid motion then perturbs the trajectories of nearby particles. When portions of a cell are repeatedly switched, a local redistribution of the pigment can occur causing noticeable changes in the appearance of the device. There are various methods for controlling the redistribution. The general effects of hydrodynamic behavior on EPID suspensions has been recently reported on by *Murau* and *Singer* [6.18].

It is also necessary that the EPID suspension exhibit a high degree of chemical stability in quiescent as well as during operating conditions. Chemical reactions in the suspension can adversely affect operating lifetime, response speed, and contrast. Therefore the choice of the suspension materials must emphasize their chemical compatibility and general inertness. Areas of major concern are: solubility of pigments in the solvents, photo- or electrodegradation of the dyes and pigments, and electrode reactions. Since the dyes and stabilizers are surface-active materials, there is the possibility of antagonistic or competitive interactions. In general, standard purification procedures are employed to remove contamination (e.g., H_2O) from all constituents. In addition, the chemical structure of the dyes, pigments, and stabilizers can be analyzed to determine and avoid possible chemical problems.

6.2.2 Device Design

As the technology of electrophoretic displays has progressed, theoretical models have been developed which extend the understanding of the various phenomena which occur in EPID devices as well as enable the design of new and improved devices.

A preliminary model of the response time of an EPID device has been proposed by *Quon* [6.20]. This analysis assumes the pigment layer to be a uniform sheet of diffuse scatterers that move through the dyed liquid. This simple approach appears to provide a good correlation with fall-time data. However, to describe the rise time the details of the deposition and packing of the pigment layers on the electrode must be understood. This packing process is rather complex and involves space-charge effects as well as possible changes in the nature of the liquid in this region of high pigment and stabilizer concentrations.

The optical properties of the pigment layer have been modelled by *Vance* [6.8], and *Singer* and *Dalisa* [6.21]. Using the Kubelka-Munk equation, *Vance* has determined the relationship between the reflectance and the pigment-layer thickness for pure dielectric particles in a transparent liquid.

Singer and *Dalisa* have calculated the amount of the incident light that reaches the pigment layer and the amount which is back-scattered to the

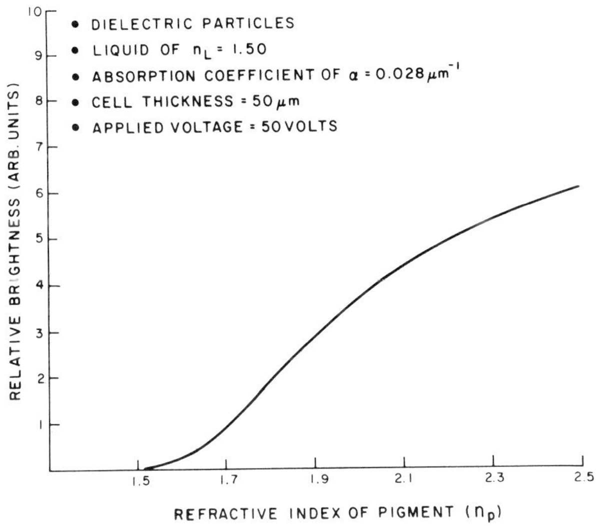

Fig. 6.9. Calculation of the brightness of ON state of a typical EPID device as a function of the index of refraction of pigment (n_p)

observer using the diffusion approximation. This model can treat absorbing as well as dielectric scatterers suspended in a dyed liquid. Figure 6.9 shows the dependence of the display brightness on the index of refraction of the dielectric pigment particles.

In EPID devices, it has been shown that the applied electric field can cause hydrodynamic instability in the fluid as discussed in Sect. 6.2.1. The origin of the fluid motion due to this instability is the nonuniform distribution of space charge (neglecting pigment) in an EPID cell. An approximate analysis has been performed to determine the critical voltage for the onset of the hydrodynamic instability and the corresponding spatial dimensions of the flow patterns in EPID cells [6.18].

For a control electrode structure, as discussed in Sect. 6.1.2, the principle of superposition and conformal mapping methods have been used to obtain a reasonably accurate three-dimensional solution for the electrostatic potential [6.9]. The electric field at various points in the potential-well structure and the voltages on the three electrodes required to operate the device, as well as the response time, have been calculated. These predicted operating parameters are in good agreement with the actual performance data discussed in Sect. 6.3.3.

6.2.3 Device Fabrication

EPID test cells and prototype devices are fabricated using commercially available indium oxide coated glass. After the glass has been cut to size, holes to be used for filling the assembled cell are drilled in diagonally opposite corners of one piece. Depending upon the design of the device an electrode pattern must be etched in at least one indium oxide layer. For example, in a seven-segment numeric device a photographic mask and standard photolithographic techniques are used to remove a thin (≈ 0.025 mm) line of the electrode layer which forms the borders of the segments and to bring the electrical contact for each segment and the background electrode to one edge of one of the glass plates. In the areas surrounding the seven segments the electrode is left intact so that an electrical field can be applied to this region to establish the background color of the display. In general, the patterned electrode plate is used as the rear side of the cell, and the viewer looks through an unpatterned indium oxide electrode. With this design the colors of the number and its surrounding background can be reversed by reversing the polarities of the voltages. Alternatively, an opaque mask which surrounds the seven segments and whose color matches the OFF state of the segments can be used; in this method the color tone cannot be reversed.

For x–y-addressed devices as discussed in Sect. 6.1.2, a control electrode structure is formed on one glass plate. The operation and fabrication procedures involved in this addressing method will be described briefly in Sect. 6.3.3.

The requirements on the seals for an EPID cell are: high resistance to the solvents in the suspension (i.e., insoluble and little or no swelling or softening), chemical inertness, electrical insulation, low permeability to moisture or solvent vapor, and good tensile and shear strength. Several types of epoxy seals have been employed with moderate success. An alternative sealing technique has been developed which uses a sheet of polymer material that simultaneously provides accurate spacing and parallelism of the two glass plates as well as excellent strength and hermeticity. The polymer is cut in the shape of a window frame that makes the perimeter seal and additional strips can be provided to insure parallelism and added strength for large panels. The polymer sheet is sandwiched between the glass plates and the seal is formed by applying moderate heat and pressure for several minutes. The resulting seal has a tensile strength of 4–5 $\mathrm{kg/cm^2}$ and a shear strength of 30 $\mathrm{kg/cm^2}$. It can withstand boiling water for over 500 h. No leaks have developed in cells that have been subjected to a temperature range of $-50\,^{\circ}\mathrm{C}$ to $+100\,^{\circ}\mathrm{C}$ as well as to repetitive thermal shocks (0 °C–100 °C).

After sealing, the cell is filled through the holes in one of the glass plates. The cell is sealed by means of an appropriately sized Teflon ball that is pressed into the fill holes. A small button of epoxy is applied over the Teflon ball to insure a leakproof seal and a slender, neat package.

6.3 Device Performance

The performance characteristics of EPID devices depend to a significant degree upon their application as well as the early state of the technology. This section will present a general description of the optical and electrical characteristics of prototype devices as well as their operating lifetime and some environmental

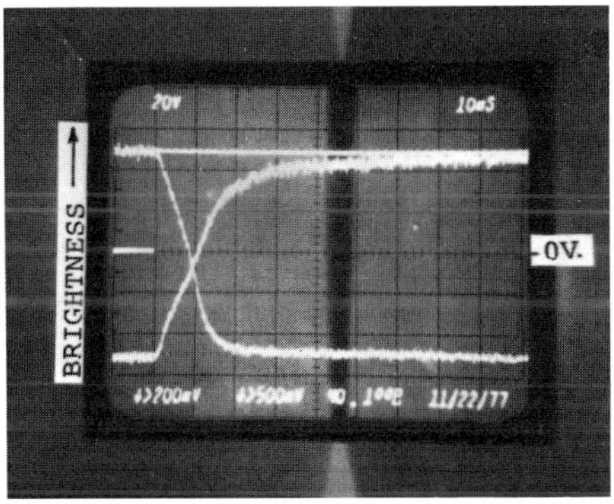

Fig. 6.10. Typical rise and fall times of the optical response of an EPID device (50 μm thick) with the applied voltage pulse. The horizontal scale is 10 ms/division

characteristics. In addition, the results obtained to date for matrix addressing EPID devices will be described.

6.3.1 Optical Characteristics

The rise time of an EPID device shall be defined as the duration required for the reflected brightness of a given side of the display cell to go from 10% to 90% of its maximum value. Conversely, the fall time is the time during which the reflected brightness goes from the 90% level to the 10% level. Typical rise and fall times along with the applied voltage pulse are shown in Fig. 6.10. The relatively long time required for the rise time to saturate is due to the deposition and packing of the pigment layer on the electrode.

After the particles have moved away from the electrode, they travel at a velocity approximately given by

$$v=\mu E=(\varepsilon\zeta E)/(6\pi\eta)\,. \tag{6.2}$$

The time necessary to transit the cell is approximately

$$t=(6\pi\eta L)/(\varepsilon\zeta E)=(6\pi\eta L^2)/(\varepsilon\zeta V)\,. \tag{6.3}$$

Figure 6.11 shows the experimental data for transit time versus reciprocal voltage.

The brightness of an EPID device is limited by the optical properties of the pigment (index of refraction, color, and size), the distribution of the pigment, and the concentration and absorption coefficient of the dye. There are also losses in incident as well as scattered light due to the glass and electrode surfaces that comprise the cell wall. An additional factor which can significantly effect the display brightness is adsorption of some of the dissolved dye onto the pigment surface. *Fitzhenry* and *Dalisa* [6.22] have determined the extent of this adsorption for some EPID suspension system and estimated its effect on the

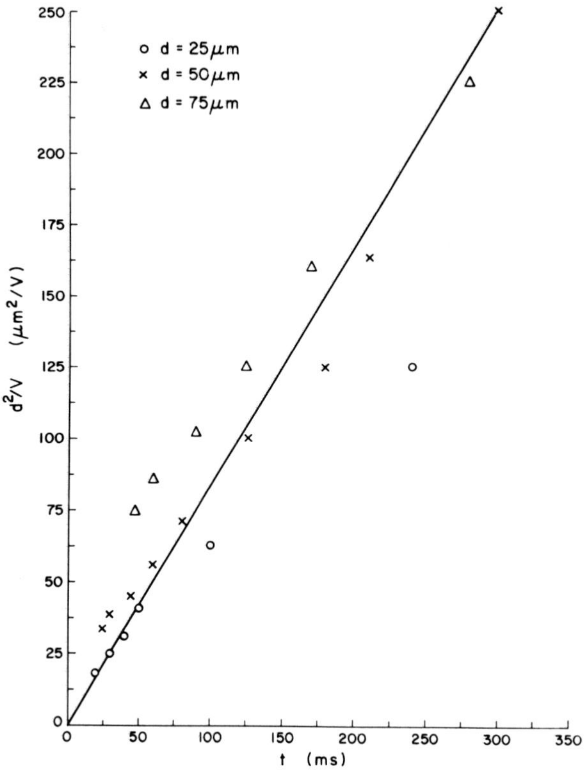

Fig. 6.11. Transit time of pigment vs reciprocal of the applied voltage and cell thickness (*d*) for a typical suspension in an EPID cell

optical properties of the device. Some criteria on the choice of suspension materials are discussed which can reduce this adsorption problem.

The brightness of the layer of pigment in an EPID device can be most easily described for the case of pure dielectric (white) particles and is expressed as a comparison with a standard white surface, e.g., $BaSO_4$, under the same illumination and measured with a detector through a photopic filter. For devices with a cell thickness of 50 μm, an applied voltage of 50 V and a standard black/white suspension employing TiO_2 pigment and black dye the relative photopic brightness is about 38%. The contrast defined as: $(B_{ON} - B_{OFF})/B_{OFF}$ is 20. For suspensions with colored pigments, e.g., Diarylide Yellow, but the same dye, the photopic brightness relative to the same standard is about 14%.

For colored systems the photopic contrast is not a good measure of the perceived contrast of the display. By using a narrow-band filter on the illumination source or on the detector, chosen such that one of the colors is rejected, a "monochromatic" contrast can be specified. This method yields numbers that are similar to a black/white contrast, but the values will of course depend upon the choice of the filter. Figure 6.12 shows the reflectance spectra for the two color states of a standard red/yellow suspension. Alternatively, the tristimulus values for each of the two color states can be measured, and a color

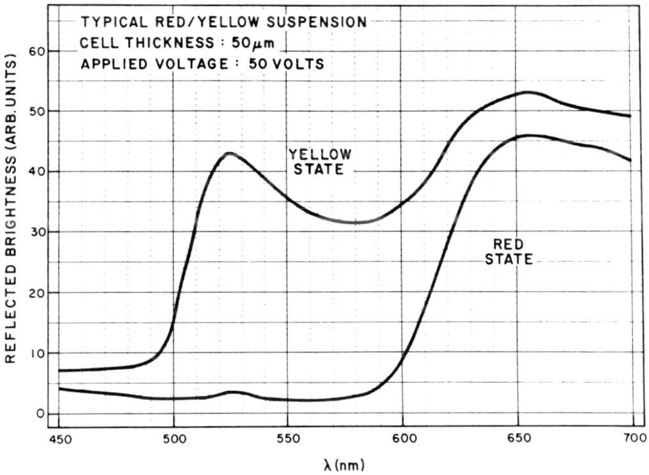

Fig. 6.12. Reflectance spectra for the ON and OFF states of a typical red/yellow suspension

difference or color contrast can be calculated (approximately 31 for the suspension shown in Fig. 6.11) using one of the standard formulae in the literature [6.23].

The pigment layer remains on the electrodes after removal of the applied voltage, providing these devices with inherent memory. The duration of this memory depends strongly upon the suspension parameters, e.g., conductivity, and composition, as well as the device configuration and the impedance of the electronic drive circuitry. With an open circuit between electrodes of a simple device and suspension resistivity of 10^{11} Ω cm the memory can be in excess of several thousand hours. The origin of the memory and its dependence upon materials has been discussed by *Ota* et al. [6.24].

6.3.2 Electrical Characteristics

The suspending liquids employed in the suspensions are highly insulating, nonaqueous liquids. Aqueous liquids are not suitable due to the unavoidable hydrolysis at the electrodes. In addition, highly insulating fluids are chosen to reduce the parasitic current which consumes power and controls electrochemical reactions.

A typical electrophoretic suspension has a current–voltage behavior that is essentially ohmic. When the potential difference across a 50 μm thick cell is 50 V dc, the corresponding current density is about 0.20 μA/cm^2, corresponding to a suspension resistivity of approximately 10^{11} Ω cm. The stabilizer and dye are most responsible for the final conductivity.

The transient current response of a simple EPID cell is shown in Fig. 6.13. The switched charge (not including the parasitic current) is the area under the

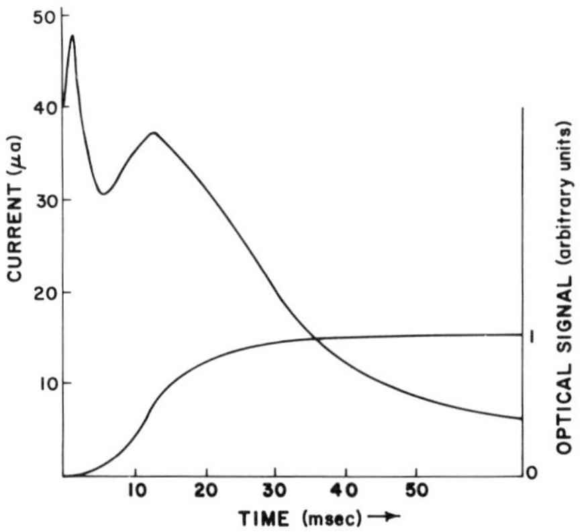

Fig. 6.13. Transient current and optical response for a typical EPID suspension

curve and for a typical suspension is about 0.1 μC/cm^2. The shape and position of the current peak in the transient current is dependent upon the applied voltage, and the composition of the suspension. The typical capacitance of a simple EPID cell is approximately 240 pF for a cell thickness of 50 μm and an area of 6 cm^2.

6.3.3 Matrix-Addressed Devices

The first attempts to matrix address electrophoretic displays employed suspensions that exhibit a voltage threshold for the removal of the pigment layer from an electrode surface, as mentioned in Sect. 6.1.2. A matrix-addressed panel reported on by *Ota* et al. [6.24] and shown in Fig. 6.14 uses this kind of threshold. It consists of a 32×184 array of elements driven in a 3/1 (60 V/20 V) voltage selection mode. It is scanned a line at a time without need for refreshing, but it requires 12 s to scan the 32 horizontal lines (375 ms/line). Slow speed and control of this chemical threshold appear to be disadvantages of this technique.

The basic concept and cell configuration of a control-electrode x–y-addressed prototype device [6.21] has been described in Sect. 6.1.2. The voltage waveforms applied to the row, column, and anode electrodes for a selected display element along with the corresponding optical response are shown in Fig. 6.15. The pigment is negatively charged, and the cell is viewed from the control-electrode side of the cell. The brightness of a half-selected display element using the control-electrode technique remains constant even after thousands of half-select operations. The addressing voltage levels on the rows and columns during the write operation need only be applied for the time necessary for the pigment to leave the potential wells and not for the entire transit time. With the present prototypes, addressing rates of 5 ms/line are possible, and substantial improvements are expected in the future.

The resolution of this display, when viewed from the control-electrode side of the cell is only limited by the uniformity of the photolithography that is used

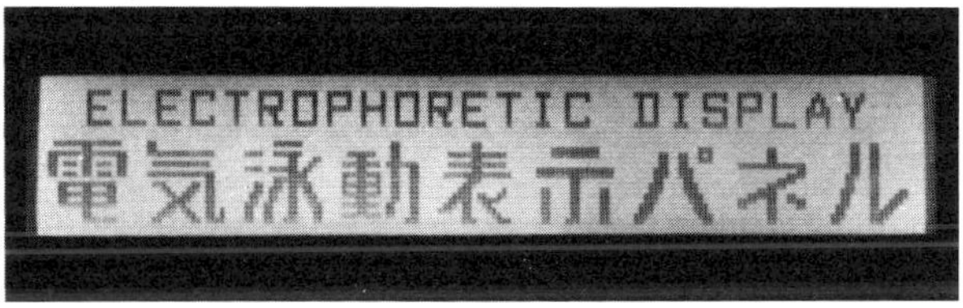

Fig. 6.14. Matrix addressed EPID panel using a suspension with a voltage threshold

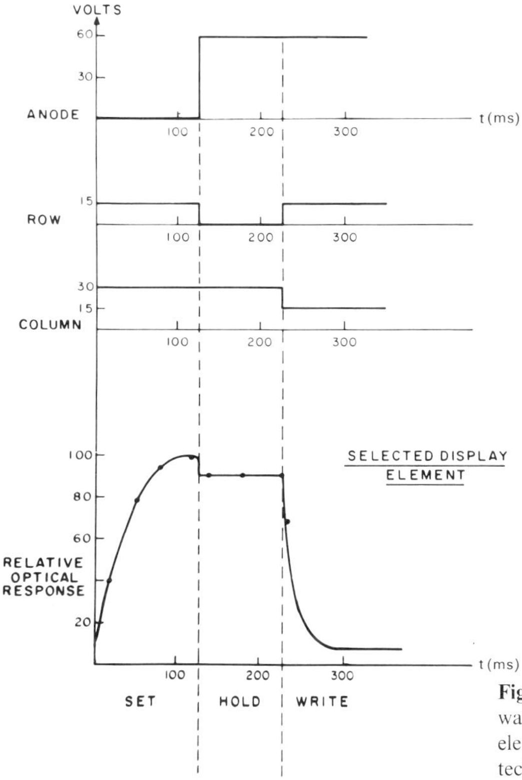

Fig. 6.15. Optical response and voltage waveforms for an x–y-addressed display element using the control-electrode technique

Fig. 6.16. Alphanumeric characters from a prototype 7×9 control-electrode-addressed EPID device

to form the wells. Presently the resolution is about 5 lines/mm and it can be extended to about 10–15 lines/mm in the future. It is also possible to erase selectively a character as well as provide a cursor to indicate any character position. Figure 6.16 shows several typical alphanumeric characters from a prototype 7×9 control-electrode device.

An alternative addressing technique which requires multiple voltage levels, simultaneously addresses all elements of an alphanumeric character and requires no threshold properties in the suspension has also been demonstrated [6.25].

It is also possible to envision the use of an active matrix such as an array of thin-film transistors (TFT) [6.26] to address an EPID panel. Whether the performance, cost, and reliability of such TFT arrays offer any advantage over the control-electrode technique for this application depends upon future developments in both technologies.

6.3.4 Lifetime and Environmental Effects

The end of the useful operating life for an EPID device is most often related to an objectionably nonuniform distribution of pigment on the transparent electrode. The principal failure modes include: agglomeration, clustering, pigment redistribution, adherence of pigment to the electrodes, and electrochemical/photochemical effects such as bleaching of the dye. In simple two-electrode test cells agglomeration and clustering are the primary causes of cell failure. In devices with patterned electrodes, clustering and pigment redistribution are usually the primary problems.

Clustering has been related to a hydrodynamic instability in the suspension, as indicated in Sect. 6.2.1. By reducing as much as possible any unnecessary charge (ions) in the suspension, this phenomenon can be effectively prevented. In most cases, the average cluster size is small (<0.025 mm) and produces no noticeable nonuniformities.

The redistribution of pigment first appears as a small area (several mm in size) of different color due to the local variation in the thickness of the pigment layer. As the operation of the cell continues, the variation in color can become increasingly evident. In certain EPID applications, such as control-electrode-addressed devices, the lateral electric fields and device structure inhibit the formation of pigment clusters and substantially reduce pigment redistribution.

In suspension-life tests [6.7] cells with standard suspensions are still in operation after more than 2×10^8 switches at ± 50 V, a frequency of 1/2 Hz and a cell thickness of 50 μm. Seven-segment numerics have been operated for greater than 5×10^7 switches without objectionable degradation, and significant improvements are still in progress.

At present, only limited tests of the effect of environmental conditions on performance and reliability have been reported. An operating temperature range of $-10\,^{\circ}$C to $100\,^{\circ}$C and a storage temperature range of $-40\,^{\circ}$C to $100\,^{\circ}$C has been reported by *Dalisa* [6.7]. *Ota* et al. [6.24] have reported that for their suspensions the response times and cell resistance are believed to be governed by the temperature dependence of the viscosity of the suspending liquids. In addition, some possibility of photochemical degradation of the dyes and suspending liquids were noted in this publication.

As this technology matures, much more detailed environmental tests will be necessary. However, the preliminary indications are that the use of electrophoretic displays, especially in commercial and consumer applications, will not be significantly limited by these factors.

6.4 Conclusions

The electrophoretic image display is a relatively young and highly promising display technology. In its present state of development it has demonstrated many desirable inherent characteristics such as distinctive appearance, low power consumption, and practical response times. It is suitable for large area displays and is expected to have a low fabrication cost. A continual improvement in the suspensions has been achieved and test devices have demonstrated an operating lifetime of better than 15,000 h. In addition, recent development work has demonstrated a practical x–y-addressing method which can use standard integrated circuit logic, greatly improved addressing rates, and high resolution.

However, due to the early stage of EPID technology, work remains to be done in more efficient charging and steric stabilization of the pigment particles, the reduction of dye–pigment interactions that can reduce brightness, as well as the improvement of photo- and electrochemical stability of all suspension components especially the dyes.

In the device technology, the development of a better understanding and control of space-charge effects, hydrodynamic behavior, and the pigment deposition and packing are areas of continued active investigation. In addition, much more extensive environmental testing is necessary. Initial tests indicate that the storage and operating temperature ranges are acceptable for most commercial applications.

Based upon the results and understanding to date, substantial progress is expected in the near future and it is realistic to expect future EPID devices to have still improved appearance, lifetimes exceeding 20,000 h, low production costs, and in matrix displays, addressing rates of about 1 ms/line with a resolution of 10 lines/mm. With their distinctive appearance and low power consumption EPID devices will be competitive in a wide variety of display applications. However, in the opinion of the author their major application area will be in the field of information systems. Here the EPID concept can provide a nonemissive, flat-panel, page-sized display with excellent resolution capable of displaying a page of textual information or graphics with a comfortable and attractive appearance, a kind of electronic paper.

Acknowledgements. The author would like to thank Ms. B. Fitzhenry, Mr. R. Liebert, and Dr. B. Singer for their assistance in the preparation of the manuscript and Mr. J. Kostelec for his support in this effort. I would also thank Dr. I. Ota, Wireless Research Laboratories, for photographs of prototype EPID devices.

References

6.1 K.A.Metcalfe, R.J.Wright: J. Sci. Instrum. **33**, 194 (1956)
6.2 K.A.Metcalfe, R.J.Wright: J. Oil and Color Chem. **39**, 845 (1956)
6.3 P.F.Evans: U.S. Patent 3, 612, 758 (1971)
6.4 I.Ota, J.Ohnishi, M.Yoshiyama: Proc. IEEE **61**, 832 (1973)
6.5 J.C.Lewis: "Electrophoretic Displays", in *Nonemissive Electrooptic Displays*, ed. by A.R.Metz, F.K.Von Willisen (Plenum Press, New York 1975)
6.6 J.Kostelec, R.Liebert: Presented at Amer. Ceramic Society, Denver, Co. (1974)
6.7 A.L.Dalisa: IEEE Trans. ED-**24**, 827 (1977)
6.8 D.Vance: Dig. Biennial Display Conf., New York (1976) p. 96
6.9 B.Singer, A.Dalisa: Proc. SID **18**, No. 3 (1977)
6.10 A.L.Dalisa, R.J.Seymour: Proc. IEEE **61**, 981 (1973)
6.11a I.Ota, T.Sato, T.Yamagami, H.Takeda: Presented at Laser 75 Opto-electronics Conf., Munich (1975)
6.11b A.L.Dalisa, B.Singer, S.Ross: Presented at Device Research Conf., Salt Lake City, Ut. (1976)
6.12 G.D.Parfitt: *Dispersion of Powders in Liquids* (Wiley, New York 1973) p. 14
6.13 E.J.Verwey, J.Overbeek: *Theory of the Stability of Lyophobic Colloids* (Elsevier, Amsterdam 1948)
6.14 B.Fitzhenry: Bull. 4th Annual Meeting Fed. of Analytical Chem. and Spectrosc. Soc., Paper # 104, Detroit, Mi. (1977)
6.15 B.V.Deryaguin, L.D.Landau: Acta Physiochem. **14**, 633 (1941)
6.16 J.Van Der Minne, P.H.Hermanie: J. Colloid Sci. **8**, 38 (1953)
6.17 W.Albers, J.Th.Overbeek: J. Colloid Sci. **14**, 510 (1959)
6.18 P.Murau, B.Singer: "Suspension Instabilities in an Electrophoretic Display", J. Appl. Phys. (1978) to be published
6.19 P.Murau, B.Fitzhenry, A.L.Dalisa, R.Liebert, B.Singer: Suppl. to Dig. of Intern. Elect. Device Meeting, Washington, D.C. (1976) p. 8
6.20 W.S.Quon: Dig. 1976 Biennial Display Conference, New York, NY. (1976) p. 92
6.21 B.Singer, A.L.Dalisa: "Optical Model of Electrophoretic Displays", to be published
6.22 B.Fitzhenry, A.L.Dalisa: Proc. 52nd Colloid and Surface Sci. Symp., Univ. of Tennessee (1978) to be published
6.23 H.Pauli: J. Opt. Soc. Am. **66**, 866 (1976)
6.24 I.Ota, M.Tsukamoto, T.Ohtsuka: Proc. SID **18**, No. 4 (1977)
6.25 W.S.Quon: IEEE Trans. ED-**24**, 1121 (1977)
6.26 T.P.Brody: "Integrated Electrooptic Displays", in *Nonemissive Electrooptic Displays*, ed. by A.R.Metz and F.K.Von Willisen (Plenum Press, New York 1975) p. 303

7. Electronic Displays

E. O. Johnson

With 4 Figures

Electronic displays have a world-wide sales volume in excess of \$ 10^9 [7.1]. This includes displays in TV receivers, calculators, computers, timepieces, and a great assortment of lesser applications. The total commercial volume will surely continue to grow as old systems expand and as electronics invades new systems, for example, automobiles. Each application has its own special requirements and optimizations, and these will continue to evolve in terms of size, speed, power requirements, color range, brightness, contrast, and all of the other possible parameters. Numerous technical approaches have appeared and will continue to do so. The technical challenge is large and extends across many different disciplines: image perception and biology, optics, physics, chemistry, and electronic circuits and systems [7.2].

7.1 Basic Functions

As indicated in Fig. 7.1, an electronic display is a "language translator" that converts time-sequential electronic signals into a spacially configured photon signal useful to a viewer. This function is carried out by two intertwined subfunctions. One is that of display-element address wherein signal electrons are appropriately routed to the various display elements. The second sub-function involves converting the signal electrons arriving at each display-

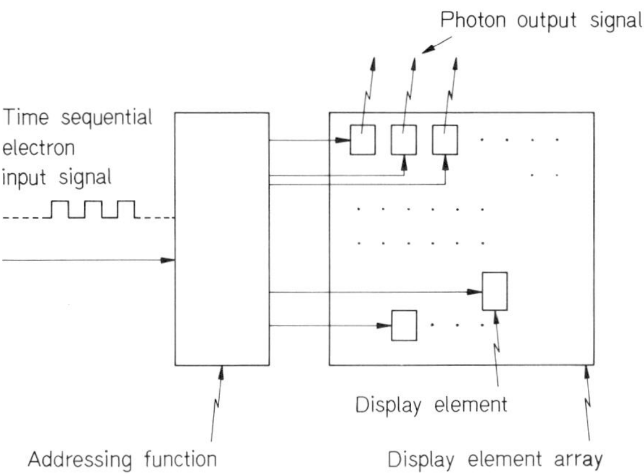

Fig. 7.1. Basic functions of an electronic display

element destination into useful signal photons. This second subfunction accomplishes the inverse of what occurs in a photon detector, or in an imaging device such as a camera tube [7.3]. As will be shown, the signal electron/photon conversion efficiency in a display element provides a convenient handle for comparing most, if not all, display devices.

The display addressing function is virtually identical to that in an electron information storage system and it, also, has been implemented by a great variety of approaches. As will be seen, the display addressing function can be discussed in terms of two parameters. One involves the allowable degree of randomness in the coupling between the input signal and the display elements. The other involves the partitioning of routing control between central and local domains, somewhat remeniscent of federal vs state control in governments.

7.2 Display-Element Classes

Most, if not all, types of display elements can be conveniently classified according to the general scheme shown in Fig. 7.2, wherein only several representative types of elements are illustrated. The class of elements at the top of the figure is distinguished by the fact that all of its elements generate their own photons internally: the bottom class of elements control ambient photons by modulation of such parameters as reflection, absorption, scattering, or optical rotation. These two major classes behave very differently in terms of their photon conversion efficiencies. The photon-generator class has a theoretical efficiency approaching unity and some elements, particularly phosphors, are not far removed from this value in practice. The ambient-photon-control class of elements have a theoretical and practical ability to greatly exceed unity conversion efficiency, but only at relatively slow switching speeds.

7.2.1 Photon-Generator Elements

The energy and electron/photon conversion capability of these is discussed with reference to Fig. 7.3 wherein the electrical power input P_i to an element is the product of the voltage V, and the current I. This can be described in terms of the electron charge e, and the number of in-flowing electrons n per unit of time t

$$P_i = VI = (eV)\frac{n}{t}. \tag{7.1}$$

The photon output power P_o is expressed in terms of the energy per photon $h\nu$, and the number of photons p being generated per unit of time t

$$P_o = (h\nu)\frac{p}{t}. \tag{7.2}$$

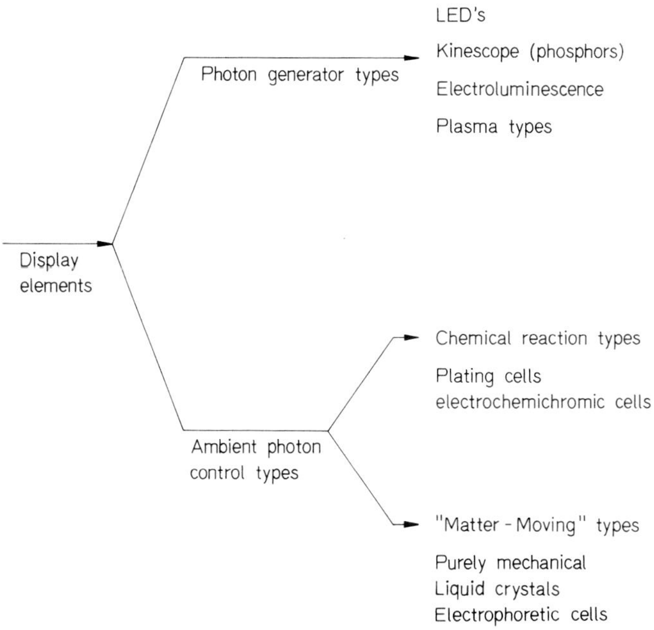

Fig. 7.2. Display-element classes

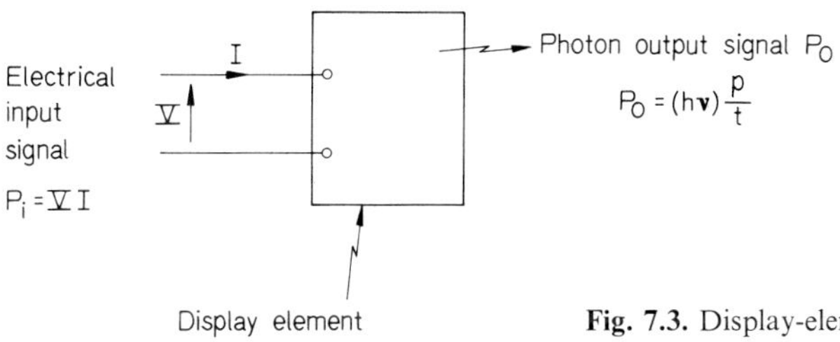

Fig. 7.3. Display-element input and output power

The ratio P_o/P_i is the power conversion efficiency η and this is identical to the energy conversion efficiency if the input and output time periods are conveniently treated as identical,

$$\eta = \left(\frac{hv}{eV}\right)\frac{p}{n} = \eta_V \eta_n . \tag{7.3}$$

The first term represented by η_V is the photon/electron energy ratio, and the term represented by η_n is the photon/electron conversion efficiency ratio.

For a solid-state semiconductor light-emitting-diode (LED) element the input voltage is approximately equal to the semiconductor band-gap energy, and this is very nearly equal to the photon energy. Thus η_V is close to unity in value. The value of η_n, however, is of the order of 10^{-2}, or less, even though the internal photon-generation efficiency can approach unity [7.4]. The problem is one of escape from the diode element because of strong absorption in the semiconductor material. This unfortunate circumstance arises because the photon energy is close to that of the band gap so that the absorption per unit path length tends to be high. Furthermore, the absorbing path length is lengthened by multiple internal reflections caused by the relatively large difference in refractive index between the semiconductor material and the outside air [7.4].

Phosphor display elements such as used in a TV kinescope provide an interestingly different story [7.5, 6]. These elements are composed of many small crystalline grains that generate photons when electrons are injected, not by a fabricated p–n junction as in the case of the LED elements, but by a freely steerable electron beam inside a vacuum space. This beam consists of relatively few high energy electrons that are remarkably efficiently transformed into a great many low-energy electrons inside the phosphor grains. These "down-converted" electrons have an average energy comparable with the phosphor band-gap energy and with the usual photon-producing centers which have a somewhat smaller energy than the band gap. In commercial phosphors these electrons generate photons with high efficiency and, since it turns out that the photon energy is significantly less than the band-gap energy, and the crystallite size is very small for useful purposes, most of the photons escape. The end result is that $\eta_n \cong 0.16$–0.4. The term $\eta_V \cong 0.3$, a value roughly comparable with unity, when calculated on the basis of the electron energies which remain after "down-conversion" of the beam electrons.

Thin-film electroluminescent elements, mostly developmental at the present time and incompletely understood, use phosphors in a solid-state sandwich structure [7.7]. Electrons are injected into the phosphor grains by the cathode bounding electrode, but generate photons with far lower efficiency, quite possibly basically so, than an electron beam in a kinescope. This low efficiency seems to arise from the high dissipative losses associated with generating electrons of appropriate numbers and energy by processes totally inside the material, itself. However, the electroluminescent type of element combines the solid-state virtues of an LED with the potentially low cost per unit area characteristic of a phosphor screen. Thus, we might expect that this type of display element will find use in applications where electrical power is not a critical problem, such as in automobiles, and where one needs relatively large and simple displays of low cost with solid-state virtues.

The plasma-type elements, such as used in plasma-panel displays [7.8], use the luminescence generated in a electrical discharge in a sealed planar-type vessel filled with a suitable gas such as argon, neon, or mixtures thereof. Like phosphor-type elements, the gas-plasma-tape has significant potential for large area displays of relatively low cost. The simple sandwich-structure cell with bounding electrodes capacitively coupled to the gas discharge exhibits various nonlinearities that simplify addressing. Attempts are being made to enhance a relatively low photon-conversion efficiency by the use of phosphors such as in

Table 7.1. Photon-generator display-elements performance summary

Element type	η	η_V	η_n
LED	$\sim 10^{-2}$	~ 1	$\sim 10^{-2}$
Kinescope phosphor	$\sim 2\times10^{-1}$	~ 1	$\sim 2\times10^{-1}$
Electroluminescent	$\sim 10^{-3}$(?)	$\sim 10^{-2}$(?)	$\sim 10^{-1}$(?)
Plasma types	$\sim 10^{-1}$	~ 1	$\sim 10^{-1}$
Theoretical limit	~ 1	~ 1	~ 1

fluorescent lamps where the photon energy conversion efficiency reaches about 20% in ordinary commercial products. The photon-conversion process in the plasma cell is somewhat comparable to that in the phosphor of a kinescope. Monoenergetic electrons are injected from the cathode electrode into the gas discharge plasma with an energy of the order of 100 V. Each injected electron is down-converted to many secondary electrons by ionizing and exciting collisions with gas atoms. These secondaries have a random energy distribution with an average energy of several volts. The secondaries excite atoms to produce photons with often good efficiency, but sometimes a large fraction of these are in the uv range and not visible. These can be converted to visible photons with high efficiency with a suitable uv/visible phosphor such as used in fluorescent lamps, which typically contain a mixture of Ar and Hg gases [7.9].

The performance of photon-generator display elements is summarized in Table 7.1. We see that η_V is comparable with unity for all elements if we consider the process after whatever energy down-conversion may exist. The value of η_n is where the big differences exist among various display elements. The theoretical maximum value of η_V and η_n would be unity, a value that is most nearly approached in the ubiquitous TV kinescope.

Light-output-conversion efficiency, while by no means the only major display parameter, has an importance well beyond simple energy conservation. It effects other parameters such as geometrical size and packing density of elements, the driving circuitry and addressing, the power-supply size and design, and these items in turn effect the cost and general acceptability of the system.

7.2.2 Ambient-Photon-Control Elements

This class of elements can be described in the previously described manner if we use (7.1) as previously noted, and (7.2) with the understanding that ambient photons are being controlled, rather than internally generated. In sunlight about 2×10^{15} visible photons fall on $1\,\mathrm{cm}^2$ of display area each 0.1 s, the approximate minimum eye-response time. If we can reflect, absorb, or otherwise completely control all of these ambient photons with 2×10^{15} circuit electrons injected into the display element, then by our definition η_n would be unity. The minimum eye-response time of 0.1 s provides the least number of ambient photons that can be usefully controlled for the viewer in a "reference" lighting situation. Obviously, for longer exposure times more photons will be involved and this will tend to increase the value of η_n in the types of devices to be described, devices whose photon-control action persists after the electrical input power has been switched off. The minimum eye-response time of approximately 0.1 s is used herein because it provides the most conservative value of η_n.

Chemical-Reaction-Type Elements

These ambient-photon-control type of elements include the types of elements shown in Fig. 7.2 [7.10–12]. All of these operate in the same basic manner in a sandwich cell that has cathode and anode bounding electrodes on each side of a solid or liquid electrolyte. Ionic species in the electrolyte are normally transparent and invisible until they are allowed to pick up or deposit a circuit electronic charge at an electrode. This transfer of charge oxidizes or reduces an ion and causes it to become more absorbing and/or reflecting of ambient light. The transformed ion may adhere to the electrode as with plating, or it may slowly diffuse away.

In this process the number of transformed ions is directly related to the number of in-flowing circuit electrons (Faraday's law). If we combine this result with Lambert's law, which relates the number of deposited particles with their optical effectiveness, we can directly determine the optical-control action of each circuit electron [7.10]. Performance is found to be surprisingly similar over a wide range of different materials: about 5×10^{17} circuit electrons are needed to block out 2×10^{15} visible photons falling on a square centimeter in 0.1 s. This gives a value of η_n of $2 \times 10^{15}/5 \times 10^{17}$, or 0.4×10^{-2}. If the deposited layer remains without requiring additional circuit charge, as is the case with several types of usable cell materials, particularly of the plating type, the useful value of η_n rises indefinitely as we lengthen the viewer observation time.

For all of these chemical-reaction-type of cells the applied element potential is in the order of a volt or two so that, again, η_V is approximately unity.

These types of display elements are promising for displays for a number of reasons. One is that their photon/electron conversion efficiency, $\sim 10^{-2}$, is roughly comparable to that of an LED (see Table 7.1) in the worse case, and can be much better than this when the above-noted storage effect can be taken advantage of, as for example when display patterns can be retained for periods longer than the minimum eye-response time. Another reason is that such elements seem intrinsically inexpensive per unit of area compared to single-crystal LEDs. Still another reason is that the elements can be made to display various colors depending upon the chemical species used [7.11, 12] and some may be used in combination with auxiliary ambients, such as uv [7.13]. The main disadvantage with chemical-reaction elements is that secondary chemical reactions are difficult to control and can greatly shorten operating life.

Table 7.2. Ideal matter-moving display-elements energy and power performance

	Inertial	Viscous	Units
Energy per element	10^{-10}	10^{-11}	joules [J]
Energy per cm^2 of array	10^{-8}	10^{-9}	joules [J]
At 10 Hz per cm^2			
Power	10^{-7}	10^{-8}	watts [W]
V	1 (nominal)	1 (nominal)	volts [V]
I	10^{-7}	10^{-8}	amperes [A]
Conditions			
$\tau = 10^{-1}$ s	switching time		
$\sigma = 1$ gm/cm^3	water density		
$\phi = 10^{-2}$ gm/cm s	water viscosity		
$l = 10^{-1}$ cm	element size and displacement		

Matter-Moving-Type Elements

This type of ambient-photon-control element comprises a wide variety of different approaches as briefly noted in Fig. 7.2. All involve the same principle. Gross assemblies of atoms or molecules are electrically caused to be distorted, or to be moved in or out of position to effect ambient-photon control. A meter needle is perhaps the simplest example. Liquid-crystal display elements are also in the same category since they operate either by a molecular twisting effect (Field-Effect Mode) or by fluid-turbulence effects (Dynamic-Scattering Mode) [7.14].

In all cases, frictional and inertial forces and energies are involved. These determine the necessary electrical-input parameters. The optical controlling action is determined by the gross size and shape of the elements that act upon the incident light. Simple physics[1] determines the mechanical inertial energy E_i involved in moving a reflecting or absorbing body of density σ [gm/cm^3] and dimensions $l \times l \times l$ a distance l [cm], in time τ [s]

$$E_i = \frac{l^5 \sigma}{\tau^2} 10^{-7}\,\mathrm{J}\,. \tag{7.4}$$

Simple physics also determines the frictional or viscous energy E_V involved when the same body is moved in a viscous fluid of viscosity ϕ [mg/cm s]

$$E_V = \frac{l^3 \phi}{\tau} 10^{-7}\,. \tag{7.5}$$

The energies per cm^2 of display area are obtained by multiplying (7.4, 5) by $1/l^2$. It is also to be noted that the kinetic energy acquired at the termination of the

1 The simple physical treatment herein was first described to the author by *Engelbrecht* [7.15].

Table 7.3. Display-element performance summary

Type of element	Photon/electron Energy ratio η_V	Photon/electron Conversion ratio η_n	Cost tendency per unit area
Photon-generating			
LED (Single xtal)	~ 1	$\sim 10^{-2}$	high
Kinescope phosphor	~ 1	$\sim 2\times 10^{-1}$	low
Electroluminescent	$\sim 10^{-2}$(?)	$\sim 10^{-1}$(?)	low
Plasma types	~ 1	$\sim 10^{-1}$	low
Theoretical limit	~ 1	~ 1	–
Ambient-controlling			
Chemical-reaction type	~ 1	$\sim 10^{-2}\rightarrow\infty$[a]	low
Ideal matter-moving	~ 1[c]	10^4-10^5[b]	–
Liquid Crystal	~ 1[c]	$\sim 10^3$[b]	low
Various others	~ 1[c]	$\sim 10^3$[b]	–

[a] Lower value at minimum eye response time of 10^{-1} s.
[b] For values noted in Table 7.2, and no storage effects.
[c] Controllable by cell design.

translation l is dumped. This is analogous to what is normally done to capacitively stored energy in a logic switching circuit.

The main points to note in (7.4, 5) are that both energies are an inverse function of time. This means that one pays an increasing energy penalty as the switching time τ is decreased, particularly when inertial effects dominate. However, if τ is of the order of 10^{-1} s, and reasonable values are used for the other quantities, such as shown in Table 7.2, then the energies and related powers for ideal devices are very small indeed: 10^{-7} W, or less, per cm^2 of display surface. Actual devices will require somewhat more energy and power because of parasitic circuit losses, electrical-coupling problems, and the parasitic mass of the supporting structure. However, in practice, the overall energy and power demands are so low that such types of elements are particularly attractive for very low power applications, for example, wrist watches.

7.2.3 Display Element Summary

A summary of display-element performance is tabulated in Table 7.3. We see that the values of η_V are all comparable with unity as one might expect in cases where the injected electrons play a direct role in the photoelectronic transition. The electron/photon conversion efficiency, however, varies greatly in the various types and classes of elements. In the photon-generator class of elements the conversion efficiency of the ideal element is unity, and this is most nearly approached by the ubiquitous kinescope phosphor. The chemical-reaction types operate in most cases with conversion efficiencies comparable to the normal efficiencies of some of the photon-generation types if the 10^{-1} s switching condition is used. However, information storage capability makes possible much better efficiency when display patterns can be retained for long times. The matter-moving display elements have a very superior conversion efficiency at slow switching speeds, both ideally and in practice, but this efficiency deteriorates monotonically as switching speed increases. This strongly contrasts with the behavior of the photon-generation types of elements.

A combination of the above properties along with the cost/unit area characteristics, also listed in Table 7.3, help identify the optimum realms of applicability for the various elements.

7.3 Addressing

This subject is no less important than that of the elements themselves. As the number of display elements increases, unambiguous address or selection of a desired combination of elements from among the many other elements becomes increasingly difficult: the required degree of addressing "marksmanship" increases. Over the years this difficult problem has spawned a great variety of approaches. Some degree of order among this diversity can be seen if addressing is discussed in terms of two characteristics: the allowable degree of randomness in the coupling between the input signal and the addressed elements and, also, the partitioning of routing control between a central location and locations in or near the elements themselves. These two characteristics are the coordinates of the diagram in Fig. 7.4. This type of diagram seems capable of encompassing almost any conceivable addressing scheme, even including those used in information-storage systems.

The regions A and B above the horizontal axis encompass addressing systems, actual or conceivable ones, wherein all addressing is centrally controlled, for example, as in a monochrome (B&W) TV kinescope. The regions AL and BL located below the horizontal axis encompass systems that have some degree of local routing in or at the elements themselves, such as in a shadowmask color kinescope [7.16] or in a LED display wherein diode nonlinearity is used for x–y-type selection. The farther a system is depicted below the horizontal axis in the figure, the greater the degree that routing is carried out locally. The extreme case of local routing control exists in the box labelled "cell-identity system", a display-addressing arrangement not presently used but nevertheless possible of implementation. Herein, each element has its own identity and responds only when its "name" is called during a general information broadcast from the input signal.

The regions A and AL existing to the left of the vertical axis encompass systems that are of the nonrandom or series-selection type. That is, the temporal order of incoming signal information to the display is maintained throughout the addressing operation and is applied to the elements in a serial or linear geometrical order such as done in a TV raster scan. This is in marked

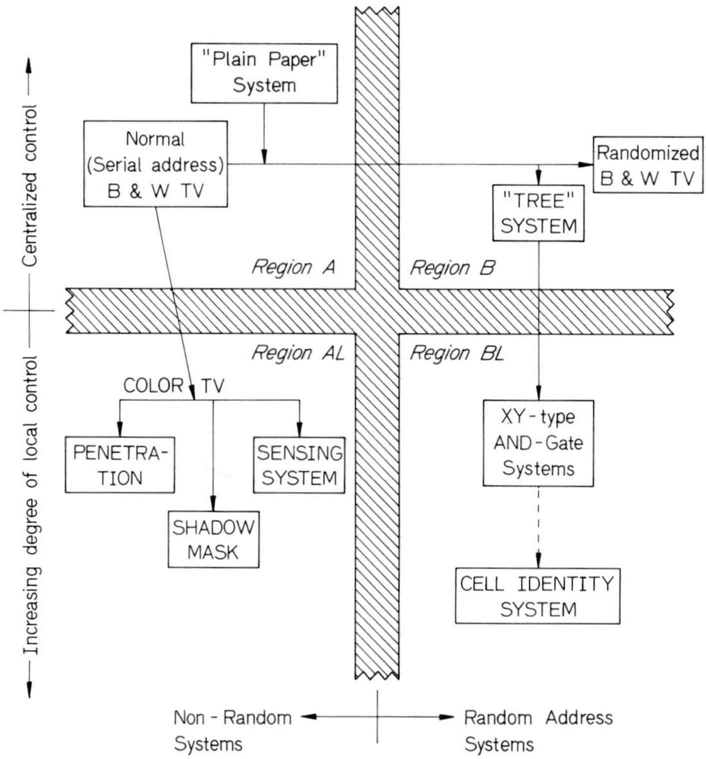

Fig. 7.4. Addressing-system domains

contrast to what happens in systems located in regions B and BL which lie to the right of the vertical axis. In these, the incoming signal information can be routed to the elements in any desired geometrical order or combination as is done, for example, in the x–y-address system of the LED display used in a hand calculator. Alternatively, the incoming signal information could be stored in a memory and randomly pulled out later in time and applied in any desired random geometrical order to the display elements.

The simplest addressing systems are located in region A, as for example, the B&W TV kinescope. All of the addressing is centrally controlled by circuitry that sweeps the steerable electron beam across the unstructured phosphor layer that constitutes the display elements of the viewing screen. The display elements themselves exercise no routing of the incoming signal information. The rigid time sequence and relative spatial location of the electron signal fed to the phosphor elements is generated in the TV camera and transmitter and is as faithfully as possible reproduced into the display elements. There is no freedom to shift bits of signal information from this disciplined time/geometry sequence, and none is normally needed. In such a situation we can conceive of each piece of signal information as existing in a time cell. The series string of these time cells and the information they contain are delivered one-by-one to the display elements in a corresponding geometrical order by virtue of the rythmic raster scan. Time and geometrical order are thus effectively made to coincide with each other and no additional ordering of information is necessary. Each piece of signal information, and each display element, is located to a precision of one part in N, where N is the total number of time cells.

Additional ordering is necessary, however, if the B&W TV display action is implemented in a random manner. In such a case the system is located in region B as shown in Fig. 7.4. Of the several different modes of possible implementation let us consider the mode wherein the time order in the above described series string of incoming information is preserved and the addressing system samples this string in a random manner and routes each sample to the appropriate display element. In such an approach the dimensional aiming precision required of the electron beam as it is randomly required to seek out a particular display element is one part in N, where N is the number of elements to be addressed. In the typical B&W TV case, N is approximately 5×10^5, and this level of precision, or geometric ordering, must be used each time a display element is selected. This geometrical ordering is an additional systems requirement, beyond that required in the previous case. It must somehow be built into the beam-deflection devices and circuitry. But that is not all of the ordering that must be added for this particular mode of random implementation. An additional amount must be built into the system to recognize the geometrical address of each random information sample, and also to provide for some degree of information storage: the information in the incoming time string must survive at least until it is sampled. Although there are other modes of random implementation that may lessen the need for some of this additional degree of ordering, in general it appears that additional geometrical ordering is required in all cases, compared to the serially addressed B&W TV system located in region A. There are other problems associated with the random approach, notably increased energy requirements and beam deflection speed, but the main point of the discussion here is that the price for randomization is that an additional amount of geometrical ordering must be built into the hardware.

We can view the comparison in another way that leads to the same conclusion. In the serial system, the location of the beam on one display element acts as a reference point for the beam in locating the adjacent element. This is quite similar to what is done when one writes by hand a sentence on a piece of paper. This situation is depicted in Fig. 7.4 by the box labelled "Plain Paper" system. In this system the location of the pen or pencil at any point, or dot bit, is the reference point for the next bit maneuver. Imagine the marksmanship required of your hand if the ink or pencil dots needed to compose alphabetic characters were set down in a purely random manner!

In the case of the familiar x–y-type AND gate systems, depicted in region BL, the geometric aiming accuracy for element address is established by the conduction matrix constructed during manufacture. In effect, the geometrical location of each element is pre-established to one part in N, where N is the total number of elements.

The above brief discussion is not intended to deprecate random-address systems. As is well known, random address in display and in information storage systems have great virtue in many cases. The above discussion emphasizes that there is a hardware price to be paid when one disturbs the temporal/spatial order of the signal information. The practical moral of this is that random address should not be used unless there is a compelling reason to do so.

As for the second addressing characteristic, the partitioning of routing control, the transition from the B&W TV kinescope to the color TV kinescope addressing system is a good example. To handle the added addressing burden imposed by the addition of color to the B&W system, it has been found much easier in practice to shift some of the routing burden locally to the region of the elements, themselves. The shadow-mask approach is one example, the penetration phosphor tube and the sensing tube, all depicted in region AL, are other examples [7.17]. In all cases, additional local structure is introduced in or near the display elements. The typical shadow-mask tube contains three adjacent electron guns, one for each phosphor color. The mask located inside the tube, closely spaced from the phosphor screen, contains several hundred thousand small holes, one for each underlying phosphor trio consisting of a red, a green, and a blue dot. The beams from the three guns are intensity modulated by color signal information and are deflected such that all three beams approximately merge at any given hole in the mask. Furthermore, these beams are made to arrive at slightly different angles ($\sim 1\%$) relative to the mask surface. This slight difference in angle allows only the emergent "red" beam to strike only the red phosphor dot of any dot trio; the "green" beam, a green dot; and the "blue" beam, a blue dot. The mask holes are smaller in size than the original beam diameters. This feature lessens beam-aiming-precision requirements as well as tailoring the size of an emergent beam to the approximate size of its respective phosphor color dot. The price for these and other advantageous features is that approximately 84% of the beam electrons strike the mask surface, rather than emerging through holes, and are thus ineffective in producing light.

The penetration-type tube introduces local structure directly into the phosphor screen itself. One version of this approach uses a multilayered screen, another uses multilayered phosphor particles. The basic principle is the same in either case. The electron-beam energy is modulated, and this modulates the beam penetration depth so that the beam deposits the majority of its energy into the appropriate color layer and thus provides the desired color of generated light. The sensing tube also introduces structure into the phosphor screen, but in a direction lateral to the beam. In one version of this, the screen is divided into red, green, and blue stripe trios. Each trio has an adjacent sensing phosphor stripe which emits nonvisible (uv) light. This light is sensed by a feedback circuit that makes appropriate adjustments to maintain appropriate landing precision of the electron beam. There are various other combinations of the above approaches, but no matter what the combination, local structure is introduced [7.17].

To accomplish the complete addressing function totally centrally in the above cases would require an impracticably high order of electron-beam-aiming precision. The slanted line was used to indicate that this local control also involves random selection of the color but only at each screen bit location.

The same general situation prevails in the random regions B and BL. Display elements can be randomly addressed centrally via an addressing tree (region B), but this becomes cumbersome if the number of elements is large. A more practical approach, well known in practice, is to use AND-gate action, mostly of the two input, or x–y-type, at each element. This greatly reduces the central routing burden without much increasing element complexity, at least when the elements, themselves, can provide the AND-gate action by means of intrinsic nonlinearity of their electrical characteristics. This general addressing approach can be carried to higher levels of local routing control, such as by using a larger number of AND-gate inputs. A typical example of this occurs in information storage systems of the magnetic-core type wherein the cores can be threaded with several control wires [7.18]. However, a higher degree of local control than exercised in the widely used x–y system is not usually easily implemented because too much is demanded of the elements or their local surrounding structure. A very slight increase in the complexity of each element of an N-element system enormously increases overall display complexity when N is a large number. This would be the case, for example, of the cell-identity system mentioned earlier and depicted in the lower part of region BL. This is an extreme case wherein local control is a maximum.

One suspects that there is an optimum allocation between central and local routing control that depends both on the state of available technology and the utilization factor of the local control mechanism. For example, there is a high utilization factor for the centralized addressing control in a kinescope system: the central system works continuously in serving all elements. In contrast, in the cell-identity system each local routing apparatus is relatively idle until that particular element cell is addressed. One might intuitively expect for displays, at least, that the optimum allocation of routing control is not far removed, if at all, from the sort of level found in the semiconductor-type x–y system, or in the color kinescope system. The variations on the theme may be a bit more subtle for information storage systems. Herein, various subdivisions between central and local control may evolve such as possibly happened in biological systems.

The same sort of approach used above for display addressing is equally applicable to information storage systems. Indeed, these can easily be fitted into Fig. 7.4. For example, audio or video tape or disk systems are of the serial-address type and would thus fit into region A, as also would an information system of the delay-line type. On the other hand, a CCD (charge coupled device) [7.19] delay-line information storage system would fit into region AL since some degree of local addressing control is used [7.20]. A variety of conceivable information storage systems of the random type would fit into region BL.

7.4 Conclusion

Simple physical principles and concepts can be of much help in providing a useful roadmap through a diverse assortment of display types. If the response time is not a critical parameter of the display, the large modulators of ambient light are the most efficient. But for speed and for observation under low ambient-light level, photon-generating devices are more attractive.

Electronic displays are of great utility and importance, and they are a continuing challenge that involves many technical disciplines.

References

7.1 Electronics **81** (1977)
7.2 L.N.Heynick, I.Reingold, A.Sobel: IEEE Trans. ED-**20**, 917 (1973)
7.3 A.Rose: *Vision – Human and Electronic* (Plenum Press, New York 1973) p. 71
7.4 J.I.Pankove: In *Electroluminescence*, ed. by J.I.Pankove, Topics Appl. Phys. **17** (Springer, Berlin, Heidelberg, New York 1977) p. 8
7.5 S.Larach, A.E.Hardy: Proc. IEEE **61**, 915 (1973)
7.6 H.W.Leverenz: *An Introduction to Luminescence of Solids* (Dover, New York 1968)
7.7 T.Inoguchi, S.Mito: In *Electroluminescence*, ed. by J.I.Pankove, Topics Appl. Phys. **17** (Springer, Berlin, Heidelberg, New York 1977) p. 197
7.8 H.G.Slottow: IEEE Trans. ED-**23**, 760 (1976)
7.9 J.D.Cobine: *Gaseous Conductors* (McGraw-Hill, New York 1941) p. 542
7.10 S.Zaromb: J. Electrochem. Soc. **109**, 903 (1962)
7.11 I.F.Chang, B.L.Gilbert, T.I.Sun: J. Electrochem. Soc. **122**, 955 (1975)
7.12 I.F.Chang, W.Howard: IEEE Trans. ED-**22**, 749 (1975)
7.13 D.P.Hamblen, J.R.Clarke: IEEE Trans. ED-**20**, 1028 (1973)
7.14 A.R.Kmetz: IEEE Trans. ED-**20**, 954 (1973)
7.15 R.Engelbrecht: RCA Zurich Laboratories (1972) (private communication)
7.16 H.B.Law: IEEE Trans. ED-**23**, 752 (1976)
7.17 A.M.Morrell, H.B.Law, E.G.Ramberg, E.W.Herold: In *Color Television Picture Tubes*, ed. by B.Kazan (Academic Press, New York 1974)
7.18 J.A.Rajchman: Proc. IRE 105 (1961)
7.19 D.F.Barbe (ed.): *Charge-Coupled Devices*, Topics Appl. Phys. **38** (Springer, Berlin, Heidelberg, New York 1979)
7.20 W.S.Boyle, G.E.Smith: IEEE Trans. ED-**23**, 661 (1976)

Additional References with Titles

Chapter 1

Special Issue on Displays and LEDs. IEEE Trans. ED-**26**, 1113–1245 (1979)

K. Okamoto, Y. Hamakawa: Bright green electroluminescence in thin-film ZnS:TbF_3. Appl. Phys. Lett. **35**, 508–511 (1979)

Chapter 3

P. Plesko: "Aging and Burn-In Behavior of AC Plasma Panels", SID Symp., Chicago, Ill., May 1979, p. 56, 57

P. Pleshko: AC Plasma Display Device Technology Overview. IEDM, Washington, DC, Dec. 1979, pp. 532–535; Proc. SID **21** (2, 3), (1980)

P. Pleshko, J.C. Greeson, Jr., K. Pearson: AC Plasma Display Panel Repair. SID Symp., San Diego, CA, April 1980

Proceedings of the Society for Information Display, Special Issue on: Plasma Displays **20** (3), (1979)

Chapter 4

J. Cheng, G.D. Boyd: The liquid crystal alignment properties of photolithographic gratings. Appl. Phys. Lett. **35**, 444 (1979)

D. Davies, W. Fischer, G. Force, K. Harrison, S. Lu: Practical liquid crystal display forms forty characters. Electronics **53**, 151 (1980)

Chapter 5

L.G. Van Uitert, G.J. Zydzik, S. Singh, I. Camlibel: Anthraquinoide red display cells. Appl. Phys. Lett. **36**, 109 (1980)

W.C. Dautremont-Smith, G. Beni, L.M. Schiavone, J.L. Shay: Solid-state electrochromic cell with anodic iridium oxide film electrodes. Appl. Phys. Lett. **35**, 565 (1979)

Subject Index